HANDBOOK

OF

MINING DETAILS

COMPILED FROM THE

ENGINEERING AND MINING JOURNAL

BY

THE EDITORIAL STAFF

Elibron Classics
www.elibron.com

Elibron Classics series.

ISBN 0-543-95708-X (paperback)
ISBN 0-543-95707-1 (hardcover)

This Elibron Classics Replica Edition is an unabridged facsimile
of the edition published in 1912 by McGraw-Hill Book Company, New York.

HANDBOOK OF MINING DETAILS

HANDBOOK

OF

MINING DETAILS

COMPILED FROM THE

ENGINEERING AND MINING JOURNAL

BY

THE EDITORIAL STAFF

FIRST EDITION

THIRD IMPRESSION

McGRAW-HILL BOOK COMPANY

239 WEST 39TH STREET, NEW YORK

6 BOUVERIE STREET, LONDON, E. C.

1912

THE · MAPLE · PRESS · YORK · PA

PREFACE

This book is a collection of articles that have appeared in the *Engineering and Mining Journal* during the last two or three years under the general head of "Details of Practical Mining," a department of the *Journal* that has been appreciated highly by its readers, many of whom have expressed the wish that a collection in book form be made, which has now been done.

In the editing of this volume, the work has been chiefly in the selection of the material and its arrangement in chapters. Now and then it has been possible to excise some paragraphs as being unessential and occasionally the phraseology of some articles has been altered a little, the requirements of preparation for the original weekly publication not always having permitted leisurely consideration, but in the main the articles now presented in this book are as they were given in the pages of the *Engineering and Mining Journal.* However, it has been necessary in a few cases to reduce the size of the engravings.

In making this collection the limitation of space necessitated the rejection of all material that did not pertain to the subjects selected for the chapters of the book, and even so it was necessary to dismiss some of the longer articles pertaining to them, which approached the character of essays rather than being the description and discussion of details. Of course a wealth of contributions pertaining to the arts of ore dressing and metallurgy had to be rejected summarily. The compilation covers the publications in the *Engineering and Mining Journal* from Aug. 7, 1909, to July 1, 1912. If all of the material that appeared in this department of the *Journal* during that period of three years had been used it would have been necessary to make a book of several times the size of this.

No claim is made that this book is a treatise, exhausting its subject, or any part of it. It is simply a handbook that is a more or less random collection of useful information, being just what passes through the pages of the *Engineering and Mining Journal* in the course of a few years. No special attempt to round out any subject has been made. Yet it will be found that some subjects are fully treated.

With regard to the authority of what is to be found in these pages: The matter in the main is merely descriptive of what is done. Nevertheless, there is frequently the injection of opinion and advice. A great technical journal is directed by its editor and is shaped by its editorial staff, but it is essentially the product of its contributors. It is a co-operative institution and its pages are a symposium of the experiences and views of many professional men. During the 18 months ending with June 30, 1912, there were 460 contributors to the

Engineering and Mining Journal, exclusive of the members of the editorial staff, and its regular coadjutors, and its news correspondents. Many of these contributors furnished articles that are now collected in this book. Their articles generally are signed. The unsigned articles are chiefly the work of members of the editorial staff of the *Journal* who have been sent into the field to study mining practice.

The heterogeneous authorship of this book naturally gives rise to some inconsistencies, some differences of opinion and some conflicts in advice. It has seemed to me best to let these stand just as in the original, since they are often merely the reflection of different conditions prevailing in different parts of the country, and if carefully read, absence of unity in this respect will not be misleading.

W. R. INGALLS.

October 1, 1912

CONTENTS

CHAPTER I

CHAPTER II

CHAPTER III

CHAPTER IV

CHAPTER V

CHAPTER VI

CHAPTER VII

CHAPTER VIII

CHAPTER IX

CHAPTER XII

LIST OF ILLUSTRATIONS

I

GENERAL NOTES

Economics of Management—Convenience and Protection of Employes and Equipment—Knots and Ties—Miscellaneous Notes

ECONOMICS OF MANAGEMENT

Checking Men In and Out of Mines.—At the Newport mine, Ironwood, Mich., where over 1000 men are employed, each man is given a brass tag with a number. Each morning as he goes to work he must appear at the timekeeper's office and present his brass tag, receiving in exchange a small cardboard check upon which is the date, his name, number and occupation. He keeps this check during the day and returns it at night with the timekeeper's notation on it showing the number of hours worked and the job or contract number, to which his time is charged that day. The brass check is returned to him when he presents his cardboard check. The time records are made up from these cards. The system entails much labor but no more than in almost any factory employing a like number of men. In order to divide the work of issuing these checks at the office windows, there are three aisles leading to three office windows; one for the surface men, one for the numbers ranging from 400 to 1000 and a third for numbers above 1000.

Mining Records (By Frederick T. Rubidge).—The collection of mining data is not always an easy matter and it not infrequently happens that men who are good mine foremen and shift bosses have little education and cannot write legibly, especially by the light of candle and with the palm of the hand as a desk. The writing of the reports gives an excuse for a half hour's leisure, and it is customary for them to retire to some convenient place in order to make out those required of them, and this usually at the end of the shift when their men should be watched and the holes are being charged for blasting. When thus writing up his report from memory the shift boss often forgets just how many men he had in this or that working place, and satisfies himself by putting figures down which will total properly. And when this report is received either by the foreman or at the main office it is scarcely intelligible, due to the combination of errors, smut and poor writing.

In endeavoring to overcome the difficulties mentioned, I have successfully replaced the pencil with the punch, the results being (1) a more accurate report, for the reason that the punching is done at the working place and without inconvenience, (2) a more legible report, being both clean and neat, in addition to being entirely printed, (3) and a report in which the duplicate is as good as the original. Incidentally the men take kindly to the innovation.

Figure 1 is a reproduction of the shift-boss report. The forms as furnished to the shift bosses were bound into booklets of 50 leaves, alternately white and light green in color, and with light fiber covers. The stopes were numbered according to the coördinate system. The coördinate numbers were also used, in connection with the level, compass point, or foot- and hanging-wall (F. W. and H. W.) to designate the location of raises and drifts, punching also the word *RAISE* or *DRIFT* as the case might be. *M D* stands for mine development and *D W* for dead work as opposed to mining proper. Under *DRILLS*, *B H* stands for block-hole drill, *H D* for hammer drill, and *P* for piston drill.

MINE OPERATING

LEV.	50	100	150	200	250	300	350	400	450	500	550	600	650	700	750	800	850	900	950	1000	1050	1100	1150	M.D.	D.W.	D'FT	PIPE	MISC.
STOPE	908	968	1020	1070	1141	1190	1239	1363	1440	1493	1550	1597	1644	1697	1747	1797	1865	1905	1983	2017	2071	2126	2226	2333	2405	RAISE	HOIST	PIL'R.
	680	622	564	500	436	360	290	229	176	123	80	4	78	139	291	348	401	451	535	612	669	716	783	812	858	FILL	TRAM.	TRIM.
JAN.	SEP.	0	DRILLS			NO. 1		NO. 2		XTRA		FUSE		CAPS		LAG'NG		SET T.		ST.	D.R.	H.	M.	TR.	T.M.	CARS		M.
FEB.	OCT.	1	BH	HD	P	100	0	100	0	100	0	100	200	100	0	100	200	100	200	1	1	1	1	1	1	100	0	GRA.
MAR.	NOV.	2	½	½	½	1	1	1	1	1	1	1	300	1	1	1	300	1	300	2	2	2	2	2	2	1	1	
APR.	DEC.	3	1	1	1	2	2	2	2	2	2	2	400	2	2	2	400	2	400	3	3	3	3	3	3	2	2	TIM.
MAY		4	2	2	2	3	3	3	3	3	3	3	500	3	3	3	500	3	500	4	4	4	4	4	4	3	3	
JUN.	1	5	3	3	3	4	4	4	4	4	4	4	600	4	4	4	600	4	600	5	5	5	5	5	5	4	4	N.
JULY	2	6	4	4	4	5	5	5	5	5	5	6	0	5	5	5	0	5	0	6	6	6	6	6	6	5	5	S.
AUG.	3	7	5	5	5	6	6	6	6	6	6	6	2	6	6	6	2	6	2	7	7	7	7	7	7	6	6	E.
1910		8	6	6	6	7	7	7	7	7	7	7	4	7	7	7	4	7	4	8	8	8	8	8	8	7	7	W.
1911	DAY	9	7	7	7	8	8	8	8	8	8	8	6	8	8	8	6	8	6	9	9	9	9	9	9	8	8	F.W.
1912	NGT.	10	8	8	8	9	9	9	9	9	9	9	8	9	9	9	8	9	8	10	10	10	10	10	10	9	9	H.W.

FIG. 1.—REPORT FORM TO BE PUNCHED BY SHIFT BOSS.

No. 1, No. 2 and *XTRA* refer to grades of powder. Other abbreviations are as follows: *S T*, stopers; *D R*, drill runners; *H*, drill helpers (or timbermen's helper if *TIM* is punched, indicating that it is the timber boss' report); *T R*, trammers; *T M*, timbermen; *GRA*, grading; *M*, mucking; *TRIM*, trimming; *PIL'R*, when punched in conjunction with a stope number is understood to be the pillar adjacent to that stope on the north. The numbers under the different classes of labor and under drills indicate shifts. For the labor it would be better to use hours instead of shifts as being more accurate but in this instance it would have made the form too large for convenient handling and punching.

For each working place a white and green sheet is punched—one operation serving for both. The punched sheets remain in the book until the end of the shift when they are torn out and left at the mine office. The green ones are retained there, filed in pigeon holes corresponding to the working places, and the white ones are sent to the main office. Each shift boss has a different punch mark. Any necessary corrections can be made by the mine foreman who has an individual punch mark.

At the end of the month the totals for each working place are made up at the main office, according to these reports, the stoping being separated from the drifting and raising. The necessary data regarding the tonnages are furnished by the surveyor and checked with the hoisting records. The surveyor also fills out the location, dimensions, inclination, etc., of the various working places. Powder is recorded in sticks but the explosives unit is necessarily the dollar. After the figures are completed copies are sent to the foremen for remarks, giving them the opportunity to explain any unusual increase in labor or explosives. If their remarks agree with the facts they are noted on the final copies which are distributed to all in authority. The punch form has also been used to advantage in recording electric haulage, hoisting and pumping records, and it is probable that it would be useful in milling and smelting.

Standard Cost Sheets.—The standardization committee of the Institution of Mining and Metallurgy on mine accounts and cost sheets, has made the following recommendations[1] in regard to the standardization of working accounts and cost sheets. These recommendations have been adopted by the council, which advises their use wherever conditions will permit.

The committee recommended that for the sake of convenient comparison, both in working accounts and cost sheets, all expenditure should be classified under the following main heads: (1) Development. (2) Extraction of ore (*i.e.*, mining). (3) Sorting at surface, preliminary crushing and transport. (4) Reduction costs (*i.e.*, ore treatment). (5) Administration charges and general charges at mine. (6) Realization charges on products. (7) Taxes and royalties of all kinds, shown separately. (8) Head-office charges.

The subdivision of these main heads into subheads must necessarily depend somewhat upon the conditions, but the advantages of adhering as closely as possible to one form, and departing from it only where necessary, are manifest. The following were suggested as desirable subdivisions:

(1) Development costs need only appear in the general cost sheet in one total, but a detailed sheet should be prepared showing the total expenditure and cost per foot in shaft sinking, driving, crosscutting, raising, winzing and plat (or station) cutting separately, as well as the proportions of these expenditures which are for labor and for materials respectively.

(2) Extraction of ore may be usefully divided into: (*a*) Stoping or breaking of ore, including under subheads: compressed-air and rock-drill costs, labor and supplies, shoveling, etc. (*b*) Timbering, filling excavations and sorting of ore in stopes (if any). (*c*) Hoisting. (*d*) Pumping. Where the cost of pumping is exceptionally heavy, it may be convenient to make this item a main head. The same remark applies to the removal of the overburden when a deposit requires to be stripped. (*e*) Underground tramming. (*f*) Sampling, assaying and surveying. (*g*) General underground maintenance. It is suggested that these subdivisions of the main head should be set out in detail in

[1] *Bull.* No. 76, I. M. M.

the general cost sheet, because mining is at once the principal item of working cost and offers the greatest scope for economies.

(4) Reduction costs should be subdivided according to the treatment undergone by the ore, *e.g.*, crushing, amalgamation, concentration, fine grinding, cyaniding sands, slimes treatment, roasting, smelting, converting, leaching, precipitation, etc.; for each of which a detail sheet should be prepared in such form as circumstances may dictate.

(5) Administration may be divided into salaries (general resident manager and clerical staff on the spot), stationery and office general expenses, traveling expenses, insurance and legal expenses, accidents, medical, sanitary and hospital expenses, stabling and sundry transport, etc.

(8) Head-office charges, besides ordinary central-office expenses, will include a great variety of items, such as agency expenses, directors', consulting engineers' and auditors' fees, bank charges, etc., also interest on debentures (if any). The individual items under (5) and (8) are likely at times to merge one into the other.

Labor Wasting and Labor Saving (By S. A. Worcester).—The persistent use of such antiquated devices as hand-trammed cars and barrows, shovels, forks, scrapers, rakes and hand-feeding operations of various kinds in large plants operated with ample capital is difficult to understand. The army of laborers necessary is always ready for a strike whenever the agitator appears, and the operation of the plant is therefore largely subject to the caprice of this element. Lack of intelligence is a most fruitful source of annoyance and of accident to the laborer and the plant.

The attitude of managers toward improvements finds, unfortunately, much support in the antiquated designs that are regularly being offered and erected by machinery builders and consulting engineers. An instance of this was the recent erection of several concentrating mills in which the concentrates are shoveled and wheeled from the table boxes, and, after drying or draining, are again shoveled and wheeled to box cars or wagons. One 100-ton mill, recently designed by a well-known engineer, requires five men per day shift and four men per night shift; another 150-ton mill doing exactly similar work, but using modern devices economically arranged, employs but three men per day shift and two men per night shift.

In the old-line mills referred to, one man per shift is usually employed as crusher feeder, whereas many well-known plants eliminate this attendance by using either a simple automatic feeder or a crusher hopper large enough to receive the mine run. Such simple and practical devices as shaking launders, conveyors, etc., for moving concentrates, are in use in a number of mills and are familiar to every progressive designer, so that there is no justification for shoveling and wheeling.

Many cyanide plants employ one or more men per shift filling sand tanks by tramming a car from a sand bin to the tanks. Automatic appliances that would save their cost in a brief period of continuous operation were available for this work when these mills were built. Well-designed excavator and conveyor systems for handling tailings and sands were proved successful a number of years ago, and the shovel and tram car are obsolete.

Many managers and some engineers contend that automatic machinery involves necessarily a large outlay, and is, therefore, prohibited except for long-running operations. While it is no doubt true that the probable life of the plant must determine the outlay within reasonable limits, yet there are many instances of labor-wasting plants which have cost more than completely automatic arrangements would have cost. Automatic devices are not necessarily cumbersome and complicated. Highly economical results can often be accomplished by simple and inexpensive apparatus, if the designer has the object fully in mind.

Mills are frequently designed so that a large sample has to be reduced or cut down daily by hand. A large sample is then taken to the assay office and recrushed and ground, the rejection being returned to the mill. In nearly all cases a laboratory crusher with proper automatic sampling apparatus can be so placed by the designer that the sample made in the mill is ready for grinding, and small enough for convenient handling, all rejections being returned by the automatic splitter to the mill. The work thus done is likely to be more accurate than hand work and the labor saved is well worth considering.

In mining operations the same disregard for economy of labor is evident. Nothing is more common than ore bins so designed that they can neither be filled to nearly the full capacity nor emptied entirely without using a shovel or scraper. Hoisting cages are still being frequently installed in metal-mine shafts, involving higher labor and power costs, greater first outlay and maintenance, greater dead weight and smaller capacity by far than the skip, which is an old and thoroughly proved appliance. In view of the present state of the art, nothing but ignorance, in most cases, can explain this blunder.

The employment of cages, with a number of laborers on each shift tramming small-capacity cars from cage to ore house or waste dump, is one of the most common labor- and power-wasting arrangements now in use. One skip-hoisting plant of my design displaced four laborers per shift, a net saving of $24 per day, by replacing cages with skips dumping into a motor car which delivers the ore to the ore house. This car is operated by a motorman. At two other plants, where the motor car is at all times visible from the engine room, the car is operated by the engineer, using a controller placed at the hoist. At two plants, where circumstances favor such an arrangement, the skip empties its load into a chute leading direct to the ore house and having deflecting doors which direct the ore to any desired bin.

At smaller hoisting plants, using the bucket, there are now several different styles of automatic bucket dumpers in use, which in many cases save one man's labor per shift. These dumpers can be placed at the top of the headframe so as to dump the ore directly into the ore house, thereby saving the wages of a top trammer. In spite of the evident economy of these simple devices, mining operators are still frequently erecting labor wasters at actually a greater outlay than would be required for uptodate rigs. The ancient hook-and-ring arrangement for dumping buckets is still being frequently erected in places where the inexpensive automatic device would save one man's wages per shift.

One favorite argument against placing the ore house close to the shaft, in order to save tramming, is that the headframe must usually be made higher for this arrangement. It is also a popular belief among hoisting engineers that the more nearly vertical rope pull of the headframe has an undesirable effect on the hoist. I am positive that a careful analysis of conditions will fully dispel this belief. Mine operators seem to have an aversion toward headframes of any considerable height, and when an extension can be no longer avoided it is made about half as much as it should be to give ample head room. As a rule it is far better economy to gain head room for ore-handling operations by a headframe of sufficient height, than to tram a car from the shaft over a dump or trestle to a point where sufficient elevation exists.

An almost invariable mistake in the design of ore houses and bins is the practice of placing the grizzly or spout which delivers ore to the bin at so low a height that the bin can be only partly filled when the rising ore chokes the spout. Then no further filling can be done until the ore is shoveled away toward the sides and corners of the bin. The spout must be so placed as to form a rock cone whose sides slope naturally from the top edges of the bin to the end of the spout, in order to fill the bin to its capacity without shoveling.

The moss-covered error of insufficient slope in bin bottoms and chutes still prevails, and it is probable that fully 50% of all wooden bins built by small operators in the Rocky mountains have bottoms so flat that they cannot be emptied without using the shovel. Horizontal bin bottoms are still being built occasionally, although careful consideration of the working of this bin will nearly always place it in deserved disfavor.

The practice of arranging bins from which ore is to be hauled by wagon, so that every pound of ore has to be shoveled into the wagon, is indefensible. Where the hauling is to be done by contractors and no lower price can be gained by using bin gates with ample height for saving shoveling, there is still the advantage that having a convenient arrangement insures better service than when all the ore must be shoveled. Teamsters' unions sometimes limit the number of loads per day per man or team, but in case of any delay the mine having spouts and gates is more likely to get its full number of loads hauled than the one without such conveniences.

In many places ore bins for loading railroad cars are still erected with spouts so low that much scraping is necessary to empty the bins, and much shoveling necessary to trim the load. If the bins and spouts are properly designed gondola cars can be loaded rapidly without the use of either shovel or scraper.

By using portable diverting spouts, box cars can be loaded and the load nicely trimmed without any shoveling. For loading fine ore, slimes and concentrates into box cars, a short portable conveyor operated by a small motor will carry the ore from the bin spout to the ends of the car and trim the load without wheeling or shoveling.

If designing engineers will adopt as a cardinal principle the elimination, as completely as possible, of all manual labor and particularly of shoveling and hand-tramming, there will be materially less encouragement and support for managers who persist in labor-wasting methods, and if the managers will exhibit, in the adaptation of modern appliances to their work, a minute fraction of the ingenuity and dogged persistence with which they at present defend and adhere to their evidently obsolete equipment, labor problems will be speedily solved and net profits increased.

Labor and Tonnage Chart as Aids in Reducing Costs (By Claude T. Rice).—To get the mining costs as low as is compatible with good mining it is essential to instill a healthy rivalry among the men and let them know that the mine superintendent, and every one in authority on the job, knows how much work they are doing. A great aid in accomplishing this at the Highland Boy mine is the posting of labor and tonnage charts where the bosses and men can see them.

The tonnage chart shows the tonnage mined by each shift, the combined tonnage of the two shifts and the tonnage sent out over the tramway (at the Highland Boy, the ore is shipped in that way from the mine), the total number of machine drills at work in the mine, the number of machines working in ore and the number working in waste. On the labor chart, the total number of men employed at the mine, the number underground, the tons mined per man employed at the mine and per man working underground are shown.

The charts are drawn on cross-section paper ruled 10 squares to the inch and a negative made from a tracing ruled with cross-section lines. From the negative a print with white background and blue lines is obtained. The scale and the headings, as well as the days of the month, are put on the original tracing cloth so that the final prints are all ready for use.

The data for the last day of the preceding month are shown as the start of each curve. The days of the month are plotted as the abscissas and the other data as the ordinates, the horizontal scale being a day to the inch, while the vertical scale varies with the different curves. The various curves are drawn in with different-colored crayons so that there is no trouble in following them.

The tonnage curves are drawn to a vertical scale of 100 tons to the inch, as at the Highland Boy mine the tonnage does not fluctuate more than 200 tons per day and this scale is ample to show with sufficient emphasis the variations in the tonnage mined from day to day. The shift tonnages are plotted from the tonnage reported by the respective shift bosses who estimate this from the number of cars dumped in the tramway bins. The tramway tonnage is reckoned from the number of buckets sent out over the line and the average weight of a loaded bucket as determined over a long period of time by checking it against the weighed ore shipped to the smeltery. The tramway curve is therefore the more accurate curve. The curves reported by the shift bosses give checks on how full the cars are loaded underground, so by comparing the curves of the tonnages mined by each shift, it is possible to see which is, in all probability, failing to load the cars properly. At the Highland Boy mine, the saving effected by correcting the practice of underloading cars, through the use of these curves has been greater than the cost of keeping them.

Below the tonnage curves, and on the same chart are plotted the machine curves. The vertical scale used on these is five machines to the inch. This scale is sufficient to give emphasis to the variations in the number of machines at work which is usually only about twenty-five. As one of the curves shows the total number of machines running on ore and another the number working on waste, and as most of the machines on development work would be working in waste, an indication is given as to whether the development work is being kept uptodate or whether it is being shirked so as to make a tonnage showing. It might be well to show the number of machines working in ore and the number on development instead of in waste; as such a curve would be more important than the waste curves unless the filling were being broken underground.

The vertical scale used on the curves representing the number of men working about the mine is 10 men to the inch. It might be well at mines where the square-set method of mining is used or where stull timbering is done, to show how many men are working at timbering, for the job with the greatest possibilities for loafing is that of timbering. It always pays to keep close track of the timbermen. On the labor chart it might also be well to plot a curve showing the number of sets or stulls put in each day so as to keep still closer track of the work of the timbermen.

On the tons-per-man curves a vertical scale of half a ton to the inch is used so as to show plainly the variations. The importance of this is evident. The drop in the labor curves shows clearly which day of the month is pay day, even if it is not marked. The tons-per-man curves also show that the best workers are not the drinking men, although this increase in the tons mined per man is due partly to the fact that less development work is done on pay day. At a mine where the stopes are being filled it would also be advisable to plot a curve, on the tonnage chart, showing the number of tons or cars of waste filling that is being dumped into the stopes. This would give a check on the progress in the filling of the stopes and the tendency to let that important element in the mining lag behind in the scramble after ore would be reduced.

The importance of these curves representing graphically the several steps in the operation of the mine is evident. They afford, in a manner that spurs the men on to do better work, a means of keeping close check on 60% of the total expenditures in the mining of the ore. The curves have been in use at the Highland Boy mine nearly a year and have been found of great aid to those in charge.

Their introduction was due to Ivan DeLashmutt, engineer at the mine. The set of curves shown in Fig. 2 are taken from the charts showing the details

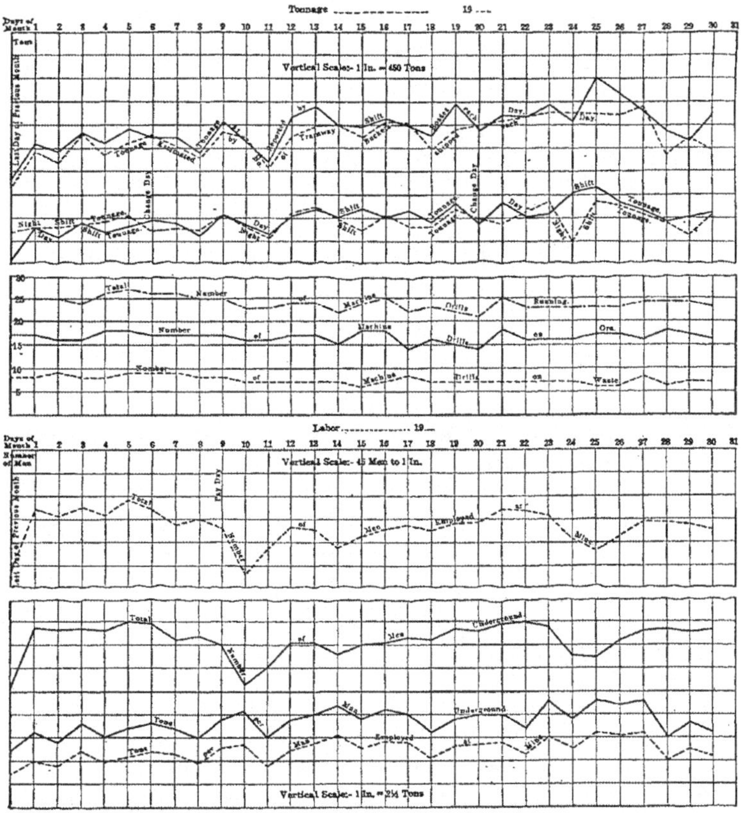

FIG. 2.—LABOR AND TONNAGE CHARTS SHOWING RECORD FOR TYPICAL MONTH AT HIGHLAND BOY MINE.

of the work for a fairly typical month. The work in keeping these charts uptodate is small. Charts 22 in. wide and 34 in. long are used.

The Automobile in Mining.—The automobile is doing much to solve the transportation problem in Mohave county, Ariz. A large auto-truck, put into use by the Dixie Queen mine recently, has been thoroughly tried out in the 25-

mile haul between the mine and Chloride and has given excellent service. Loads as high as four tons have been successfully hauled to the mine in less than half the time required by teams.

[Throughout the Western desert regions the automobile has proved an important factor in the development of the mining industry. The outlying, and formerly inaccessible, districts have been brought into easy communication with the centers of population, thus stimulating and facilitating operations. The use of automobile trucks has also become extensive in the mining industry all over the world.—EDITOR.]

CONVENIENCE AND PROTECTION OF EMPLOYES AND EQUIPMENT

Acetylene Lamps.—The Republic Iron Co. has been using acetylene lamps in its mines for about one year. When the lamps were first introduced, they were sold to the miners at the same price as the ordinary lamps in which "sunshine" was used. After all the men had been supplied, the additional lamps were charged out at 75 cents each, which is the actual cost. With ordinary care one lamp will last six months. The require 1/2 lb. of carbide per day. That quantity is given to the miner at the beginning of each shift, at which time he returns an empty can which is filled during the day and thus is ready for the next morning. Each lamp burns about two hours before it is necessary to recharge. The lamp is small and light, and is worn on the hat, the same as the ordinary oil lamp. The great advantage in the use of these lamps is cleanliness. There is no oil and no soot. They also give a better light and do not foul the air to such an extent as some other illuminants. The acutal cost of using acetylene lamps is about one-fourth that of candles, and one-half that of "sunshine" or oil.

Some superintendents have tried to introduce the acetylene lamp but have failed; some of the excuses are that the miner does not like the lamp; that it is too much bother to charge it with carbide twice a day underground, and that the lamps get out of order too easily. The lamp as usually constructed is too frail for rough usage. The reflector causes some trouble, and the thread which connects the carbide chamber with the water compartment wears out easily. These are mechanical difficulties that the manufacturers can readily overcome. With proper care the lamp, as constructed, will last a year.

The Penn Iron Mining Co. has been using Baldwin acetylene lamps about two years at its mines near Vulcan and Republic, Mich. After a thorough trial, the lamps have proved to be cheaper than either candles or sunshine. They are much cleaner than candles or oil, give a better light and burn better in poor air.

Candle Tests (By Claude T. Rice).—Candle tests are run frequently at mines, but generally candles of the same make are tried—the Granite candle, a soft one, against the hard wax candles, the Goodwin or the Schneider. There are several factors affecting the life of a candle, the most important being the condition of the air. Air currents, poor ventilation, and hot workings cause candles to burn dimly and unevenly, so all the material is not consumed. The size of the candle has an important bearing upon the cost per man, and often a cent per man can be saved by changing the weight.

The accompanying table, made from the results of a test at a fairly well ventilated mine of about the average temperature for a deep mine, is interesting as showing the merits of some new candles or rather candles that are not so well known to western miners, while it also brings out the importance of the weight of the candle used. Since the tests were made the cost of candles has been reduced at this mine to 3.203 cents per man, using 12-oz. "Sunlight" candles.

CANDLE TESTS

Make of candle	Standard oil "granite"	Perry (hard wax)	Perry (hard wax)	Perry (hard wax)	Werks (hard wax)	Schneiders (hard wax)
Weight, ounces	12	14	12	13	14	14
Length of test, days	31	15½	17	26	23	29
Number of men working per day	345¼	306	322½	396¾	366	380
Price per box at mine	$3.45	$4.64	$4.08	$4.36	$5.057	$4.956
Candles used per man per day	4.35	3.06	3.94	3.58	3.30	3.7
Cost per man per day	$0.062	$0.071	$0.067	$0.065	$0.0697	$0.0778

CANDLE TESTS (Continued)

Make of candle	Goodwins (hard wax)	Standard oil, No. 1 hard	Sunlight (soft candle)	Sunlight	Sunlight
Weight, ounces	14	12	13	13	12
Length of test, days	14	16	30	31	30
Number of men working per day	388¾	383¾	429¼	375½	377½
Price per box at mine	$5.15	$4.35	$2.93	$2.93	$2.75
Candles used per man per day	3.5	3.7	3.44	3.546	3.538
Cost per man per day	$0.749	$0.064	$0.0421	$0.0433	$0.0403

Underground Repair Shops.—At the middle working-level stations of the shafts of the Tonopah Mining Co., and also in the Goldfield Consolidated mines, repair shops are maintained for fixing all air drills underground. A machinist is employed on day shift to make these repairs, and no machines are sent to surface unless there are some unusual repairs to be made. At the Tonopah mines the shop is equipped with a drill press which is especially useful in boring out broken stud bolts on the air chests, the only other power-driven machine being a grinding wheel. Power is supplied by a 1-h.p. electric motor.

Abandoned Shafts and Open Cuts.—Abandoned shafts and open cuts should be kept securely covered or fenced to prevent accidents due to persons falling into them. This, however, is not always a simple matter, due to the fact that vandals steal the material with which the openings are covered or fenced. Only recently in South Africa an old shaft was covered with timbers too heavy to move, and they were actually chopped in pieces in order that they might be carried away. In another case a wire netting had been placed around an open pit and someone had deliberately taken the wire down and stolen it. In practically every mining camp there are abandoned shafts that are not covered or even marked. Too much care cannot be taken to make these places safe. In the vicinity of Nome, Alaska, there are dozens of open shafts on the tundra, with nothing to mark their situation save the small pile of dirt that has been taken out. During the winter the snow drifts over these, completely covering them, thus making an excellent trap for the cross-country traveler.

Speaking Tubes in Mines.—Speaking tubes are used for communicating from one part of the mine to another in the Allouez mine in the Michigan copper country. They can be used up to distances of 1000 ft., beyond which the service is not satisfactory. In the Allouez mine the tubes are 1 1/2-in. pipes. A tee is inserted in the line wherever it is desired to establish a point from which communications can be sent. The mouthpiece is made by screwing a short nipple into the tee. The nipple is closed by a wooden plug, attached to a string or chain hung from the pipe to prevent loss. The nipples are kept closed when not in use. To call a station a knocker, consisting of a 2-in. nipple that encircles the 1 1/2-in. pipe above the tee, is used. Lifting and dropping the knocker on the tee causes a sound that is clearly transmitted by the pipe. For greater distances than 800 ft., 2-in. pipe should be used.

Safety Appliances in Mines (By Lee L. Wilcox).—It is the purpose of this article to describe the various safety devices used at the mines of the Republic Iron & Steel Co. They naturally divide themselves into two classes; those tending to prevent accidents and minimize the dangers to which workmen are subjected, and those which relieve suffering and lessen the attendant dangers after an accident has occurred.

The first class, being preventive, has occupied most of the time and has included every phase of work about the mines. It was started by the appointment of a committee from among the mining captains and mechanics, whose duty it is to make regular monthly inspection trips, to report on the condition of the various mines and to recommend means for improving the conditions where in their judgment it seemed necessary. This committee performs its work faithfully and efficiently, and the result has been the extensive adoption of safety appliances.

This can be best described by citing some of the devices employed. In the shops all pulleys, belts, gears, and such equipment are enclosed with strong wire-cloth guards. Line shafts and jack shafts are provided with overhead walks for the oilers and repair men. Hoisting engines, compressors, and other machines are protected by substantial handrails wherever moving parts may endanger the employees.

In the matter of underground protection the company is very particular. All openings, such as shafts or raises, are thoroughly protected by gates, railings or bars, making it practically impossible for a man to step into them accidentally. In addition to this ore chutes are equipped with grates, and ladderways are provided with permanent platforms at intervals of about 12 ft., through which are left openings only large enough to allow a man to pass. Ladderways in hoisting shafts and air shafts are electric lighted wherever possible, and no one is allowed to carry a lighted candle in them, thus insuring protection against fire.

The shafts are equipped with sprinkling devices for use in case of fire. These are made by running a 2-in. pipe the entire depth of the shaft. To this pipeline are connected, at 10-ft. intervals, pipes which extend horizontally around the shaft and which are drilled with 1/4 holes at intervals of 4 in. The stations and pump-rooms in the mines, where there is little water, are protected in the same manner. The places so protected can be thoroughly sprinkled in a short time.

In dealing with the second class of safety appliances, which have to do with conditions after an accident, the company has installed two complete sets of the Draeger rescue apparatus, 1910 type, one pulmotor and the necessary secondary equipment to keep this apparatus in perfect working order.

Rescue parties consisting of from four to eight men, who are entirely familiar with the underground workings, have been organized at each mine. These parties are drilled regularly in the use of the apparatus. The drills consist of climbing ladders, using picks and shovels, carrying men and other such work. They are varied from time to time as the instructor may direct. Together with the rescue parties first-aid parties are organized. The company has a physician at these meetings to instruct how to handle injured men and how to apply bandages.

Improvements in Mine Bunks.—Proper sanitary conditions and comfort for employees should be seriously considered by operators in every line of business. Health and satisfaction among those employed at a mine is practically conducive to an increased and steady production. This has been given consideration at the Sunnyside mine, in Eureka gulch, San Juan county, Colo. Here the mine has provided reading rooms, baths and every modern convenience. The question of sanitary sleeping quarters within limited space has been solved by the installation of a bunk, patented and manufactured by Charles Scheer, of Silverton, Colo. It is made up of piping with appropriate coupling and joints. As shown in Fig. 3 it can be used with any coil spring. It is provided with side rails for the protection of the occupant. The parts are entirely separable and can be readily transported. The healthful atmosphere that prevails in the sleeping quarters at the Sunnyside mine is ample proof that it fulfils the desired object. In one feature alone, the elimination of the bed bug, the installation has repaid the company.

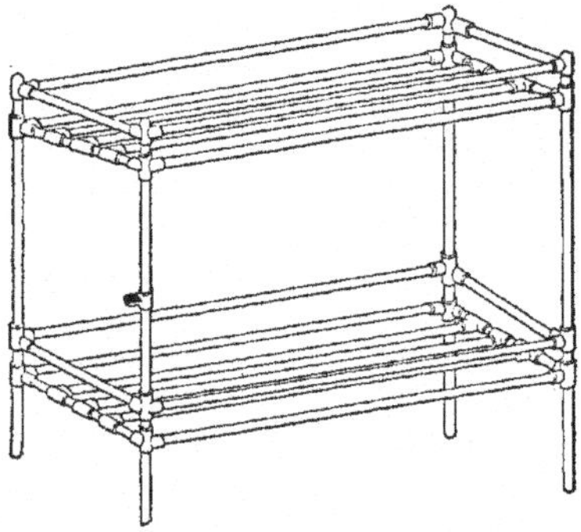

FIG. 3.—SANITARY MINE BUNK BUILT OF PIPE AND FITTINGS.

Portable Houses.—One of the continual problems of the prospector and miner is that of his cabin. To a certain extent this is being answered by the builders of portable houses. These range from 7×9 ft. to about 18×30 ft. in floor space, which means from one to six rooms. These houses can be set up or taken down without any tools in three hours, or less, are weather proof, and the material is also guaranteed against mildew or rot. They are usually screened, and completely proof against insects, a matter of importance in tropical or mosquito-infested districts. The weight of the houses per square foot of floor area varies from about 4.4 lb. in the smallest size down to 3.2 lb. in the larger sizes, while the prices range from about 80 cents per square foot of floor area for the smallest down to about 65 cents per square foot for the largest size. From the above data a miner knowing about the size of house he desires, can closely approximate its weight and cost.

A Sanitary Underground Latrine.—The latrine, the details of which are shown in Fig. 4, is in use at the Goldfield Consolidated mines, and after a test of several months has been found to be absolutely sanitary. Moreover, as the box closes tight, there is no odor of lime scenting up the whole level, as is the case when open boxes are used. The latrine is constructed of No. 10 sheet iron and can easily be made by any blacksmith. The top and bottom are riveted to the body, which is 26 in. high and 16 in. in diameter. The top is made by flanging over a form, as this is the easiest way when there are several to be made. This form is made from a plate of 3/4-in. iron, and the inside edge of the ring is turned in a lathe to an easy curve. The outer edge of the sheet forming the top piece of the box is flanged 2 in., so as to stiffen the sides. The

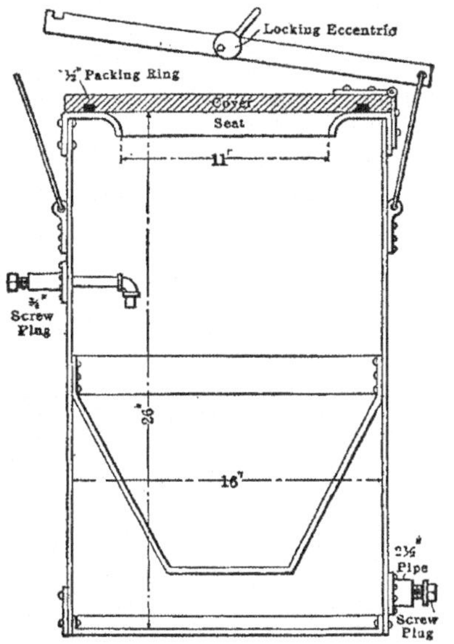

FIG. 4.—SANITARY LATRINE USED UNDERGROUND AT GOLDFIELD.

bottom piece is also flanged over the same form. A flange for a 2 1/2-in. pipe is riveted to the cylindrical body at the bottom, into which a pipe is screwed for an outlet. The outlet is closed with a screw plug, as a valve would be in the way and liable to be broken off when the latrine is taken to the surface to be emptied. A hole is drilled through the screw plug, and a pin put in it so it can be removed without touching it with the hands. Near the top, on the opposite side from the discharge, a flange is riveted for a 3/4-in. pipe, also closed by a plug. A short piece of pipe extends from the inside of the flange, ending in an elbow that is fastened to a short nipple pointing down to the bottom of the box. The cover piece is a circular plate of 1/2-in. iron in which a groove is cut about 3/8 in. deep to hold a ring of 1/2-in. engine packing, to make a tight joint between the cover and the seat of the latrine when the cover is locked in place. The cover is hinged to the body of the latrine, and is locked in place by means of a bar, fastened at one end of the body by a link so that when the latrine is in use underground it hangs straight down by the side, entirely out of the way. On the other end in the top of the locking bar is a notch over which slips a link fastened to the opposite side. A small eccentric with a short lever arm is carried on a bar for locking the device.

After using the latrine, a little lime is thrown in and then the cover closed down, but not locked, as the weight of the lid is enough to keep the box fairly air-tight. The latrine is placed in a small room partitioned off for the purpose. Before removing the latrine the top is locked in place. It is then loaded on a truck and hoisted to the surface. There the discharge plug is removed by means of a wrench, and connection made between the washing-out jet at the top of the latrine and the water line at a pressure of about 120 lb. per square inch. The seat is also washed by a hose.

Inspection Department of the Goldfield Consolidated.—The inspection department of this company is in charge of the fire-fighting corps. Its members devote the entire 8-hour day to inspecting the mine workings, hoisting engines, hoisting ropes, cages and passages and ladders used by men. The miner who neglects keeping his working place safe is discharged. Records of all accidents are kept, the cause being specified after an investigation. The men are impressed with the importance of attending to minor injuries, especially cuts, in order to avoid septicemic infection. Since the organization of the department in May, 1910, there have been but two fatal accidents. The records show that the majority of the accidents were due to falls of roof, and falls down ladderways and chutes. Most of the ladderway falls were caused by the miners having one hand employed in carrying tools. The miners are encouraged in every way to report dangerous conditions, and are commended for their interest even though an investigation shows the report to be not well founded.

Goldfield Consolidated Fire Equipment (By Claude T. Rice).—Since the fire in April, 1910, did such serious damage to the Goldfield Consolidated mill, the company has spent much money in installing a comprehensive fire-fighting equipment. As the square-set method of timbering is used, it is necessary to keep a supply of timber at the sawmill. This timber is protected by several hydraulic monitors, mounted on platforms, high enough for the operator to get a good view of the yard, in a manner similar to that adopted at most of the big sawmills on the Pacific coast. The monitor used, made by A. J. Morse & Son, is much cheaper and less elaborate than most monitors put on the market; it costs only $15, while others sell for about $150. It has a 2-in. inlet and 1-in. discharge pipe. It discharges about 350 gallons per minute at a pressure of 80 lb. at the orifice, and has a range of about 150 ft.

At the shaft and buildings there are hydrants with lines of 2-in. hose coiled nearby. At each mine and at the mill are fire stations, where are kept hose carts, ladders, a 40-gallon chemical barge, picks and fire lanterns, all sealed in place to prevent tampering. All hydrants are kept sealed and are inspected about once a week, for if the water were turned off at one of these places, it might take several minutes to discover the difficulty, during which time the fire would be making headway.

As the floor and much of the interior construction of the mill are of wood, 16 connections for hose are provided. Under the ore bins, where there is a possibility of a fire starting and gaining headway before being discovered, a sprinkler fire-extinguishing apparatus is installed. The main water line is carried on the trestles of the railroad that connects the mines with the mill, and there are numerous hydrants for hose connections at each trestle. The reservoir feeding this system is on top of Columbia mountain, and has a capacity of 146,000 gallons. There is a pump supplying the reservoir at the rate of 120 gallons per minute from a tank holding 30,000 gallons. The tank is replenished at the rate of 100 gallons per minute. In the fire-prevention equipment there are 11,500 ft. of 5-in. pipe, 5000 ft. of 4-in. pipe and 10,000 ft. of 2 1/2- and 1-in. pipe, a total of 26,000 feet.

Comparatively speaking, the amount of water available for underground fires is small. In contending against such fires, reliance would have to be placed upon bulkheads, so placed as to keep the fire within a restricted area. The mines are inspected after each shift for the discovery of incipient fires. At each mine there are two Dräger helmets.

KNOTS AND TIES

The Diamond Hitch (By W. H. Storms).—There are many things that prospectors should know in addition to a knowledge of minerals and formations and how to hit a drill, and one of the most useful of the others is how to "throw the diamond hitch." The diamond hitch is used to secure a pack on the back of an animal in such manner that the rope at the top of the pack forms a diamond, and is the most satisfactory way to fasten a pack that has been devised.

The first essential is a good pack saddle, to which are attached back and breast straps to keep the load from shifting on steep hills. The next essential is a strong rope. A 5/8-in. manila is good, but a rawhide lariat is better if it is not too new and stiff. First secure the pack saddle on the animal, and be sure that it is secure, for pack animals soon become tricky and will swell up by holding their breath while the saddle is being cinched. To beat this draw the straps tight and pretend to make them fast, watching the animal, and when the breath is exhaled, quickly take up the slack.

When the pack saddle is securely cinched the various articles of the pack may be placed on the back of the animal, being held in place by preliminary lashing, good enough to hold temporarily. After all is in place you are ready for the diamond hitch. This requires two men, one on each side of the animal.

He on the left is the thrower; he on the right of the animal, the cincher. The thrower first ties one end of the pack rope securely to a ring attached to a broad leathern cinch, as at *A* in Fig. 5. The other end of this cinch is equipped with a hook *B*. Having made the rope fast, the thrower, holding the rope and ring end of the cinch in his left hand, tosses the hook end of the cinch under the animal's belly, where it is caught by the cincher with his left hand. The latter now stands ready to use his right hand when his partner throws him the loop *C* over the back of the animal. The loop is quickly caught in the hook, and the thrower adjusts the rope so as best to secure the various articles of the pack,

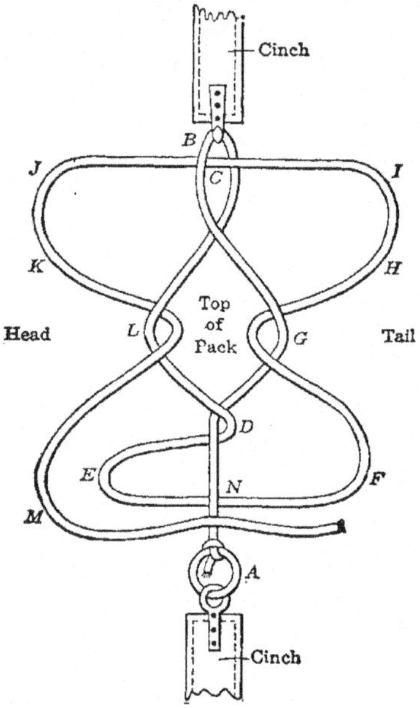

FIG. 5.—THE DIAMOND HITCH.

while following the lines shown in the sketch. The cincher now draws the long end of the rope tight, and throws the loose end forward over the back of the animal to the thrower, and together they adjust the rope to the pack, while keeping the loose end taut. A turn is taken beneath the rope at *D*, and it is then carried forward and around the lower corner of the pack at *E*, and backward along the lower side of it toward *F*, where it is taken upward to the top and a second turn taken at *G*. From there it is passed around the back and upper corner at *H*, down to *I* and around that corner, and forward along the lower right-hand side of the pack to *J*, upward and around the corner at *K* and looped under at *L;* thence forward and downward to *M*, and along the lower left-hand side, where a turn can be taken in the crossing rope at *N*, and the loose end secured in the diamond at the top of the pack.

Throughout this entire performance the rope must be kept as tight as possible, and the end secured so it will not slip. The diamond hitch is quickly undone, once the end is loosened, for there is no portion of it that binds, or works into hard knots.

Splicing Wire Rope.—Wire rope is susceptible of almost perfect splicing and the operation is so simple, states F. L. Johnson in *Power*, that it may be learned in an hour by any mechanic who is at all skillful in the use of ordinary tools. For all kinds of transmission rope the long splice is used and should not be less than 16 ft. in length for 1/2-in. rope and increasing to 30 ft. for the larger sizes. Where the splicing must be done in position, rope blocks are used to draw the wire taut, as in Fig. 1, care being taken to make fast far enough from the ends to leave plenty of room for the splice and the men who make it. If possible, it is better to hold the rope taut, mark the splice on both ends, by securely winding with No. 20 annealed-iron wire, throw it off the sheaves and make the splice on the floor or staging, as may be most convenient. The strands of both ends are unlaid, back to the points wound with wire, the hemp core cut off and the ends of the rope brought together with the strands interlaced, as shown in Fig. 2. Any strand, as *a*, Fig. 3, is now unlaid and closely followed by the corresponding strand 1 of the other end of the rope which is pressed closely into the groove left by the unlaid strand. The unwinding of one strand and the inwinding of the other are continued until all but about 12 in. of strand 1 is laid in, when *a* is cut off at the same length and both strands securely tied with cord. Strands 4 and *d* are next treated in the same way and the process is repeated with each pair of strands until all are laid and cut, the projecting ends being tied as shown in Fig. 4 to prevent unwinding. When this has been done the splice is bent and worked in all directions until the tension in all the strands is equal and the rope as flexible there as elsewhere. If this is not done and there is more tension in some of the strands than in others when a stress is put on the rope, these strands will pull into the rope, making a bad looking and weak splice. Next, the open or free ends of the 12 strands are carefully trimmed and wound with fine wire. Two rope-and-stick clamps, Fig. 5, are now secured to the rope, one on each side of an end crossing, as in Fig. 8, for the purpose of untwisting the rope to allow tucking the strand ends into the middle of the rope. There are two ways of tucking in these ends. They are first straightened with a mallet; the long ends of the rope-clamp handles are twisted in opposite directions, separating the strands and exposing the hemp core, which is cut off and pulled out between the points to which the tucked-in strands will reach and the ends forced into the place formerly occupied by the core. This is most easily done with the aid of a marlin spike, which is passed over the strand that is to be tucked and under two strands of the rope, Fig. 6, and moved along the rope spirally following the lay and forcing the free end, as shown in

Fig. 7, into the core space. In the other method the strands are more widely separated by untwisting the rope with the clamps, Fig. 9, slipping the free end between the strands and correcting slight kinks by the use of a mallet. The order in which the ends are tucked in is immaterial. Some operators prefer to

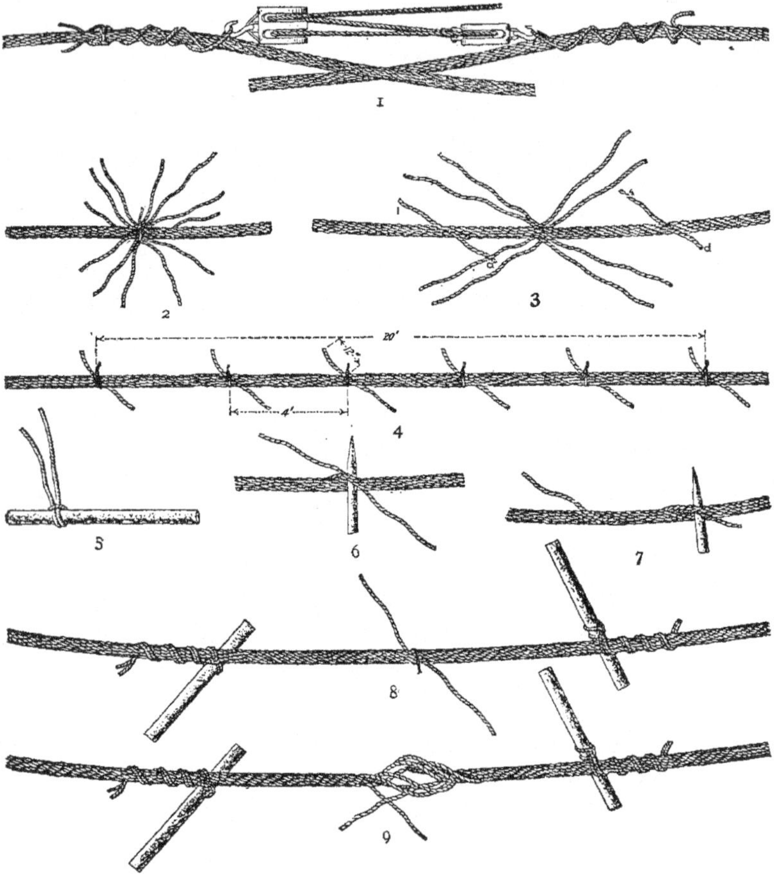

FIG. 6.—METHODS OF SPLICING WIRE ROPE.

tuck all the ends pointing in one direction before any of those pointing the opposite way, while others finish each pair of ends in series. If the foregoing directions are intelligently followed the splice will be uniform with the rest of the rope, of nearly equal strength throughout, and after a few hours' use it will be almost impossible to detect the splice. Four-strand hemp ropes can be spliced in the same way, this splice being known among riggers as the long splice.

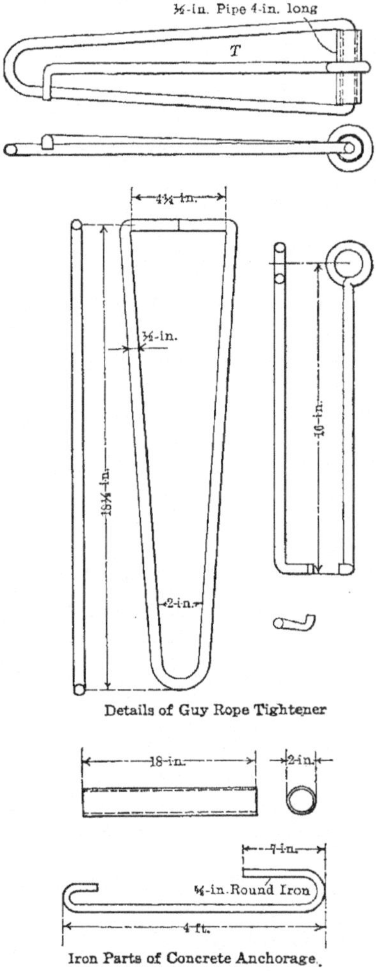

FIG. 7.—GUY-ROPE TIGHTENER AND ANCHORAGES.

A Guy-rope Tightener.—In Fig. 7 are shown the details of a guy-rope tightening device which is used in the southwestern Missouri zinc district. The tightening device resembles a buckle in construction. It is made of 1/2-in.

wrought iron, with a piece of pipe over the broad end of the buckle-shaped link so as to allow the tongue to rotate readily without much friction. This tightener is fastened to the anchor of 5/8-in. wrought iron after the guy rope has been secured to the tongue of the tightener. By using the tongue as a lever, the guy line is wound around the broad end of the buckle until the proper tension has been obtained. Then the tongue is locked by hooking its end over the buckle part of the tightener. The iron parts for a concrete anchorage in which a pipe cross piece is used are also shown.

MISCELLANEOUS NOTES

Disposal of Waste.—In the stripping of the large ore deposits on the Mesabi range, the disposal of the overburden is one of the large problems that has to be solved. In most cases this waste material is hauled from 1 to 2 miles in side-dump cars and emptied over the side of a dump, 50 to 100 ft. high. The shifting of the standard-gage track to keep up with the growth of the dump is a matter of considerable expense. In nearly all cases this shifting is accomplished by hand labor. At one or two mines it is done with the wrecking car by simply attaching four chains to 30-ft. sections of track, lifting them bodily and setting them over the required distance. At one mine a track shifter is being tried. Where there is ample room and no danger of damage suits arising, the latest device is to use water to wash the material from the railroad tracks. The necessary requirements are a high bank upon which to build a 15- or 20-ft. trestle, and a deep gulch in which to deposit the dirt. A substantial trestle is constructed and this is filled up to the ties with dirt from the dump cars. Just beneath the ties and on the outside row of posts is placed a 6-in. water pipe with 1/2-in. holes in one side, 18 in. apart. These holes all open out from the track. The bottom of the trestle is buried sufficiently deep to hold it in place and there is little danger of its washing out. The material is dumped over the side of the trestle from the cars and the water plays upon it, washing it down the slope to the low ground. In this way the angle of repose for the dirt will be 10 or 15° instead of approximately 40°. In the case under consideration the track is about 75 ft. above the side of a small lake. This flat alluvial fan will extend out a long distance and add largely to the capacity of the dump. In this way a solid track can be maintained and it need only be moved once in many months, whereas, in the usual scheme it is necessary to shift the track almost daily. The shifting operation is expensive and in addition the track is always in bad condition.

Mine Tailings for Filling (By Lucius L. Wittich).—While the use of mine tailings as railroad ballast is not new, the method employed by the St. Louis & San Francisco R. R. Co., west of Joplin, Mo., is, I believe, novel. The tailings were introduced into abandoned and dangerous zinc and lead workings directly beneath the railroad right-of-way through four 8-in. drill holes put down from the surface to penetrate the drifts, 140 ft. beneath. Hopper-shaped excavations were made at the top of each drill hole, the holes having been sunk between the rails of the track about 20 ft. apart. More than 16,000 cu. yd. of mine gravel were used in filling the old drifts, the greatest volume put through a single hole having been in excess of 9000 cu. yd. That the tailings might be equitably distributed and made compact, streams of water from two 2-in. pumps were poured into the hoppers, the water carrying the tailings through the drill holes and spreading them through the drifts. As a precautionary measure, to prevent the tailings spreading over too wide an area, the ground of the old workings, at certain points where it connected with other drifts, was shot down and these barriers acted as walls which stopped the promiscuous distribution of the waste chat, but which were porous enough to permit the big volume of water escaping into other drifts, from which it was pumped again through the pumps that had been stationed in the mine.

Strength of a Mine Dam.—The total pressure exerted by water on a dam is found by multiplying the wetted area of the face of the dam, expressed in square feet, by 62 1/2, the weight in pounds of 1 cu. ft. of water, and this product by the vertical height of the surface of the water above the center of gravity of the wetted area, says *Coal Age*. To calculate the required thickness of a dam, let

t = Thickness of dam (in.);
r = Shorter radius of dam (in.);
w = Width of opening or span (in.);
p = Pressure of water at dam (lb. per square inch);
S = Compressive strength of material (lb. per square inch).

For an arched dam

$$t = \frac{pw\sqrt{4r-1}}{4S}$$

The shorter radius of the dam should be from one-fourth to one-third greater than the clear span or width of opening. For practical reasons, it would not be advisable to build a dam less than 20 in. thick.

II

EXPLOSIVES

Blasting and Handling of Dynamite—Storage of Explosives—Frozen Dynamite.

BLASTING AND HANDLING OF EXPLOSIVES

"Don'ts" in Using Explosives.—The following "Don'ts" were embodied in a lecture on explosives, delivered recently by A. E. Anderson before the mining men of Telluride, Colo. While many of them are by no means new, occasional repetition is not undesirable; if properly observed, fewer dispatches headed "Blown up While Thawing Frozen Dynamite," etc., would be seen.

Don't forget the nature of explosives, but remember that with proper care they can be handled with comparative safety.

Don't smoke while handling explosives, and don't handle explosives near an open light.

Don't shoot into explosives with a rifle or pistol, either in or out of a magazine.

Don't leave explosives in a field or any place where cattle can get at them. Cattle like the taste of the soda and saltpeter in explosives, but the other ingredients would probably make them sick or kill them.

Don't handle or store explosives in or near a residence.

Don't leave explosives in a wet or damp place. They should be kept in a suitable, dry place, under lock and key, and where children or irresponsible persons cannot get at them.

Don't explode a charge to chamber a bore hole and then immediately reload it, as the bore hole will be hot and the second charge may explode prematurely.

Don't do tamping with iron or steel bars or tools. Use only a wooden tamping stick with no metal parts.

Don't force a primer into a bore hole.

Don't explode a charge before everyone is well beyond the danger zone and protected from flying débris. Protect the supply of explosives also from danger from this source.

Don't hurry in seeking an explanation for the failure of a charge to explode.

Don't drill, bore or pick out a charge which has failed to explode. Drill and charge another bore hole at least 2 ft. from the missed one.

Don't use two kinds of explosives in the same bore hole except where one is used as a primer to detonate the other, as where dynamite is used to detonate Judson powder. The quicker explosive may open cracks in the rock and permit the slower to blow out through these cracks, doing little or no work.

Don't use blasting powder, permissible explosives or high explosives in the same bore hole in coal mines.

Don't use frozen or chilled explosives. Most dynamite, except Red Cross, freezes at a temperature between 45° F. and 50° F.

Don't thaw dynamite on heated stoves, rocks, sand, bricks or metal or in an oven, and don't thaw dynamite in front of, near or over a steam boiler or fire of any kind.

Don't take dynamite into or near a blacksmith shop or near a forge on open work.

Don't put dynamite on shelves or anything else directly over steam or hot water pipes or other heated metal surface.

Don't cut or break a dynamite cartridge while it is frozen, and don't rub a cartridge of dynamite in the hands to complete thawing.

Don't heat a thawing house with pipes containing steam under pressure.

Don't place a hot-water thawer over a fire, and never put dynamite into hot water or allow it to come in contact with steam.

Don't allow thawed dynamite to remain exposed to low temperature before using it. If it freezes again before it is used it must be thawed again.

Don't allow priming, the placing of a blasting cap or electric fuse in dynamite, to be done in a thawing house or magazine.

Don't prime dynamite cartridges or charge or connect bore holes for electric firing during the immediate approach or progress of a thunder storm.

Don't carry blasting caps or electric fuses in wearing clothes.

Don't tap or otherwise investigate a blasting cap or electric fuse.

Don't attempt to take blasting caps from the box by inserting a wire, nail or other sharp instrument.

Don't try to withdraw the wires from an electric fuse.

Don't fasten a blasting cap to the fuse with the teeth or by flattening it with a knife; use a cap crimper.

Don't keep electric fuses, blasting machines or blasting caps in a damp place.

Don't attempt to use electric fuses with the regular insulation in unusually wet work. For this purpose secure special waterproof fuses.

Don't worry along with old, broken leading wire or connecting wire. A new supply won't cost much and will pay for itself many times over.

Don't handle fuse carelessly in cold weather, for when cold it is stiff and breaks easily.

Don't store or transport blasting caps or electric fuses with high explosives.

Don't store fuse in a hot place, as this may dry it out so that uncoiling will break it.

Don't "lace" fuse through dynamite cartridges. This practice is frequently responsible for the burning of the charge.

Don't operate blasting machines half-heartedly. They are built to be operated with full force. They must be kept clean and dry.

Don't cut the fuse short to save time. It is dangerous economy.

Don't expect explosives to do good work if it is attempted to explode them with a detonator of insufficient strength.

Preparations for Blasting (By M. T. Hoster).—In almost every one of the various mines and mining districts of the country some peculiarity in the procedure of the miner in firing his round of holes may be observed. The necessary operations are discussed under the subheads following, and the best practice outlined.

Cutting the Fuse.—Fuse should never be less than 5 ft. in length and often 6 or 7 ft., depending on the number of holes, whether in a shaft, raise, drift or stope. It should always be cut straight across (not slanting) with a sharp knife or cleaver, for if the fuse is not so cut, some of the powder may fall out or the pointed end may bend over and seal the powder from the fulminate when the fuse is pushed into the cap.

Crimping.—Crimping the cap on the fuse is usually done with a crimper, the teeth or a knife. Crimping caps with a knife or the teeth is not only dangerous but ineffective and is often responsible for expensive misfires. With the thin-jawed crimper, the cap is grooved on the fuse, the great disadvantage of this being that such a sharp groove may often squeeze the shell into the fuse so closely that the powder train is choked or cut, causing a missed hole. Two or more such grooves may be lightly pressed on each cap, but water may leak in as the cap does not fit the fuse closely.

The broad-jawed, or California crimper is far better, as the metal is not grooved and still fits the fuse closely. With the California crimper, the cap should be pressed several times, revolving the fuse and cap slowly each time the jaws open, squeezing lightly at first and stronger as the cap contracts about the fuse. This crimp gives a water-tight fit. Many missed holes are unquestionably due to imperfect crimping; hence this operation should be left to a reliable man and not merely to the miner.

The Primer.—At times, one will see a miner loading his holes by tamping in a stick or two of powder, then inserting the fuse and tamping in more powder, the fuse lying between the wall of the hole and the powder. This is inefficient. A primer should always be prepared beforehand. The best two methods of making a primer are: (1) Open up one end of the cartridge, extract a small amount of powder, press a hole into the center of the stick of powder and insert the fuse. Then twist the paper cartridge about the fuse and tie it tightly with cord. For wet holes the space at the top of the cartridge is often filled with grease. (2) Simply stick a hole into the cartridge and insert the fuse. This method does not require as much time and is not as good as the first, but will give satisfactory results. This primer must be handled with care to prevent the fuse from coming out and should always be inserted into the hole fuse-end first, which necessitates a bend in the fuse and there is a possibility of this nicking the fuse and causing a misfire.

Greasing.—For wet holes the fuse is often greased with axle grease or crude oil. Some miners rub grease into the entire fuse, but this is not necessary, as any good fuse will withstand water for a few hours at least. If grease is used at all it should be rubbed on the fuse only where it enters the cap so as to prevent water from soaking into the cap at that point.

Loading the Holes.—For dry holes, all of the paper cartridges, excepting the primer cartridge, should be slit with a knife before placing in the hole. The object of this is to give the powder a chance to spread out when tamped and so fill the entire hole. For wet holes it is best to slit the first two cartridges only, as the water will fill up any space not taken up by the remaining powder. The most effective place to have the powder is at the bottom of the hole, hence the object of slitting at least the first two sticks. In loading a hole with, for example, five sticks of powder, put in the first two sticks, tamping each solidly into place. Then put in the primer, tamping but lightly, and finally the last two sticks of powder, which should be pressed in well. Some miners say to load the primer last, but this hardly seems advisable. Never load the primer first, and always use a wooden tamping stick.

Tamping with Clay or Mud.—In some districts soft clay or mud is tamped into the hole after the powder has been loaded but whether there is any good resulting from this tamping is doubtful. It is often advantageous to force a little earthy matter into a hole in order to prevent the sparks of some nearby hole from igniting the powder or to prevent the powder in a dry hole from falling out, but otherwise the force of the explosion can hardly be affected by this tamping. If a hole is tamped at all it should never be filled to the collar, as this makes it difficult to discover a missed hole.

Splitting and Spitting the Fuse.—Each fuse to be spit should be cut open at or near the end so as to expose sufficient fuse powder for igniting. The three general ways in which this may be done are: Slice the last inch of the fuse in half; cut a slit in the side of the fuse; fork open the end of the fuse. The first is a poor way, as the powder exposed may all fall out before it is time to spit. The second method is much used as there is the least chance of the powder falling out and it requires little time. The chief disadvantage to this method is that when spitting, the fire may travel away from instead of toward the cap, and the miner, being in a hurry, may not notice this. The last method is undoubtedly the best, for although some of the fuse powder may fall out, enough will always be left where the arms of the fork meet. For very wet places the second method has an advantage over the third in that the powder is not as likely to become wet.

The old and still much used method of spitting by use of candles or miners' lamps is slow, inefficient for a large number of holes, and dangerous. Spitting one fuse by candle, the second fuse from the first, the third from the second, etc., is good but can be used only when the holes are close together. The advantage of this method is that, when through spitting, the miner is certain that all the fuse was spit properly. A more common, and far better, scheme is to use an extra piece of fuse about 18 in. long, which is slit or notched along one side, the notches being about 1 in. apart. When ready, light one end of this fuse, and as the fire travels along the fuse it will spit out at each notch in succession. As the fire spits from the first notch, ignite the first fuse, with the second notch the second fuse, etc., the advantages being that there is the same interval between the lighting of each fuse and the method is rapid, safe and easy.

Table for Cutting Fuse.—The accompanying drawing, Fig. 8, shows the design of a table for conveniently cutting fuse. It is used at the Blackberry and

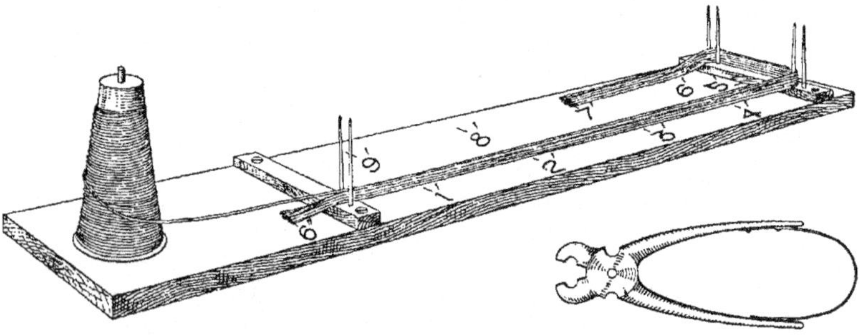

FIG. 8.—FUSE TABLE AND CAP CRIMPER USED IN JOPLIN DISTRICT.

Montana mines, near Joplin, Mo. The coil of fuse is carried on a conical spindle at the end of the table, and from this spindle the fuse is unwound as it is measured. The fuse is held between a series of two 60-d. nails driven into two cleats of 1×2-in. wood that are screwed to the table. The cutting is done 6 in. ahead of the first pair of nails, so that there are no long, loose ends in the way. On the table are marks which are used in measuring the several lengths of fuse usually required.

In cutting the fuse, the loose end from the coil is run through the first pair of nails, taken around the turning nails, passed through the last pair of nails that hold the loose end when the fuse is being cut, and the end is pulled along to the mark indicating the length desired. Then with the fuse cutters the fuse is cut off at the other end, and another piece of fuse measured until all the fuses have been cut, the different pieces being left securely held between the nails until the miner is ready to cap them. To aid in the cutting, the fuse cutters, which are ordinary crimpers, are made with a spring riveted to one of the handles, so as to keep the jaws apart. This increases the speed of cutting the fuse somewhat, and makes the use of the crimpers more convenient.

Blasting in Wet Shafts (By E. M. Weston).—It frequently happens, especially when shaft sinking with machines is in progress in wet shafts, that the pumps have to be withdrawn before blasting. This may mean that the water has risen to a height of several feet above the bottom of the shaft before the holes can be fired.

If fuses are used they should be well greased just where the detonator has been placed on one end and inserted in the primer. It is true economy to use two fuses and two detonators in the leading or cut holes of the round thereby lessening any risk of the whole round being hung up by misfires. All fuses should be of the same length, from 6 to 12 ft., depending upon the number of holes to be fired. In large shafts on the Rand 12-ft. fuses are used. As soon as the holes are charged the ends of the fuses are tied to a small plank or a wedge that has been used to rig the base for the machines, 6 in. or more of the end of the fuses projecting above the plank and as the water rises all fuse ends float.

The miners work in pairs lighting the fuses from the back of the shaft toward the bucket or skip in the center. "Cheesa sticks," as they are called locally are made by splitting blasting gelatine and wrapping it around pine sticks about 18 in. long. In a wet shaft the fumes given off, though poisonous, are rapidly absorbed by the water and the use of such sticks in shaft sinking is justified. One miner carries the lighted stick and the other cuts each of the back rows of fuses about 1 in. from its end and rapidly bends it back, while the other miner applies the stick before the composition becomes damp. When the fuse is seen to spit the next one of the row is lighted. If the fuse does not take fire after the first cut has been made, another cut is made, and no fuse is left until it is seen to spit. The next rows are dealt with in the same way, except that the fuses are cut several inches further down so that each row nearer the center of shaft will explode before those first lighted.

If considered more convenient the reverse order of lighting may be adopted, but the advantage of lighting the back rows first is that the men are near the bucket when all the fuses have been lighted; they are not obliged to walk over lighted fuses to reach the bucket. In this way fuses can be lighted with certainty even in the wettest shafts and it is worth while at mines where lower-grade explosives are used to have some blasting gelatine on hand with which to make these lighting sticks, as gelignite and dynamite are too granular and brittle for the purpose. Substitutes for "cheesa sticks" made of chemical combinations and designed to give off no carbon monoxide or nitrous oxide are being introduced with more or less success for general mine work on the Rand. I have not however, heard that they have been used successfully in wet-shaft sinking.

Blasting in Wet Ground. Where a blast is to be fired in wet ground, soap or tallow should be smeared over the safety fuse at the place where it enters the blasting cap in order to keep the charge in the latter perfectly dry. Oil or grease should never be used for this purpose as they are likely to soak into the fuse and destroy the efficiency of the powder which it contains.

The Calumet System of Lighting Fuse—There are many different ways of lighting blasting fuse, but one of the best, when there are not too many to be lighted, is the method used in mines of the Calumet & Hecla company at Lake Superior. This method consists in taking the end of the fuse sticking out of the hole and plastering it to the wall with a piece of moist clay, a spot being chosen so that in case there is a draft the candle used to light the fuse will be sheltered from the air current. A snuff of a candle about an inch long and having a wick about 1/2 in. in length is then placed in the clay so that the flame will reach the fuse. The candle snuff is then lighted and allowed to burn for about 1/2 minute, or long enough for the tar in the fuse covering to begin to bubble out. The candle is then blown out and the fuse to the next hole is prepared in a similar manner. This fuse is heated, the candle in turn is put out and the procedure is repeated until all the fuses are ready. Then the miner relights the snuffs and hurries away. As it takes a minute or so for the candle snuffs to burn through the fuse far enough to reach the powder train, the miner has ample time to get to safety, while there can be no failure of the fuse to light and there is no necessity for the miner to linger at the breast or face after the holes are spit, as sometimes a miner must do when the powder has become dislodged from a split fuse, and it becomes necessary to cut it again. Many miners lose their lives by lingering in just this manner. The fuses can be cut so that the holes will explode in any order desired, just as when the fuse is lighted in some other manner. The time required for the flame to burn through the wrapping seldom varies enough to cause the holes to explode out of the order desired. There is a limit to the number of holes that can be fired by one man in this way. It is perfectly safe for a man to light as many as 10 fuses in this manner, but as there is a certain time consumed in relighting the snuffs, it is debatable whether it is not safer to use some other method when more than that number of holes has to be fired by one man.

The method is good for firing shots in wet ground, provided that the snuffs can be protected from the drip of the water. When that cannot be done the same principle of firing—burning through the covering to light the powder train in the fuse—can be utilized by putting the fuses, several in a bunch, on piles of oily waste and lighting the waste.

However, when spitting his shots in the ordinary manner, a miner has had to recut the fuses, he should never, as a last resort, hold the fuse in the flame of his light in the hope that by setting fire to the end of the fuse he can avoid a misfire, for if he has to run before the powder train catches, the covering may smolder for many minutes before the powder is reached. The miner, thinking that the fuse did not ignite, returns to the face. In this way many men are injured in blasting. The snuff method of firing the holes avoids all such dangers, as the flame of the snuff soon burns through the covering to the powder. The important point in the Calumet system of firing is that there can be nothing that will cause the miner to linger at the face after once he has started to light his shots.

Prevention of Drilling into Misfired Holes (By John T. Fuller).—A common cause of accident in shaft sinking is drilling into a misfired hole or striking a pick into a misfire while mucking. Such accidents are due frequently to lack of information as to the exact situation of the holes of the blasted round. There are several ways of lessening the danger. One is to have each shift charge and blast the holes drilled by the preceding shift. This second shift then mucks out and drills the holes for the next shift to blast. A second way is to have each shift drill, blast and muck its own round, leaving the shaft clean

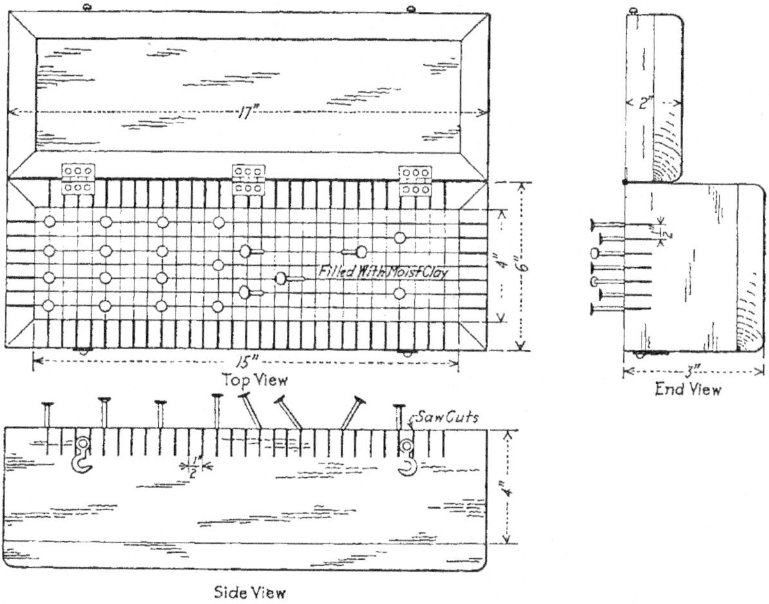

FIG. 9.—CLAY-FILLED BOX WITH NAILS, SHOWING POSITION OF DRILL HOLES.

for the next shift. Both of these methods, however, involve more or less of an upset of the usual order of doing things. The usual order is to have the shift coming on muck the rock broken by the preceding shift, then drill and blast a round of holes, leaving the muck in turn for the succeeding shift.

In a large shaft, sunk under my direction, the device herein described was used to aid in locating the holes drilled by the previous shift. A box was made of the same shape as the shaft and built to a convenient scale, as shown in Fig. 9. It was about 4 in. deep, and along the edges saw cuts about 1 in. deep were made, the distance between each representing 1 ft. of length or width of shaft. A substantial cover with hooks to fasten it securely shut was also provided. This box was filled with moist clay, leveled off even with the edges and lines scribed across the surface of the clay by a pocket knife and a straight-edge, using the saw cuts as guides. The box of clay thus represented the shaft bottom, subdivided into 1-ft. squares. When the miner went down to charge the holes he carried this box with him; also a pocketful of 20-d. wire nails. Before charging a hole, he would thrust his tamping stick to the bottom, which enabled him to gage the direction of the hole. He would then thrust a wire nail into his clay model as closely as possible to the direction and position of the hole in the shaft bottom. The squares greatly aided properly locating the holes in the model. Thus, when the miner had completed charging and had lighted the round, he returned to the surface with an almost exact reproduction of the layout of holes.

The box, with its record, was turned over to the miner in charge of the succeeding shift, with such other information as could be gathered from listening for, and counting the reports, etc. When the men went down to muck out after the smoke had cleared, they took the box with them, using it as a guide in directing the mucking and in locating any possible misfires after the shaft had been cleaned out and before work was started on the new round of holes. It was then a simple matter to withdraw the nails, smooth over the clay and prepare the box for the next record. Not a single accident due to drilling or mucking into misfired holes occurred in this shaft, although it was about 30×8 ft. and was sunk to the 1000-ft. level.

Cartridges for Tamping.—The importance of tamping dynamite in drill holes, especially the lower grades, is rapidly becoming recognized in the mining practice of the United States. In many instances the powder companies furnish paper cartridges at small cost for incasing the tamping, and in some mines such casings are used. This, however, is not common practice. The general practice is to use fine material from the vein for tamping. This is far better than using no tamping at all, but the sharp pieces of rock that such material usually contains are likely to cut the fuse. In the copper mines of Lake Superior, the miners engaged in drilling and breaking work by contract, and pay for the powder they use. They have an excellent method of making tamping cartridges. The tamping is clay sent down from the surface that is moistened and mixed with fines from the lode, so that the mass is coherent when tightly packed.

The casing is made from ordinary newspaper, according to two methods, in one of which the newspaper is sealed with candle grease, while in the other no sealing is used. By either method a good cartridge is obtained, but the better method is that in which the paper of the casing is sealed with candle grease. In both methods the newspaper, in single thicknesses, is wrapped tightly around a piece of old shovel-handle. When candle grease is used, the paper of the cartridge ends in a straight line running lengthwise with the piece of shovel-handle, about which it has been wrapped five or six times. The outside edge of the paper is sealed by dropping candle grease on it and pressing the paper together while the grease is hot. About 1 or 1 1/2 in. of paper extends below the piece of shovel-handle. This is folded up and sealed back on itself by hammering it on a piece of wood after some candle grease has been dropped on it. The shovel handle is then removed, the cartridge filled with tamping, and the other end sealed in a similar way after the tamping has been compacted by tapping the bottom end of the cartridge lightly on a board. Fig. 10 shows the method of making the cartridge. In the other system no candle grease or other sealing

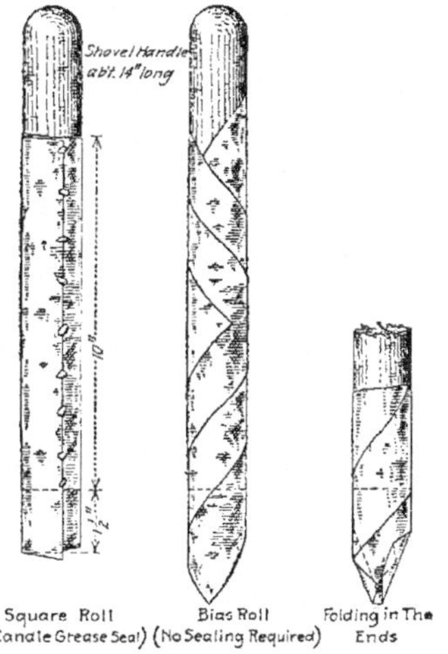

FIG. 10.—METHOD OF MAKING TAMPING CARTRIDGES.

substance is used. The cartridge is not so strong; yet is plenty good enough for the purpose. More paper has to be used, as the piece of newspaper is wrapped about the shovel-handle 8 or 10 times. The lower end of the cartridge is sealed by folding the paper back upon itself and hammering it vigorously on a piece of board so as to mat it together. The seal is not good, but the cartridge can be removed, tamping put into it and compacted by tapping the bottom on a board. Then the top end is sealed by folding it in several times on itself and compacting the paper of the end by pounding it against a board. Such cartridges are strong enough to stand any ordinary usage in loading holes.

Another device is illustrated in Fig. 11. It is made by threading one end of a 9-in. piece of brass tubing of suitable diameter. The bore of the tube is fitted with a plunger tapped to take a 10-in. rod about 1/4 in. diameter. The other end of the rod is threaded and attached to a handle by two flat nuts as shown.

The threaded end of the brass tube is closed by a cap through which the plunger rod passes as shown in the sketch. A hole 1/4 in. diameter is drilled in the side of the brass tube 1 in. below the cap. This tube, a small planed board, a few newspapers, and a batch of moist, fine dirt are all that is required to make neat cartridges that will give no trouble in tamping a drill hole. The newspapers are cut into strips 10 in. wide by 8 in. long. The plunger of the molding tube is withdrawn to the cap end and the bore is packed with moist dirt, the 1/4-in. hole at the cap end permitting the air to escape. When the bore has been filled the tube is stood upright upon the planed board and the plunger is pressed down to compact the dirt into a cylinder; the tube is then laid upon its side and the cylinder removed from the tube by pushing on the plunger handle, so that the ejected cylinder of dirt lies upon a piece of the newspaper laid smoothly upon the planed board. To complete the cartridge it is only necessary to roll it up in the paper and fold over the ends. These tamping cartridges should be carefully placed side by side in an empty powder box as soon as made, and

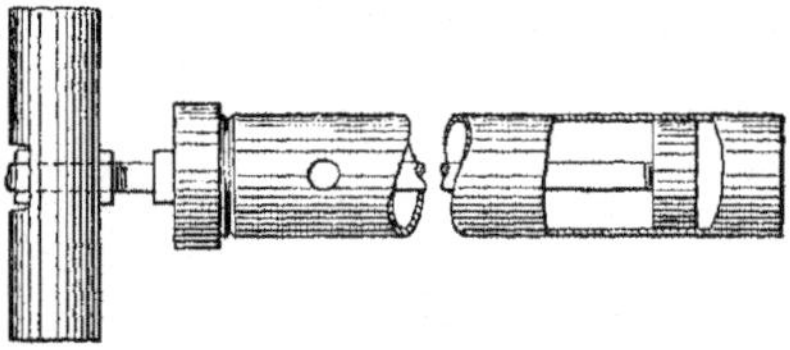

FIG. 11.—DEVICE FOR MOLDING TAMPING CARTRIDGES.

when needed, the box of cartridges is carried to the place where blasting is to be done. As all the cartridges are of uniform size no trouble will be experienced in placing them in the drill holes if the tube is of the proper diameter for the size of hole in which the cartridges are to be used.

In the diamond mines at Kimberley, South Africa, according to John T. Fuller, tamping cartridges are made by native boys delegated for this purpose. The cartridges are made of cylinders of mud covered with the paraffin paper in which the bundles of dynamite sticks are originally wrapped.

Use of High Explosives.—At a great many mines and quarries a prejudice exists among the miners against the use of dynamite of a higher grade than 40 %, this regardless of the hardness of the rock mined. In mining a moderately hard and easily fractured material 40 % explosive gives good results, and has the advantage of being comparatively safe. However, even 40 % dynamite is not a substance with which anybody can afford to be careless. If a man is careful and uses ordinary intelligence in handling dynamite, 60, 80, or even 95 % blasting gelatin can be handled with a reasonable degree of security. In many of the mines of the West where ore with a hard, tough quartz gangue is encountered, 60 and even 80 % dynamite has proved far more satisfactory than the lower grade material, giving a broken product which can be easily shoveled in the cars without previous cobbing or popping. In several instances, nitroglycerin and even guncotton have been used with success.

The pyrrhotite ore encountered in the mines of the Tennessee Copper Co. is extremely hard and tough in many places. In the Eureka open-cut mine, where the ore is very tough and cemented with a blue quartz gangue, the company is experimenting with 95% blasting gelatin manufactured by the Du Pont Powder Co. As far as can be seen from the work already done, this explosive is giving excellent results. Owing to the prejudice of the miners against high explosives, it was introduced at first without telling the miners of its grade. Now, however, after the men have become accustomed to using it, they feel no more fear of it, seemingly, than they do of the ordinary 40 or 60% grade dynamite. It has been found to shatter the rock much more than a lower grade powder, and about eight or nine sticks of the gelatin will do as much work or break down as much rock as nearly double that many sticks of 40% dynamite.

In the underground workings, objection is made to the use of gelatin even where the men are not afraid of it, as they find it will not stay in upcast holes. In tamping it acts like so much rubber and unless some paper is stuffed in the mouth of the hole it is, very likely to roll out, this of course being a source of danger.

The decision as to the grade of dynamite to be used should depend upon the results obtained. In a locality where labor is expensive, it may be more economic to use high-grade explosive on account of the labor saved from blocking or sledging the ore shot down, whereas, with similar ore, in another region where cheap labor is available it may prove cheaper to use a lower grade of dynamite and break the large blocks of ore by hand. Each superintendent must determine the grade of explosive best adapted to his conditions.

Joplin Scraper and Loading Stick.—The ore in the Joplin mines is a mineralization of stratified beds. The roof is good and in stoping a system of drilling is used, in which a bench is taken up by a series of flat holes after the ore next the roof has been blasted out in a heading. In this bench work some of the deepest drilling in underground metal mining in the United States is done. Holes 18 ft. deep are not uncommon, 16-ft. holes are common, and 12- and 14-ft. holes are typical practice in the bench portion, which is called the stope.

To break the heavy burdens, it is necessary to chamber the bottom of the flat holes, and because of the rapid wearing of the bits, it is often necessary to blast the hole for gage, so that one bit may follow another. This springing of the hole is called "squibbing." The squibbing makes the bore of the hole ragged, therefore difficult to load. This and the occurrence of cavities in the ore, through which some of the holes pass, makes "railroading" of the powder to the bottom of the hole ahead of the loading stick impossible. A pointed tamping stick is therefore used. Some sticks are pointed with a piece of copper wire; others with a copper nail. The ground is flinty, so that the use of a steel wire nail in the tamping stick would almost surely be attended by sparking. A copper nail is sometimes used, but this is not considered good practice by the powder companies. The best practice is to use a tamping stick with a wooden point. In some of the sticks a point is made by sharpening the tamping stick itself, but it is preferable to insert a round piece of hardwood in an auger hole in the end of the stick and to sharpen this inserted piece of wood. The stick of powder is impaled on the point, so that it can be made to pass any irregularities in the bore and deposited at the bottom of the hole. This is important, especially in squibbing, as part of the hole may be lost if the squibbing-dynamite is not at the bottom of the hole when fired. The other end of the stick is used in tamping.

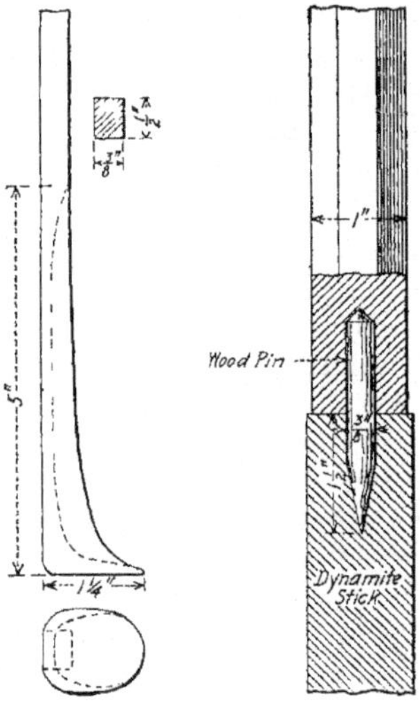

FIG. 12.—SCRAPER AND LOADING TOOL USED AT JOPLIN.

On account of the crumbling of the ore, scraping out the deep holes is a difficult operation, and it is, therefore, important that the machine man be able to keep the spoon of the scraper in its proper position in respect to the bottom of the hole. Instead of using round iron for making the scraper, a piece of $3/8 \times 1/2$-in. steel is used for the shorter, and $5/8 \times 1/2$-in. for the longer scrapers. The spoon is made about 5 in. long, with a nose nearly the diameter of the steel used in the hole. When the wide side of the scraper rod is horizontal, the nose of the scraper is turned either up or down in the hole. It is, therefore, possible to pull the scraper with its load out of a hole, while with a round handle, it is not uncommon to lose the load. The advantage of a handle of rectangular section is evident. Drawings of the scraper and loading tool used at Joplin are shown in Fig. 12.

Breaking Ground for Steam Shovels.—Two systems of blasting the ground to be excavated by steam shovels are used in the Mesabi district, Minn. The overburden to be blasted is glacial drift, while the ore is soft hematite. Where the bench to be broken does not exceed 20 or 25 ft. in height, the usual method is to drill holes 15 to 18 ft. from the edge and about 15 ft. apart, extending along the entire bench to be broken. A handle bar is fastened to the drill steel and two men operate the drill by means of this handle. The drill is lifted about 2 ft. and dropped by its own weight. The men walk around in a circle 3 ft. in diameter and in this way turn the drill. When the hole is the required depth (about 20 ft.) one or two sticks of 7/8-in. powder are lowered and discharged to spring the hole. The hole is again opened and 5 to 15 sticks of dynamite are placed in the bottom of the hole and discharged to spring the hole further, so that it will contain 10 to 15 kegs of black powder. When loading the hole with black powder, a stick of dynamite, in which two caps and fuses are inserted, or an electric fuse attached, is lowered in the hole and powder is filled around it. After the powder is in place, the hole is firmly tamped with sand. Each hole is discharged separately. During cold winter weather only one hole is discharged at a time, and the loosened material is moved before it has an opportunity to freeze. In the summer often 15 or 20 holes will be fired, one after the other, yielding enough loose material to last the steam shovel several days.

The other system of loading, called "gophering," is used when the ground is so dry or sandy that the vertical hole cannot be kept open, or when the bench is too high to drill from the top. In this method holes are drilled in from the side. The total depth is 20 to 25 ft. A pointed 1 1/4-in. drill bar is used. After it has been driven a few feet with hammers, the bar is withdrawn, and sticks of 7/8-in. dynamite are placed end to end in the hole and discharged. This loosens the ground and the dirt is then taken out by means of a long-handled shovel. In this way a hole 10 or 12 in. in diameter is opened. The hole is inclined at an angle of 15 to 20°. The lower end of the hole is sprung with 8 or 10 sticks of 7/8-in. powder and cleaned out. Ten to 15 kegs of black powder are then used. The black powder is fed into the hole by means of a box 3×3×15 in., nailed to a 22-ft. pole. Another method of loading is by means of a V-shaped trough made of 3/4×4-in. boards. The powder is poured into the upper end, and the trough given a backward and forward motion, and in this way the powder soon finds its way to the bottom of the hole. Two men can usually put in two of these holes per shift. If the ground is loose enough to work with the shovel, it is not necessary to spring the hole before the bottom is reached. When boulders are encountered, they are broken by discharging dynamite on their face. In the case of a boulder too large to break in this way, it is often necessary to start another hole.

The Necessity for Strong Detonators.—In detonating high explosives, the stronger or sharper the initial shock the quicker and more thorough is the detonation of the charge. If the detonation is slow and incomplete a greater quantity of explosive is required to do the same work, and large volumes of poisonous gas are evolved—a matter of serious consequence in underground work. Quick and complete detonation results in a minimum of flame, a point of first importance with those explosives intended for use in the presence of inflammable gas or coal dust. Electric fuses or blasting caps too weak to detonate a charge of high explosives frequently generate sufficient heat to ignite it. The effect of a detonator on a charge of high explosives in a bore hole is by no means infinite, but decreases with distance. It is, therefore, easy to understand the necessity for using detonators sufficiently strong for the effect of the detonator to extend as far as possible through the charge. It should not, however, be understood that the detonator should be placed in the center of the charge, for numerous tests have shown that the greatest effect of a detonator is straight away from its loaded end, and in a line with its long axis, *i.e.*, a detonator will explode a cartridge of dynamite farther away from it, if it is lying with the loaded end pointed toward the cartridge, than it will if it is lying parallel to the cartridge. It may be impossible to explain this, but it is known to be a fact. In deep bore holes loaded with long charges, it is well to place caps in cartridges of explosives at intervals of at least 5 ft. throughout the charge, so that the effect of the explosive material which they contain will extend the entire length of the charge.

Priming With Electric Fuse.—To prime a high-explosive cartridge for electric blasting the fuse cap should be inserted into the center of one end of the cartridge and pointed directly toward the opposite end. The two lead wires should then be brought together up one side of the cartridge and tied in place with string at points an inch or two from either end of the cartridge. The common practice of inserting the cap diagonally into the side of a cartridge and then looping the wires about the cartridge in several half hitches is to be condemned. In looping the wires, the insulation is likely to be broken, causing short circuiting or leakage of current in wet work; the wires may even be broken. The common practice, when the cap is pointed diagonally toward the end of the cartridge, is to place the cartridge so that the end of the cap will be nearest the outside or top of the charge. Any pull on the lead wires tends to swing the cap in a position more at right angles to the long axis of the cartridge. Thus the end of the cap may easily be swung entirely out of the explosive. In blasting, the principal part of the detonating charge should be placed in the center of the cartridge of explosive, and not to one side or entirely outside, against the paper. A great many missed shots are doubtless caused by improper priming.

Device for Clearing a Hung-up Chute (By J. Bowie Wilson).—All underground managers have at some time been worried by ore chutes choking and hanging up out of reach of the trammer's bar. When the material consists of fine clayey stuff the only remedy is practically to dig it out. If the material consists of rock, even if it contains a proportion of clay, and the block is due to the large pieces keying together and arching over in the chute, the following method of clearing the chute will be found better and certainly safer than attempting to shoot down the pass by tying pieces of dynamite to a tamping stick. This scheme is used at the Mount Morgan mine in Queensland, and consists in firing a wooden plug from a small cannon placed in the bottom of the chute. The cannon is made from a length of steel shafting, the center being bored out in a lathe and a touch-hole drilled large enough to take the ordinary fuse in use at the mine. The end of the shafting is turned down to fit into a hole in the top of a short length of 9×9-in. hardwood timber of such shape that when the device is laid on the floor of the chute and resting against the chute door, the cannon will point up the center of the raise. Fig. 13 shows the construction of the cannon. It is charged with ordinary black blasting powder and a plug of hardwood timber is tapped home in its mouth. A fuse is then inserted into the

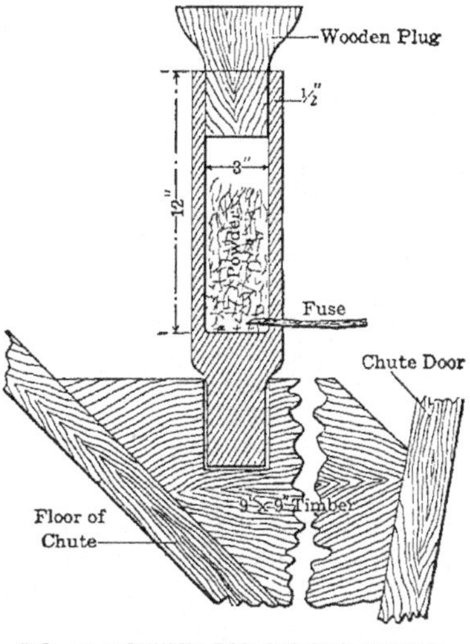

FIG. 13.—CANNON FOR OPENING CHUTES.

touch-hole to explode the powder. In action the cannon is a big popgun, and on the powder exploding, the wooden plug hits the keyed material with a good, sudden blow. If the shot is successful the material falls upon the gun and its carriage, but these, being of a simple design, are in no way hurt and can be recovered. A knock against the side of the truck suffices to clear the cannon of any material and it is thrown down in the tunnel beside the chute until required again. If the chute is not freed by the first shot it will generally be found to open after several. A great advantage of this method is that the men are in no danger when using it.

STORAGE OF EXPLOSIVES

Powder Magazines.—The mining law of the Transvaal is strict with respect to the construction and management of magazines for explosives. Some of the provisions are the following:

Care shall be taken that explosives magazines are erected only at such places where there is a suitable depth of soil or sand. In no case shall an explosives magazine be erected upon rocky ground. In the construction and erection of the magazines only the lightest and most suitable material available shall be used for walls and roofs. Solid arched roofs are particularly prohibited.

The magazine shall be at least 6 ft. 6 in. in the clear from floor to ceiling, and shall comprise at least two compartments, one called the lobby, which is accessible from the outside, and used for the reception and delivery of explosives; the other called the storage room, which is accessible only from the lobby and is used for the storage of explosives.

The outer door to the lobby must open outward, and be faced with sheet iron about 1/4 in. thick, and fitted with a good lock, for which purpose a padlock will not be deemed to be sufficient.

The compartments of a magazine shall be properly ventilated by cowled ventilators in the roof, or properly protected ventilating channels in the gables. The highest temperature allowed in the storage room shall not exceed 95° F.

The ceiling of a magazine shall be of wood, and its inner sides shall be wood-lined, the lining to be at least 3 in. distant from the walls and the intervening space filled with some uninflammable non-heat-conducting material. The floor shall be of wood, of sufficient strength, and shall be well ventilated beneath; it shall also be provided with a proper drain for insuring the dryness of the magazine. Roofs of galvanized iron shall have wood lining immediately against the iron. All nails, fastenings, locks, keys and fittings inside the magazine shall be made of wood, brass or copper.

The magazine shall be fitted with a reliable lightning conductor, supported on a vertical post standing clear of the building, but not more than 18 in. from one of the walls, and rising at least 6 ft. above the highest point of the magazine. This lightning conductor shall be carried to a properly laid earth plate.

A surface or sub-surface magazine shall be surrounded by an outer earth wall, and the bottom of the inner slope of the same shall not be less than 3 ft. from the sides of the building. This earth wall shall have a natural slope on either side, and be 3 ft. wide at the top, and as high as the highest point of the roof. The approach to the precincts of the magazine, through the outer earth wall, shall have a strongly built gate, which shall be kept locked. The entrance to the magazine shall be either in a broken line, or the door shall be protected by an outer protecting earth wall entirely shielding the entrance.

A reliable self-registering thermometer shall be kept in the storage room of every explosives magazine. At least one pair of magazine shoes shall be kept in the lobby of every magazine, and no person shall enter the storage room of any magazine except when wearing such shoes or when barefooted.

Powder House with Concrete Roof (By Claude T. Rice).—A powder house with a concrete roof, such as is used by the Copper Range company's mines in Michigan, is shown in Fig. 14. In designing the roof the weight of 150 lb. per cubic foot of concrete and 31 lb. pressure per square foot for wind, snow and other loads were taken, and the ordinary force diagram for designing

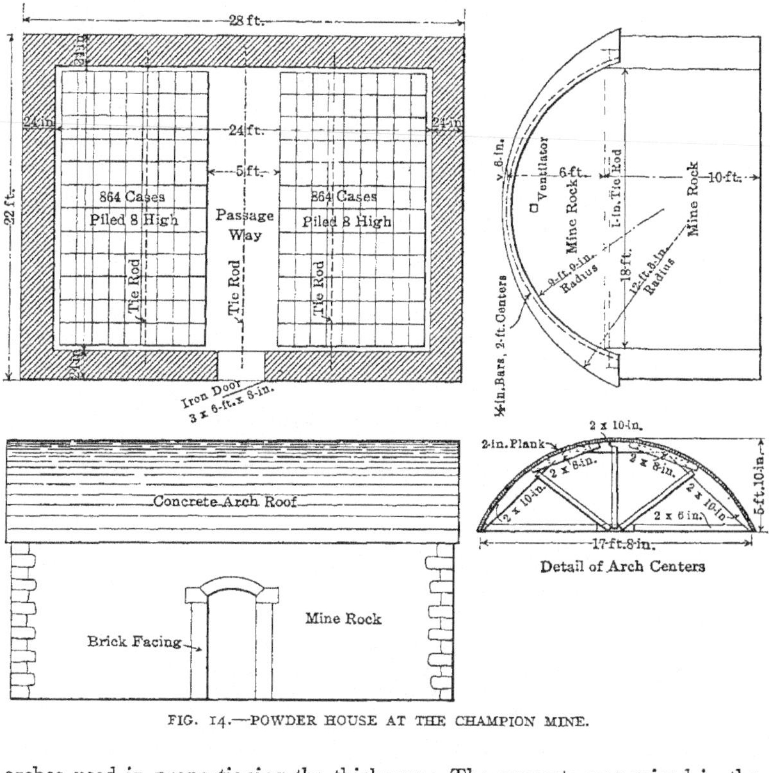

FIG. 14.—POWDER HOUSE AT THE CHAMPION MINE.

arches used in proportioning the thickness. The concrete was mixed in the proportions 1 : 2 : 4 to 1 : 3 : 5. No waterproofing mixture was added, but instead the whole roof was covered with a thin, smooth layer of 1 : 1 cement. The concrete is reinforced with rods of 1/2-in. iron at 24-in. centers. The side walls are of masonry built of waste rock from the mine, and are tied together at the top crosswise by three 1-in. rods. A small ventilating hole is left in the end walls near the top. The building is not for the thawing of the dynamite, but merely for its safe storage at the surface.

Concrete Powder House.—At the properties of one large Eastern mining company the powder houses are about 6×12 ft., built entirely of concrete. The door and door frame are wood, but covered with sheet iron. The floor is cemented. These houses are used for the storage of only a few boxes of powder for immediate use at the mine. There are a number of these buildings, and each shipment of powder is distributed among them. This avoids the storage of large quantities at one place. However, the design of these magazines is not one that is to be recommended.

Powder Storage Underground.—At the Leonard mine, Chisholm, Minn., only one box of powder is taken underground for each working face. This powder is kept under lock and key in a box, 2×2×4 ft. In this box is also kept one box of candles and whatever fuse and caps are necessary for each face. The day shift and the night shift each have a key to the box. These boxes are so distributed that they are not less than 75 ft. apart and at a safe distance from the working face. This distribution of powder prevents any serious explosions, such as may occur when many boxes are kept in one magazine.

[However, it is not good practice to keep caps and fuse and powder in the same box.—EDITOR.]

FROZEN DYNAMITE

The nitroglycerin in dynamite freezes at 42 to 46° F., and when frozen is insensible to detonation but explodes readily by friction or by breaking or cutting the cartridge. It should be thawed by putting it in a watertight case and immersing in warm water, or by laying it in a warm room. It should not be thawed by putting the sticks in warm water, as a certain amount of the nitroglycerin is lost through its coming out of the cartridge and sinking to the bottom. This water may perchance be heated again, in which case the collected nitroglycerin is likely to explode. It should also be noted that in case any water does contain nitroglycerin mingled with it, that the water should be poured out carefully and not thrown out violently on the ground. It is also dangerous to hold a stick of frozen dynamite over a hot object, as one drop of the nitroglycerin may ooze out, fall, explode and set off the stick. It is hardly necessary to state that it is imprudent to thaw dynamite by carrying it down one's boot leg or inside one's shirt. There are only two safe ways and those are the ones given above, and those should not be pursued too enthusiastically. Make haste slowly.

Thawing Dynamite.—A dynamite thawer, described by R. E. Tilden, of Winnemucca, Nev., is effective, safe, and has the advantage that the smallest mine can afford one; its cost is nothing. To thaw the dynamite use a large bottle, fill with warm water, then place the cartridges around the bottle in layers, wrapping them in place with a woolen rag or cloth, tied by a string. The size of bottle selected should be governed by the amount of powder to be thawed for use in one place in the mine and by the time that will elapse, after being taken underground, until the powder is used. The bottles may be conveniently carried in pails. Powder taken into the mine in the morning will be thawed and remain so until evening.

A simple, economical and perfectly safe method of thawing dynamite is that which has been used for over a year at the Van Roi mine, near Silverton, Slocan district, B. C. Douglas Lay, superintendent of the mine, describes this as being merely an adaptation of the widely known principle of heating by hot-water coils. The thawing house is a building occupying a floor space of about 8 × 10 ft. Placed in it is the water-supply tank—a barrel—kept full, or nearly full, of water, which is heated by means of pipes passing to a coil in a stove in another building about 300 ft. distant from the thawing house, and at a considerably

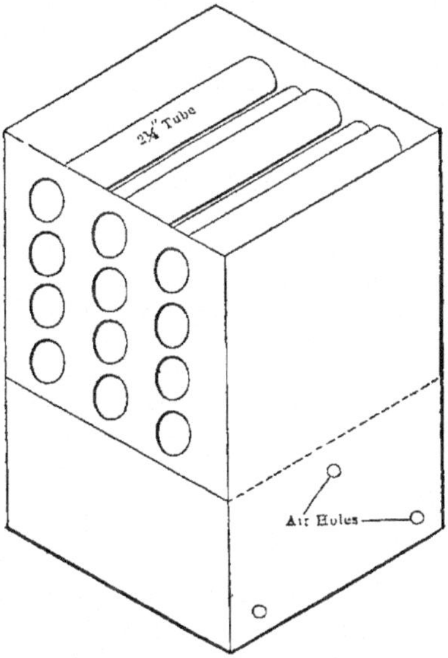

FIG. 15.—THAWER FOR DYNAMITE.

lower level. The pipes from the barrel, which connect with the stove coil, are placed in a box buried in the ground, to insure insulation. In the thawing house there is a box 5 ft. long, 2 ft. wide, and 2 ft. deep, with a lid. A large pipe coil, connected with the hot-water column, is in the bottom of the box; it is covered with sawdust, and the sticks of dynamite are placed in the box. The sawdust serves to absorb any exuded nitroglycerin, and is renewed frequently. The building containing the heating stove and coil is erected near the entrance to one of the mine adits and is used as a dry room by the miners. As already mentioned, it is remote from the thawing house, so that the latter would be absolutely safe, even if, from any cause, the former should catch fire.

In Fig. 15 is shown a type of thawer used by the Oliver Mining Co., Hibbing, Mich., which has given excellent satisfaction. It is made of galvanized iron, about 10×10×16 in. high. Twelve 2 1/4-in. tubes are soldered in, giving the appearance of a tubular boiler. Six inches from the bottom is a water-tight partition, above which is water to cover the tubes. The ends of the tubes are open for the insertion of the sticks of dynamite. Two or three short pieces of candles placed below furnish heat for warming the water, unless hot water is available. A metal cover is placed over the box. The candles are thoroughly inclosed beneath and obtain an ample supply of air through three ventilation holes on each side.

At the Traders' mine, Iron Mountain, Mich., a small house has been constructed in which to thaw all the powder used in the mine. The building is 10×12 ft., built of 12×12-in. timbers placed close together and is on a low piece of ground sheltered by a high bank. Exhaust steam from the boiler house is

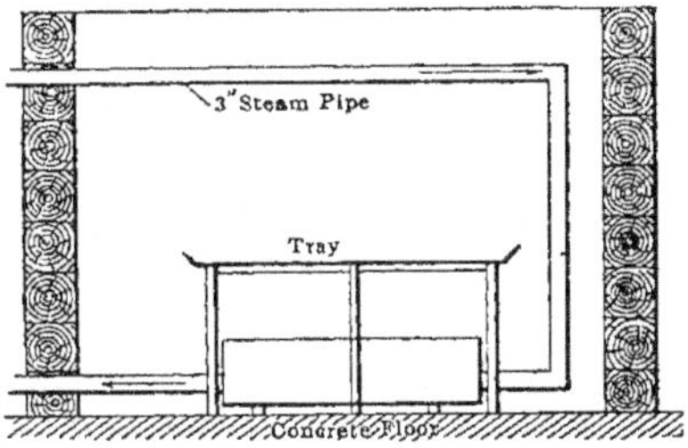

FIG. 16.—THAWING HOUSE AT TRADERS MINE.

used for heating the building. It enters through a 3-in. pipe above the tray upon which the powder is placed, and then passes down and into a cylinder 12 in. in diameter and 5 ft. long which is beneath the tray. Fig. 16 shows the arrangement. The tray itself is made of 1/2-in. iron bars placed about 1/2 in. apart. It is 3×6 ft. and affords ample room for four or five boxes of powder. This has been in use 14 years. Only one day's supply of powder is kept in this building. The fuse and caps are in another building 50 ft. away. The main powder storage house, where carload shipments are kept, is 1/4 mile distant.

Thawing Dynamite by Electricity.—A method of thawing dynamite used by the Vermont Copper Co. consists in heating the thawing box by electricity. The box referred to is about 16×40×36 in., and stands on edge. It is made of wood with a number of trays that will slide in and out. Each tray is made of small strips of wood, and it is upon these trays that the powder is placed. Doors are made as nearly air-tight as possible. The lower 8 in. of the box has wire coils through which the current is passed, and the heat from the coils warms the box. Above the coils is a thin sheet of galvanized iron, to prevent any powder or nitroglycerin from coming in contact with them. The sheet iron is covered with asbestos sheet. The box is usually thoroughly heated before placing the dynamite on the trays, and then the current is turned off. The thawer is placed in some remote part of the mine, at the end of a drift which can be closed from draft, and is thoroughly inclosed in a small house.

III

ROCK DRILLS

Bits and Drill Parts—Pointers on Operation—Economics of Practice.

BITS AND DRILL PARTS

Air-hammer Drilling in Sticky Ground (By George E. Addy).—The sticking of a drill steel in soft, clayey ground may be overcome by welding a piece of ribbed drill steel, *A* in Fig. 17, to a bull pick *B*. The drill so made can be driven with a hammer drill. If a set of three drills is made, the shorter steels having the larger points, a hole 3 ft. deep can be drilled. These pick-pointed drills will be found to be time savers as compared with the usual types of drills, or with driving a bull pick with a hammer.

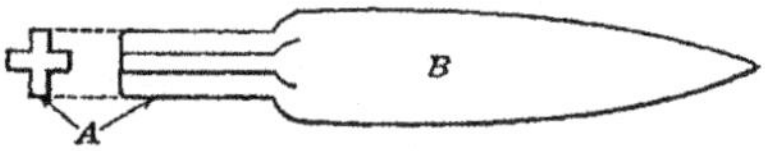

FIG. 17.—A PICK-POINTED DRILL FOR SOFT GROUND.

A Drill for Soft Ground.—The vanadium deposits of San Miguel county, Colo., occur in sandstone which is generally soft and often moist. It is difficult to drill this sandstone because the moist sand sticks in the holes. To obviate this the miners use a hand drill made of steel tubing, in which saw teeth are cut on the end, set out for clearance as shown in Fig. 18. With this drill, upper holes are regularly drilled, the cuttings coming out of the bit through the hollow center. The device is satisfactory within the limits for drilling shallow holes in soft wet ground.

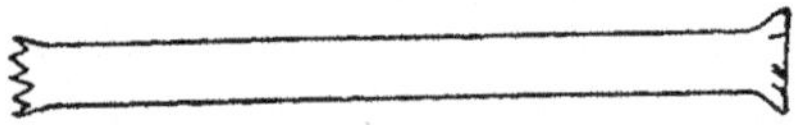

FIG. 18.—DRILL MADE OF STEEL TUBING.

Design of Drill Bits (By Ward Blackburn).—The efficiency of a rock-drilling plant is determined by the amount of power delivered to the bit and converted into cutting power. No matter how efficient the compressing plant and the rock drills, or how careful the drill operator, if the bit is dull or of incorrect shape, the rock is not cut, good energy is wasted, and money and time are lost in reaming the bore hole or in pulverizing the rock, and the real object of the work is defeated.

At most mines little attention is paid to the drill bits by anyone except the blacksmith. The drill runner is satisfied if the bits run out a change and hold gage well for the next bit to follow. The fewer changes, the better he likes it. The same is true of the blacksmith. The longer a bit is in service the less frequently must it be sharpened. The bit is designed for use as long as possible without resharpening and the cutting properties take second place.

In designing a good bit both cutting and staying properties should be considered, but the day of hand-sharpening and of chuck bolts is rapidly passing, and now the efficiency sacrificed to staying properties is determined largely by the cost of transportation and handling of bits and permissible loss of gage. It is not influenced to the same extent as formerly by the labor of sharpening or trouble of changing bits. The cost of transportation and handling of bits amounts to a considerable item which increases as the lasting qualities of the bit are replaced by cutting qualities, so that it often governs the extent to which the design may be changed.

Light, compact mechanical sharpeners can now be obtained at a reasonable price, and as these machines reduce the cost of sharpening to a low figure and enable the blacksmith to make several hundred bits per day, shop considerations offer no serious reason for retaining merely staying qualities in the design of the bit. Nearly all rock drills can be provided with a wedge chuck, or some means of holding the bit behind chuck keys so that bits can be changed in a fraction of a minute. This largely overcomes the objection to short bits. In several mines the cost of transportation and handling of bits is greatly reduced by installing the power sharpener with oil or coke furnaces underground.

Allowable loss of gage is an important factor, for the greater the loss the larger must be the diameter of the shorter bits to insure a certain diameter of hole at a fixed depth. This larger diameter means a loss of power, so this point can never be lost sight of in determining the bit to be used. Operators who believe the gage should be retained irrespective of the amount of power used, run the edges of the wings of the bit back about 1 in. in a line parallel to the axis, making the bit a reaming tool. If this is done, as soon as the cutting edges wear a little the shoulders begin to bind and stick against the sides of the hole, and it must be reamed out to a size which will allow the bit to move forward. When such a bit is removed from the hole it will show greater wear on the edges of the wings than on the cutting edges. As the bit is apparently as hard 1/2 in. back, as on the cutting edge it is reasonable to suppose that there has been as much work done in wearing the steel of the wings as in wearing an equal amount at the cutting edge, that is, if the reaming edges show more wear than the cutting edges, it is to be supposed that they have done at least as much work as the latter. In other words, as much if not more power has been expended in reaming the hole as in cutting it; half the power has been used in cutting an area of say 7 sq. in., that of a 3-in. circle, and half the power has been used in reaming an area of a little more than 1/2 sq. in., the area of the 1/16-in.

rings reamed out. Furthermore, the rock drill is designed to cut rock by means of a blow. It is not a reaming machine, and as soon as a bit is cutting or reaming the side of the hole it is retarding the action of the drill. This strains the rotation and has a tendency to twist and break the bit. Power expended in reaming is power lost. From this it is apparent that because of the rapid loss of power and the wear and tear on the rock drill, little of the efficiency of the bit should be sacrificed to the reaming qualities.

It is sometimes claimed that unless the bit has reaming edges, the hole will become rifled. This was true to some extent when the bits were forged by hand, but it was due to the difficulty of getting each corner or wing equidistant from the center of the bit. With most power sharpeners this difficulty is removed, and with the correctly formed bits there is little trouble from rifling.

While it is true that one bit cannot be selected as a standard for any and all classes of rock, still the power-sharpener manufacturers, especially those who have gone into the study of drill bits, are gradually working toward a few standard types. The consensus of opinion is that under average conditions, a good bit for power-drilling work should embody the following points:

(1) It must take full advantage of the chipping and fracturing of the rock. In a bore hole there is a certain depth to which rock will fracture when struck by a sharp tool. If the tool is driven deeper than this it will not fracture the rock; it will crush it. If the cutting edge of the bit is blunt it will not get full advantage of the fracture, and considerable of the force will be expended in crushing or pulverizing. The bit acts as a wedge.

(2) The wings of the bits should be as thin as is consistent with standing-up quality to allow for the ejection of the cuttings. If the wings are left heavy there is little space for the escape of cuttings and consequently they are held in front of the bit and continually churned and ground, the bit does not easily reach solid rock and the blow loses a large proportion of its cutting power before it reaches the rock.

(3) The bit must be perfectly free in the hole at all times; because of the tendency of the rock to fracture, the drill hole will be a trifle larger than the cutting edge. If the bit is so designed that the cutting edge is its greatest diameter and will practically remain so until dull, it will always remain free in the hole.

(4) The bit must allow equal wear on all corners. If the bit is not symmetrical, that is if the ends of all the cutting edges are not the same distance from the center of the bit, the longest end will cut a groove in the side of the bore hole and a rifled hole will result. Furthermore, the wear is unequal and the extra strains in the steel often break it. The rotation of the drill is impeded and the parts subjected to excessive wear.

(5) The bit must be dressed in a manner consistent with the treatment of good steel. It must not be overheated, or worked while too hot. Light rapid blows should be used in forging. The bit must be tempered properly. The proper temper should be determined by experiment and rigidly adhered to. The

bit should be allowed to cool thoroughly after forging and should then be reheated for tempering. It should never be sharpened and tempered on the same heat.

In Fig. 19 are shown bits designed to fill these requirements. An angle of about 90° for the cutting edge is generally accepted as the correct angle. If the angle is greater the bit has a tendency to crush rather than fracture the rock, and as a rule the bit cuts slowly. An angle less than 90° gives a bit which will cut fast, but it has a tendency to break off and wear rapidly. There is also a tendency for it to enter the rock, especially soft rock, past the point of fracture, and expend energy in crushing or wedging out the rock. The "mudding" powers are greatly diminished. The thickness of the wings may range from 3/4 in. on a large steel, to 3/8 in. or even less on a stoper and hammer-drill bit. The wing must not be so thin that it breaks off, still, the less stock necessary the better the clearance for the cuttings.

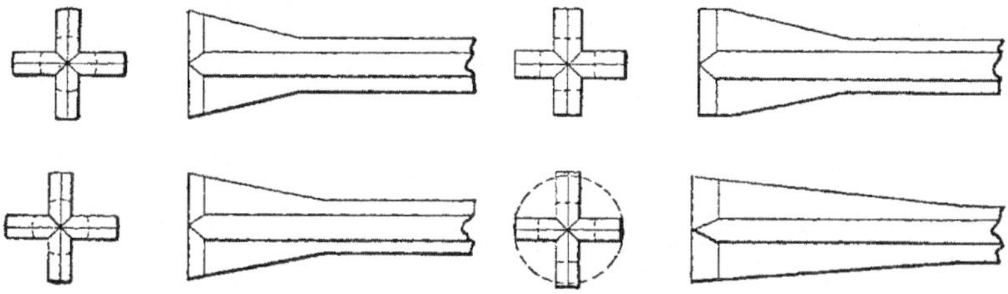

CORRECTLY MADE BITS. INCORRECTLY MADE BITS.

FIG. 19.—DESIGNS OF DRILL BITS

The third point is probably the hardest to comply with. When the bits shown in Fig. 19 are first placed in the hole the cutting edge is the largest diameter, and as it may be safely assumed that the rock will fracture 1/16 in. beyond the bit, the latter must be free in the hole. The wings of the bit taper in the ratio of one to four, or one to five, and as the cutting edge wears, a shoulder or reaming edge forms on the wing. When it wears between 1/16 and 1/8 in. from each wing, a shoulder 1/4 to 5/8 in. forms and there is a point between these where the cutting edge is fracturing the rock in the side of the hole to a diameter equal to the diameter of the bit through the shoulders. In other words, the shoulders of the bit are just touching the sides of the hole. If the bits run past this point, the cutting edge will not cut the hole to a diameter large enough to admit the bit and the shoulders immediately become reaming surfaces.

The taper of 1 : 4 or 1 : 5 is determined upon as a medium. If less taper is used, say 1 : 7, the shoulders become more prominent and the cutting edge has less allowable wearing surface before the bit binds in the hole causing the shoulders to become reaming surfaces. If the wings have a greater taper, say 1 : 2, the wear is excessive, and the loss of gage prohibitive. The sharp taper leaves the ends of the cutting edge in a weakened condition and they easily break up and chip off.

The necessity for having the drill bit concentric is obvious. In hand sharpening and even with some power sharpeners it is not uncommon to find one of the ends of the cutting edges protruding 1/32 to 1/16 in. beyond the others. Some sharpener manufacturers have effectively overcome this difficulty by entirely inclosing the bit under a heavy pressure while it is being forged. When this is done there can be no question as to the corners falling within a circle.

The tempering and manner of heating and treating the bit is generally left to the discretion of the blacksmith. This is a mistake, as a little attention and study given the subject by the manager or superintendent will often result in marked improvements. Many kinds of steel now on the market, and the scientific methods of treatment are entirely new to the average smith.

A word should be said about mechanical sharpeners. These machines have been greatly improved in the last few years, and now extremely light, simple designs of remarkably high capacity and efficiency are within the reach of all. The bits dressed in the sharpener are always perfect in form and gage and everything considered, will give from 25 to 50% greater efficiency than if sharpened by hand. In addition to the increased speed of drilling there is a noticeable saving in the wear and breakage of machine parts and the bit itself lasts longer and wears much better.

When installing a power sharpener one should not lose sight of its main advantage, increasing the efficiency of the drill bit. It should be installed with the idea that because of the marked reduction in the cost and time of sharpening one can afford to furnish the drill with better steel more often. It is impossible to make a hard and fast rule as to when a power machine should be installed, but it is certain that a sharpener will effect a material reduction in cost per steel dressed, lessen the machine repair parts required, save steel and steel breakage, and greatly increase the drilling speed of machines.

Rand Drill Steel and Bits (By E. M. Weston).—Tests have recently been made by Robert Allen at the Robinson Deep mine to determine the most suitable steel for making drills for use in the Rand mines. Many varieties of steel have been tested in the following way. From 40 to 60 drills made from each brand of steel were sharpened by hand, weighed, measured by a micrometer gage, then sent to the mine where they were used in 2 3/4-in. machines to drill the hardest rock. The depth of holes drilled and time taken were noted. The drills were then sent to the surface where they were again weighed and measured and the quality of the steel compared by the loss in gage. It is difficult to convince the Rand miners that the proper heating and tempering of the steel has an important bearing on the efficiency of the drills, so it was necessary to recommend steel of such carbon content as would permit of direct plunging in the type of tank giving a limited depth of immersion for cross bits. The bits as finally adopted, are shown in Fig. 20. They are used in 2 1/2- and 2 3/4-in.

machines. The starting bit is used to drill 15 in. of hole, the second 21 in., and the third and fourth 24 in. each. With higher air pressure I believe that the third and fourth bits should be made of 1-in. steel, as even with 70 lb. air pressure a 2 3/4-in. machine will bend some of the 7/8-in. drills. A carbon content of from 0.7 to 0.75% is the highest that will permit of steel being satisfactorily welded and which will temper without cracking on direct plunging.

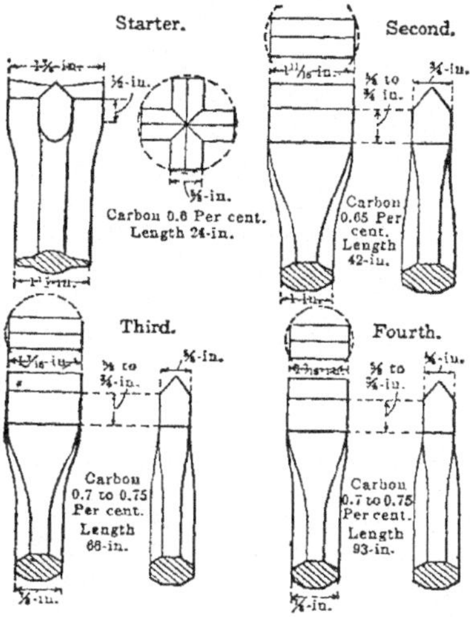

FIG. 20.—DRILL BITS USED ON THE RAND.

Ejecting Sludge from Drill Holes (By E. M. Weston).—A method of ejecting mud and cuttings from a drill hole has been devised by Mr. Tippet, an Australian, that may result in making it possible to use a one-man drill in the mines of the Rand. He conceived the idea of withdrawing the sludge through, instead of forcing water down hollow steel. He intended to bring the sludge right through the machine by means of a suction device which was to be worked

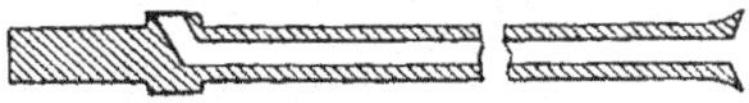

FIG. 21.—HOLLOW-STEEL BIT WITH SIDE OPENING.

by the exhaust air. While using a rose bit, he noted that when the suction was not operating, the drillings were being vigorously ejected. The sludge evidently entered the hollow in the steel with considerable velocity on the down stroke, the inertia of which was not entirely overcome during the period of the return stroke. All that is necessary to take advantage of this effect is to employ hollow steel and forge a collar near the shank, then to drill a transverse hole to connect with the central channel at a point just below the chuck. Steel of this construction has been found to hold its gage remarkably well. The steel used with piston drills is 1 1/4-in. diameter, through which passes a 3/4-in. channel. The central channel becomes filled with sludge but does not choke with pieces of rock (granite) even when as large as a ten-cent piece. Fig. 21 shows the transverse and longitudinal passages of the improved drill steel.

Improved Chuck for Piston Drills.—In the North Star mines, at Grass Valley, Calif., a special type of chuck designed by Messrs. Paynter and Bastian, employees of the company, is used on the piston-machine drills. The peculiarity

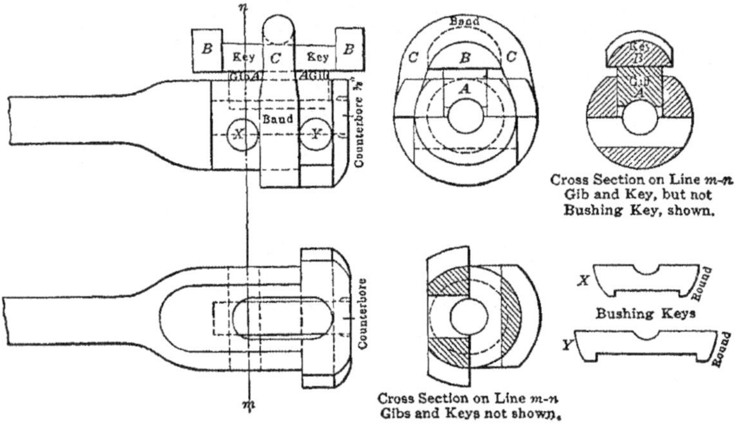

FIG. 22.—NORTH STAR BOLTLESS CHUCK FOR PISTON DRILLS.

of the chuck is that it includes no bolts, and hence does not require the use of a wrench for tightening the grip upon the drill shank. The working drawing, Fig. 22, shows the details of the chuck and clamping arrangement. The chuck is drilled as usual to receive the shank of the drill steel. A slot, above and parallel to the shank of the steel, is cut in the chuck to receive a gib *A* that bears against the shank of the drill. Below the drill socket and perpendicular to the axis of the chuck two holes are cut to receive bushing keys *X* and *Y*, that bear against either end of the lower part of the drill shank and take up wear from the chuck. A strap or band *C* fits around the chuck and over a tapered key *B* that bears on the gib *A*. The key *B* is tapered away from the end of the chuck so that as every impact of the drill against rock drives it further under the strap *C*, the gib is forced more tightly against the drill shank. There is, hence, no tendency of the drill to become loose in the chuck. On the other hand, it is held more securely at each stroke. The key *B* is made with a heavy head at either end. To fasten the drill in the chuck the key is driven tight by a blow upon the head at the larger end. A blow on the other end of the key serves to loosen it and allows the drill to be removed. This type of chuck has been used for several years in the North Star mines and has proved entirely satisfactory. Its advantage over the ordinary type where bolts have to be drawn tight every few minutes should be evident. The construction embodies no particular difficulties.

Shaping Chuck Bolts (By H. Lawrence Brown).—The accompanying sketch, Fig. 23, shows a device for making chuck bolts for machine drills. It consists of the plate *B*, made of 1-in. iron, bolted to a work bench by 3/4-in. countersunk bolts *A*. To the plate are riveted the pieces *C* and *D*, of 1-in.

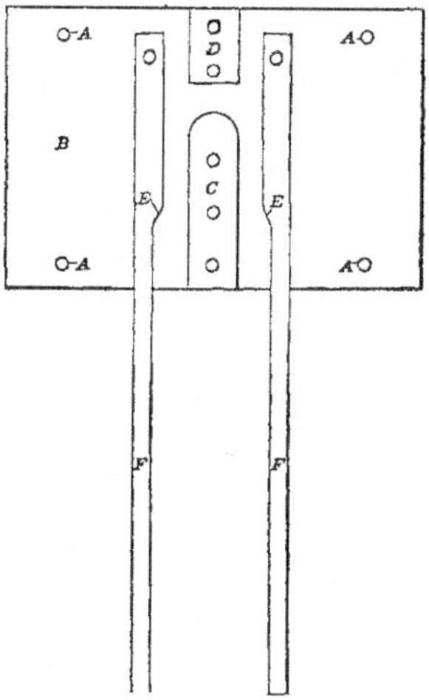

FIG. 23.—DEVICE FOR SHAPING CHUCK BOLTS.

material, with a curve in the piece *C* corresponding with the bend desired in the chuck bolt. The levers *F* are about 2 ft. long and are offset as shown at *E*, where the thread on the bolt begins, in order not to injure the threads while the bolt is being shaped. The threaded bolt is heated and placed between *C* and *D*, while the levers are open. Closing the levers to the position shown bends the bolt to the proper shape.

Improved Drill Post Collar (By Albert Mendelsohn).—An improved post collar now being used in some of the copper mines of Lake Superior is shown in Fig. 24. It does away entirely with the two bolts of the collar at present in general use, and can be loosened or tightened by a single blow of the miner's wrench. It consists of a cast-steel band *A* of diameter slightly greater than that of the post. A tool-steel gib *B* fits into a slot in the band, and directly over the gib a wedge-shaped key *C* of the same material is driven. The inner face of the gib is shaped to an arc of the same radius as the post and when the wedge is struck on its wide end the gib is forced tightly against the post. An advantage of this collar over the one at present in general use is that time and work are saved, because there are no bolts to tighten and loosen. This time may not amount to much on a two-man machine, but with the introduction of the one-man "butterfly" drill and the attention to details necessary in running it, any device that will save two or three minutes per hole drilled is of importance. At present the miner using a one-man machine has to adjust nine bolts; the elimination of two is, therefore, no small item. Another advantage, and one which is of importance on one-man machines, is the fact that this improved

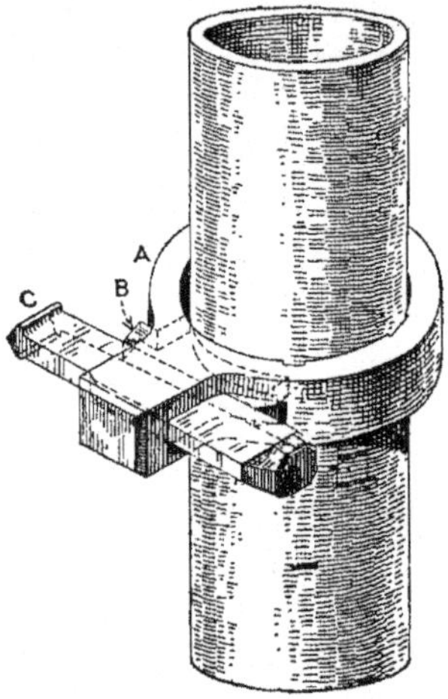

FIG. 24.—DRILL-POST COLLAR WITHOUT BOLTS.

collar can be rapidly loosened and tightened, and if necessary with one hand. In raising the machine on the post, if the post is wet or the machine too far in on the arm, the arm will not catch on the post. This means that the machine and arm must be held in the elevated position while the collar is loosened, slid up the post under the arm and tightened. In any case, speed is desirable under these conditions, and with a one-man machine it is imperative. Incidentally, it might be mentioned that these collars were used with the two one-man machines with which four men recently drove 285 ft. of 6×7-ft. drift in one month; exceptional drifting for the copper country.

Drill Post with Removable Screw.—In the Copper Range, Hancock and Quincy mines single-screw posts, the jack screws of which are removable, are used whenever a single-screw post is required in making a cross-bar setup with a machine. The details of the screw are shown in Fig. 25. The device has even been used in shaft sinking and is said to have proved as satisfactory for that work as an ordinary post, while it possesses the advantage that owing to the fact that the screw feeds ahead instead of the jacking nut, as is the case in the ordinary single-screw post, the screw can be stuck into a hole in the wall, and the bar jacked without any trouble arising from projections from the wall interfering with jacking. The separate screw and nut also possess the advantage that such a single-screw jack can be used with several different lengths of bars.

The nut of the jacking device is octagonal in shape with four steel-bushed holes in its sides for receiving the jacking bar, and one end of the nut is made big enough to go over the end of the post as a collar. The screw has a blunt end that goes against the ground, while in it is a keyway that goes over three

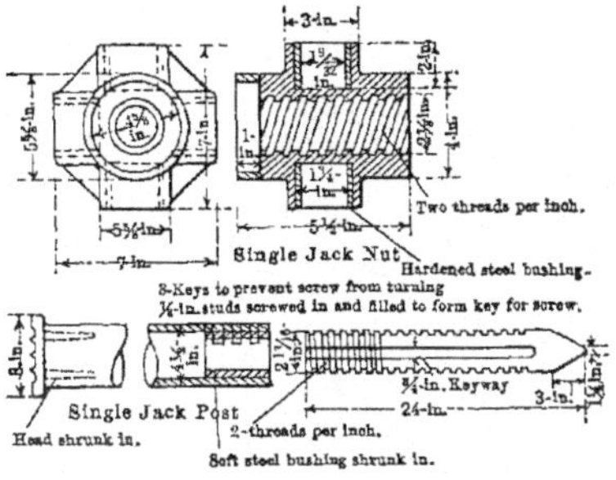

FIG. 25.—DETAILS OF DRILL COLUMN WITH REMOVABLE SCREW.

lugs in the end of the post so as to keep the screw from twisting with respect to the post as the jacking nut is turned. The post has at one end the ordinary toothed head shrunk on it, while at the other there is a soft-steel bushing or guide shrunk in for the screw. Through this end of the bar three holes are drilled and tapped for 7/8-in. studs. These studs are screwed in tightly and then filed flat to serve as the keys to go into the keyway in the jacking screw, as well as to hold in the guide bushing. The barrel of the post is a piece of ordinary 4-in. gas pipe.

Jack for Machine Drill Columns.—Machine-drill jacks become quite heavy as the length of the column increases, especially when a column with two jack screws is used. In the Michigan copper country, machines have to be set up with columns as long as 13 ft. It therefore becomes important to make them as light as possible, especially now that one-man machines are being introduced. In driving drifts it is often desirable to have two lengths of column; in such a case the Osceola type of machine column, in which the jack proper is in one piece and the post in another that can be lifted out of its socket in the jack, is convenient. When the post has to be moved, the weight to be lifted is divided in two as the two parts are moved separately. At the Osceola and the Calumet & Hecla mines, the post consists of an ordinary 4-in. gas pipe for both two-man and one-man machines, with a light cast-iron cap fastened to the upper end. In the bottom of this pipe a notch is cut to fit over the lug in the bottom of the socket hole of the jack. Fig. 26 shows the design of this screw part, which is cast in one piece with socket holes, into which the nuts of the jack screws are fastened. The lug on the bottom of the post socket is now cast square, but in the future it is probable that it will be cast with a V-section, so that

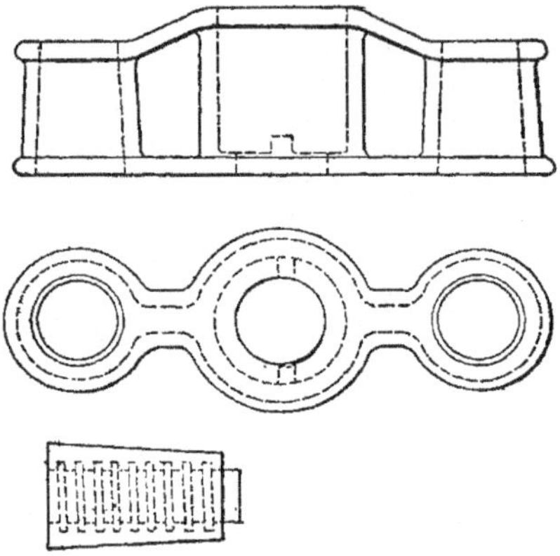

FIG. 26.—CAST-IRON JACK FOR A DRILL COLUMN.

the notches in the bottom end of the post can be cut to fit tight, for often at present these notches in the post are cut too wide, allowing the post to turn slightly when the machine is started and throwing the hole slightly out of alignment. Such a machine jack can be made cheaply, as there is no machine work on the casting, while its use is a great advantage where a two-post jack is required for one-man drills.

POINTERS ON OPERATION

Removing Stuck Drills.—In drilling deep holes by hand with 15- or 20-ft. steel, the steel often sticks; or, in some cases, a drill is actually driven in soft ground as one would drive a stake, and then it becomes necessary to resort to some means to remove it. This may be done easily by taking a piece of steel 1 1/4 or 1 1/2 in. thick, 6 in. square, with a hole in the center about twice the size of the drill. This is placed over the head of the drill and two steel wedges driven in, to fasten the piece of steel on the drill, thus forming a good shoulder against which to hammer. A few blows on this will soon loosen the drill. It is much more effective than a chain and lever, or pulley block.

Wrench for Removing Stuck Drills (By Claude T. Rice).—A miner often loses an hour or two through the sticking of a drill in a deep hole. The fault is usually the miner's. A small quantity of sticky drillings will wedge a drill so that, unless there is a good wrench for twisting it out through the drillings, it cannot be removed, and a new hole must be drilled. Cruciform steel has less tendency to stick in damp holes than octagonal or hexagonal steel, as the ribs act like a conveyor to eject the drilling, provided that the hole is not too flat and the drillings are not too sticky. E. M. Weston has advised the use

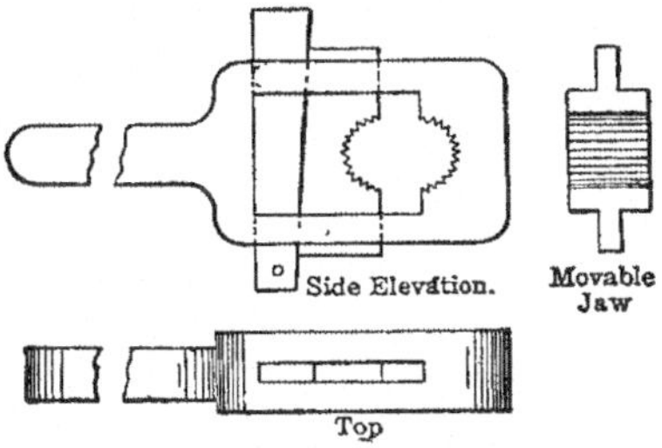

FIG. 27.—A DRILL-TWISTING WRENCH.

of drill steel with a rolled screw to overcome this difficulty and on the Fort Wayne electric drill, bits with twisted shanks are used. Similar steel was used a few years ago with good results in piston drills in certain of the New York iron mines. Twisted drill steel does not stick in a hole, I understand, but it is expensive and probably not so satisfactory as octagonal or cruciform steel.

The most satisfactory wrench that I have seen for twisting drill steel from a hole is that in use in the Calumet & Hecla mines and shown in Fig. 27. It is made of 3/4-in. iron and consists of a frame to which one of the toothed jaws is

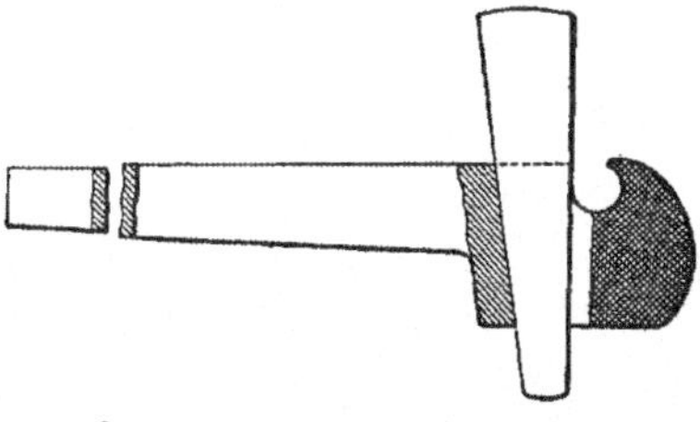

FIG. 28.—COTTER WRENCH FOR STUCK DRILLS.

fastened; the other slides loosely in the slot of the frame, being prevented from slipping out by shoulders. Back of the loose jaw is a key, prevented from falling out by a pin through its small end. This, driven downward by a hammer, forces the teeth of the jaws into the drill steel so as to grip securely. The extension of the frame forms a handle 18 in. long, but in case this is not long enough, a piece of pipe can be slipped over it. The wrench is simply and strongly made, so that it can be hammered, twisted, and jerked without injury. It has been in use for several years at the Calumet & Hecla mines and has proved satisfactory.

In Fig. 28 is shown the design of the wrench that is used at the Mohawk mine in Michigan to aid in loosening drills that become jammed in bore holes. The wrench is quite similar to the one that is used at the Calumet & Hecla mines. The grip on the drill is obtained by driving in a key or cotter that grips the drill between itself and the jaw of the wrench. The wrench is much heavier, and does not seem to be quite as handy as the one used at the Calumet & Hecla mines. In working the drill out a chain is fastened to the drill, thence it is passed around the machine post. One man pulls on the chain while the other twists the drill with the wrench.

Wrinkle for Piston Drill.—It is frequently necessary for a machine-man to release his hold on the crank to throw water into a hole or attend to the many little details that are constantly requiring his attention, and a freely feeding drill will crank itself back too rapidly to permit its being left. Miners often hang a wrench on the crank or lean a piece of steel against it or twist the hose around it. The last two expedients are unhandy and inconvenient; any weight hung on the crank will slip off unless the drill is inclined steeply downward. A hole drilled through the crank at the base of the handle, through which a wire may be slipped to hold the weight will expedite the miner's work considerably and do away with the temptation to allow the feed to work stiffly. A couple of 7/8- or 1-in. nuts bound closely to the crank in this manner will not slip around or get tangled with the crosshead.

Cleaning Drill Holes (By J. H. Forell).—In drilling upward slanting back holes in dry ground, the drillings are often removed by a squirt gun, an improved

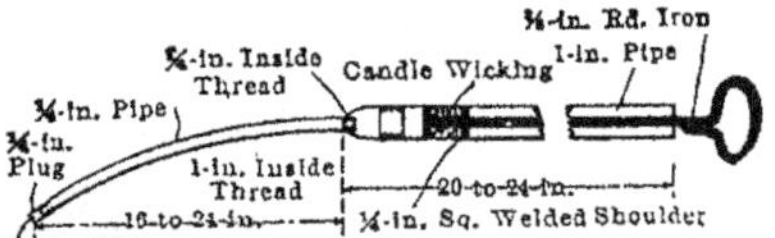

FIG. 29.—SQUIRT GUN FOR CLEANING DRILL HOLES.

form of which is shown in Fig. 29. A piece of 1-in. black pipe, 20 to 24 in. long, is threaded at one end. A cap, slightly cone-shaped at one end, is turned on a lathe and threaded at the large end to fit the 1-in. pipe; the small end is tapped to receive a 1/4-in. pipe 16 to 24 in. long. The outer end of the 1/4-in. pipe is threaded internally to hold a plug through which passes a 1/16-in. hole. The 1/4-in. pipe is curved to enable the operator to introduce it into the hole without danger of being struck by the chuck of the machine. The plunger is made of 3/8-in. round iron and is packed with cotton candle-wicking held between two shoulders of 1/4-in. square iron about 4 in. apart. The advantage of this squirt gun is in the long, narrow pipe, which can be introduced into the hole a foot or more, giving a greater pressure at the cutting end of the hole, whereas, with the old-style gun the greatest pressure was near the collar.

Preventing Freezing of Air Exhaust.—Alcohol is used to prevent freezing at the exhaust of the air-operated shot drills in use at the site of the new station in New York of the New York Central railroad. This company has had to do an enormous amount of core drilling within the last 15 years, and in this work several types of drills have been used. One of the most interesting outcomes of the work is the freezing-prevention appliance that can be readily adapted for use on any small air-operated machine where trouble is experienced from freezing at the exhaust. Alcohol is admitted to a vertical part of the air-supply pipe and close to the machine. The device used for feeding the alcohol is attached to the feed pipe as is a lubricator. It consists of a piece of 1 1/2-in. pipe about 12 in. long, fitted with bushings at the top to take a piece of 1/2-in. pipe 6 in. long. The small pipe is closed by a valve. The other end of the 1 1/2-in. pipe is fitted with bushings to take a 3-in. piece of 1/2-in. pipe, to which a valve and an elbow are attached and into the elbow another short piece of 1/2-in. pipe is screwed. The threaded end of the last-mentioned pipe is screwed into a hole tapped into the air pipe so that when screwed tight the 1 1/2-in. pipe is parallel to the air pipe. By opening the top valve, the 1 1/2-in. pipe can be filled with alcohol. That valve is then closed, and, when the machine is operating, the lower valve is opened just enough to permit drops to pass slowly. The admixture of alcohol vapor in the air effectually prevents freezing at the exhaust. The 1 1/2-in. pipe, which has a capacity of about 1 pint, is usually filled with alcohol twice per shift.

Cutting Timber by Small Hammer Drills.—Small hammer drills are used with a chisel bit in the Hecla mine at Burke, Ida., for cutting off the crushed ends of timbers. In the stopes the greatest pressure is from the squeeze of the walls, and as the ends of the stulls and caps become splintered and crushed, it is necessary to cut them off and put in new blocking. It is often difficult to get at the crushed timbers, and, even when accessible, it is not an easy job to chisel or saw the wet and twisted fibers by hand. The cutting of the wood is rendered quite easy with the drills, and by using sufficiently long bits, almost any desired place can be reached. The drills and chisels may also be used to great advantage for chiseling wall plates when an extra shaft compartment must be added, or in cutting off posts to ease up drift sets in heavy ground. In a number of mines, the blocking on drift caps is shot out when it is necessary to ease up on the sets; this work can much more safely and surely be accomplished with the drill.

An Air Moil for Cutting Timber Hitches (By S. H. Hill).—In the Lake Superior district it has been customary to cut the hitches required in timbering by hand, usually with a moil. However, since a great number of first-class air-hammer drills have come upon the market the use of an air moil for this work has met with favor upon the grounds of economy and speed. The air moil can, of course, only be used in headings that are piped for air. A reducer can be used on the end of the pipe and air for the hand tool taken from the nipple used for heading machines. However, this necessitates doing the hitch cutting or squaring when the heading machines are not in use or while one of them has been purposely stopped. The introduction of a manifold on the end of air pipe, having one opening especially for the hand tool is more satisfactory. There is also a possibility of using the air moil in sampling breasts, etc.

Boring Flat Holes with Air Hammer Drills (By Clarence C. Semple).—In the smaller mines of the Cripple Creek district, air-hammer drills are used to bore flat holes, some of which are almost horizontal. The cuttings are removed from the hole by a blowpipe, an instrument designed for the purpose by H. E. Harris, the local agent for the Waugh drill. The blowpipe consists of a brass tube 1/8 in. in diameter and as long as the deepest hole drilled. One end of the tube is attached to a short piece of 1/2-in. hose by a 1/2- to 1/8-in. bushing and a hose coupling; the other end of the hose is fitted with a coupling, and nipple connecting with a 1/2-in. valve. The air hose from the mains connect with the air inlet of the drill by a tee, to the third branch of which the blowpipe hose and valve is attached. Cruciform steel is used and the blowpipe tube is extended into the hole between two of the lugs of the steel, where it turns easily with the steel without catching on the walls of the hole as the drill is rotated. The quantity or air necessary to blow the cuttings from the hole is regulated by the valve of the 1/2-in. hose. The blowpipe makes a dusty working face if the ground is dry, so the miners usually wear respirators or throw a little water into the hole with a can. It is also possible to connect the blowpipe by a second 1/2-in. hose and a three-way valve, with a water supply so that water can be introduced into the hole to allay the dust, and air then used only to blow out the mud, the three-way valve being used to control the flow of water or air.

[The adaptation of air-hammer drills to use in raises and drifts is brought out in articles in Chapters IV and V.—EDITOR.]

ECONOMICS OF PRACTICE

Drilling with Double Screw Columns (By P. B. McDonald).—A single-screw column or bar, rigged horizontally, is of course invaluable for supporting a machine drill over a pile of muck in a drift; in some cases, it is more easily handled, due to its length coinciding more closely with the width of the working than the double-screw column with the height. Operators often purchase single-screw columns for occasional work and the miners get into the habit of using them generally around the mine because they permit of more time being wasted in clearing away muck.

Under ordinary circumstances better results can be obtained if the muck is cleaned out promptly, and the drilling done on double-screw columns rigged vertically with an arm. Due to the extra movement permitted by the arm, the holes can be placed with greater accuracy and convenience. When men can conveniently place holes where their judgment directs, there is less chance of the cut failing to break. For the same reason it may be possible to break the cut with one or two holes less than would be required with the horizontal bar, which does not allow much leeway in the movement of the drill.

When a horizontal bar is used, the setup has sometimes to be made a few inches too far back; if the sides of the drift are uneven, or the miners misjudge the distance from the face a few inches. This results in shortening the length of the hole that can be cut with any one drill. Holes shortened four inches, make a loss of 7% on the time spent in rigging, blasting, changing drills, etc., which operations consume approximately 50% of the miner's time, making a net loss of 3 1/2% on the total work on the cut. This case is especially marked with drill machines having too short a feed screw, as many of them have.

The face of the drift is usually uneven, and the horizontal bar has to be rigged far enough back to allow room for a "starter" on the farthest projecting surface. Holes drilled at an angle from a horizontal bar, such as cutting-in holes, will not penetrate so far in the line of the drift as the other holes, making an uneven space when the round is blasted. Of course, this can be corrected by shortening the straight hole, or by using an extra drill on the angle holes, but both of these operations mean a loss of time. In general, a cut drilled from a single-screw column will be shorter than one drilled from a double-screw column, so that the time sepnt in rigging, changing drills and blasting is not utilized to the best advantage.

It is always necessary to rig the horizontal bar twice, the second occasion being for the purpose of drilling the bottom holes. This consumes from 10 to 30 minutes. With the double-screw column it is often possible to drill an entire cut from one rigging.

The rigging of the double-screw column vertically is usually accomplished in less time than is required to rig a single-screw bar horizontally, especially where much blocking has to be done, because the blocks that rest on the top will slide out from between the single-screw bar and the side of the drift. Such details may affect the day's work of a single miner but a small amount per day, but the aggregate difference in work done by a force of men in a year is large.

Bundling Drill Steel.—At the Hamilton shaft of the Chapin mine, where a small number of men are at work as contract miners, the steel is delivered at the surface from the shop. As the miners come out for their dinner, each drill gang selects its drills for the next shift. These are all bundled together and tied with a wire. While the men are out at noon, the steel is lowered and sent to a common distributing center near where the men are working.

The majority of the work is by contract, and each contractor's tools are numbered. This saves the trouble of sorting the steel underground where the light is usually none too good. It also facilitates the distribution of the steel. The steel is furnished by the company while the powder, fuse and other tools are charged directly to the miner. At the Ludington shaft where most of the work is now being done, one man is employed whose only work is to look after, collect, and deliver the drill steel to the miner.

Handling Drill Steel at Champion Mine.—The best way of handling drill steel at a mine where there are several working shafts is often a serious question. There are several places at some mines where the steel is sharpened on being brought to the surface. At the larger mines it is generally better to sharpen the steel at some central place. The reason that the smaller mines favor the sharpening of the steel at each shaft is that the cost of gathering the steel is great unless the quantity is large.

This problem of gathering and handling the steel has been solved in a simple manner at the Champion mine of the Copper Range company in Michigan. At this mine there are four working shafts from each of which steel is gathered. Shaft *B* sends up about 325 drills, shaft *C* 250, shaft *D* 200, and shaft *E* 180 drills per day. The drills are sent up loose on the cages, and at the surface are loaded directly into tank-steel boxes that are carried in the bed of a wagon in summer or sleigh in winter. These boxes are made in three pieces, a bottom and two hinged sides to which rings are fastened for receiving the hoops of the chains used to lift the boxes out of the wagon. While in the wagon the sides of the boxes are held up by the sides of the wagon or sleigh.

In loading these drills they are sorted according to size so that they can be sharpened at the shop without any needless increase in the handling. Each shaft has its own wagon or sleigh. When it has been loaded with dull drills the team which does the hauling for the shafts comes for the wagon in the morning and hauls it to the blacksmith-shop with its load of drills. While the team goes after another load from another shaft, the drills are unloaded by means of an air lift attached to a trolley that travels on an I-beam carried in the frame of the shop.

Picking the boxes up with their load of drills at *A*, in Fig. 30 by hooking chains into the rings on their sides, the 2 1/2-ton air-lift *B* traveling on the overhead trolley takes its load to rack *C* in back of the heating furnace and on the same side as the drill sharpener so that the drills can be readily placed in the furnace. There the chains are unhooked from the box and the sides allowed to fall flat on the drill rack *C*. When all of the drills have been taken from the box during the progress of sharpening, the box is lifted with the air lift and taken over to rack *D* where it is put down between some pegs that will hold up its sides.

As fast as the drills are sharpened they are sent over to the other side of the heating furnace to be tempered. On the tempering side is a cooling rack where the drills are held for the proper color to come and then are plunged into the cooling tank which is at the side of rack *D*. From the cooling tanks the drills are put in the boxes resting on rack *D*, being placed in the boxes of the mine to which the drills belong. As soon as all the drills for one shaft have

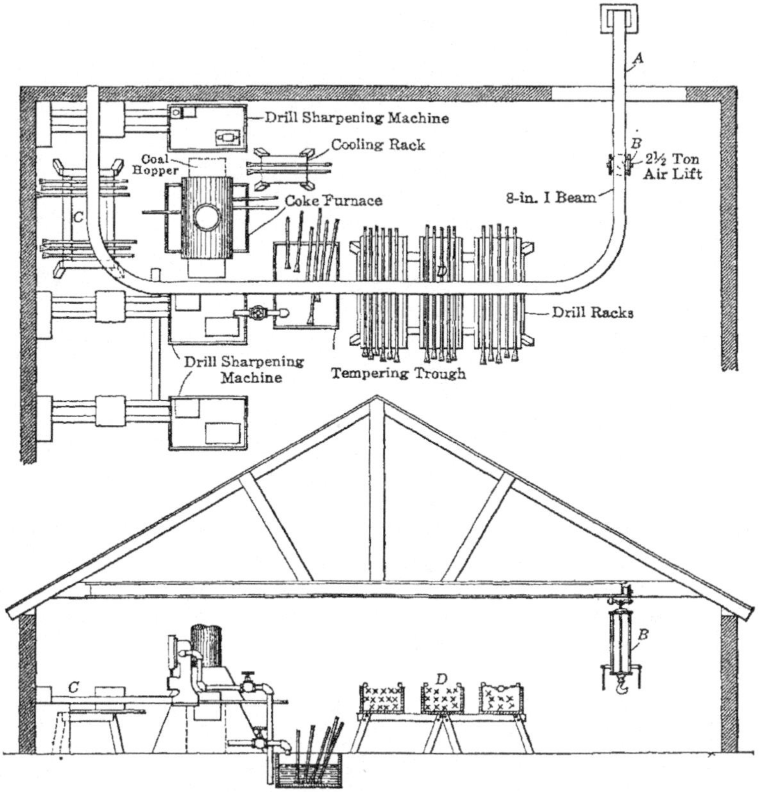

FIG. 30.—DRILL SHARPENING PLANT AT CHAMPION MINE.

been tempered, the boxes are picked up by the traveling air lift, taken back and loaded into the wagon that is standing under the loading place. Then the drills are hauled back to the shaft, where the boys sort out the drills belonging to each machine by the numbers that are cut in their shanks and load them on the cage, those for one machine pointing up, those for the next machine on that level down, thus each machine gets back as many drills as were sent up.

In this way the handling of the drills is reduced to a minimum, and yet all the advantages of central sharpening are retained for scattered shafts. The only investment other than would be required ordinarily is the providing of a wagon costing $50 or $60 to serve each shaft. The hauling is done by one of the teams that is maintained to do general hauling about the mine. This system of handling the steel has been in use for a long time at the Champion mine and is satisfactory.

Mine Dust Prevention on the Rand (By E. M. Weston).—There are two types of dust arresters in use on the Rand, one being merely an arrangement to trap the dust in a wet sack and the other adopting the old suction principle. That designed by Doctor Aymard belongs to the first class and the other is known as Pursers'. The government, the mine owners and to a certain extent the miners have begun to realize the terrible loss to the community and to the industry in the destruction of the skilled-laborer supply and the life, moral nature and efficiency of the miners that is caused by unhealthy conditions in mining. Of these the worst effects as regards health have been caused by the gases produced by explosives and more particularly by the dust produced by blasting, by rock drilling in upper holes and by shoveling.

With regard to dust from drilling uppers with either piston or hammer drills, most miners refuse to use a water jet to kill the dust in the hole, but lately I have met a few miners here who are using water jets in raises. The general complaint is that owing to the splash and drip, it is pleasanter to die of phthisis than of rheumatism. Sprays affixed to the machine have never been popular here, as they increase the humidity of the air greatly and do not lay more than 70% of the dust. Leyner drills or any hammer drills working with hollow steel are not in use, though I think the improved Leyner drill using hollow steel bits in one piece may yet find a place here for certain work. When a fair sized mine like the Nourse blunts 27,000 drill bits a week the question of maintenance of hollow steel is a serious one. I can say that at present there is no serious attempt being made on this field to destroy the dust in the hole itself when boring dry holes.

Many attempts have been made to design an apparatus that will collect all the dangerous dust at the mouth of the hole. Mr. Remeaux of France has recently done some work on the subject. It is comparatively easy to design a device that will collect the dust, but the trouble is to design something that will not detract from the efficiency of the work and will not hinder the men, or take too long to adjust. Pursers' dust arrester, shown in Fig. 31, consists of a short length of piping of such a size at one end and perhaps split so that it can be driven into the mouth of the hole formed by the starter drill bit. To the outer end is fixed a T-piece with the opening pointing downward and to this opening is fixed a reducing piece and an air cock and small air jet to act as an injector and on the end is a wet bag or a pipe opening under water. The idea being that after the hole is started the pipe is driven in and the air connection made and the suction of air will draw in the dust. There are several disadvantages connected with this device: (1) It involves the use of a certain amount of air and, as the ordinary 3/4-in. hose in use does not supply a 3 1/4-in. drill with sufficient air as it is, it thus hinders drilling. (2) It could not be used in a steep hole as the dust would fall past the opening. (3) It involves the use

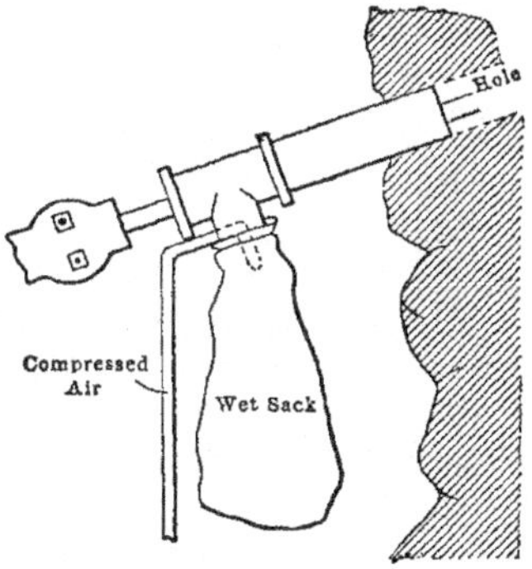

FIG. 31.—PURSERS' DUST ARRESTER.

of an air hose about the feet of the drill tender, a "spanner boy," where it is sure to get damaged or to be in the way in removing drills from the chuck.

The construction of Doctor Aymard's device and its use with hammer and piston drills is shown in Fig. 32. A conical ring, which can be made by riveting sheet iron (or better by a drop forging out of a piece of weldless tubing) is hung by two trunnions in the yoke of wrought iron which is connected with a bar

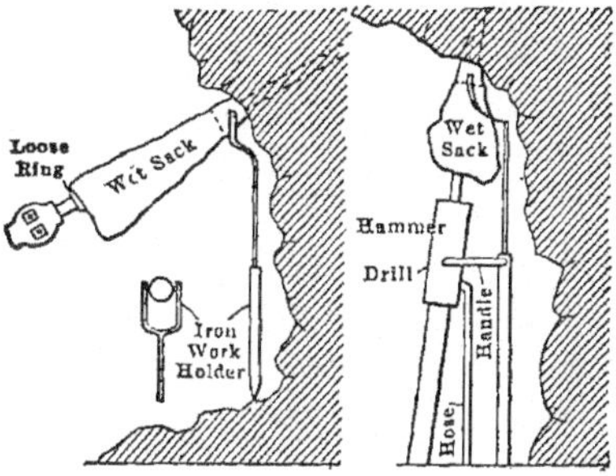

FIG. 32.—AYMARD'S DUST COLLECTOR.

sliding in a tube which can be clamped in any position by the set screw shown. Over this ring is sewn a bag of jute sacking of the shape shown and with a piston drill the other end of the sack is supported by a loose ring which slides on the drill bit. With a hammer drill the bottom of the bag is tied round the drill steel. This apparatus requires some small trouble in fixing, but I believe it to be a practical device that might be adopted in many American mines with advantage where the hammer drill has already earned the unenviable name of "widow maker." The miner if he wishes to, can make a good use of this device which costs only £1 to make, and it will pay miners to install it. The sacking is, of course, kept damp to settle and collect the dust. I have heard it said that the bag is liable to catch the chuck of the machine but this is easily guarded against with care. On the mines of the Rand, the only real objection to the use of dust arresters with piston drills is that where three drills are employed in one face, there is no room for their use; but on one of the large mines the other day quite a number of these devices were shown that had been wilfully damaged by the miners to avoid using them. The new miners' phthisis compensation act will have one good result as it makes it in the interest of the mine owners to secure as many convictions as possible against their workmen for breaches of the act, as three convictions render a miner ineligible for compensation.

The Dwyer Dust Arrester.—The Dwyer dust arrester is intended for use with rock-drilling machines that can be so operated that the exhaust air will pass down the tubular steel bit to the bottom of the hole being drilled, to blow the dust and cuttings made by the bit through the annular space between the bit and walls of the hole. The device is a receptacle that permits escape of the air, but retains the cuttings. In the accompanying illustration of the device,

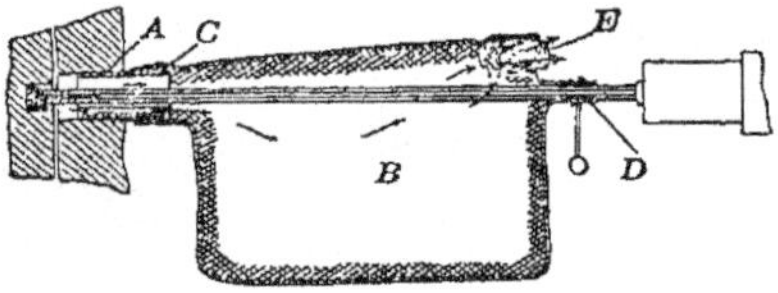

FIG. 33.—DWYER DUST-COLLECTING DEVICE.

Fig. 33, *A* is a sheet of metal rolled into a cylinder, the edges of which are free to overlap as much as may be necessary to introduce it into the mouth of the hole cut to sufficient depth by a starting bit, to permit introduction of the collar far enough into the hole to obtain a tight fit. To the collar is tied a bag *B*, which has an opening at *C* for slipping over the collar. The bag is attached to the collar by an elastic band. At *D* there is an opening in the bag through which the drill passes. To the neck of the bag at the point *D* is attached a cord carrying a weight at the end. This cord is wrapped about the neck of the bag to hold it to the drill; the weight making tying unnecessary. The bag *B* may be made of some light open fabric through which the air will pass, but which will filter the dust, or heavy material, such as leather, may be used, in which case a large opening *E* is made in the bag in which a sponge is held, through which the air escapes while the dust is retained. The device was invented by William E. Dwyer, of Leadville, Colo., and is patented.

Water Blast for Allaying Dust.—The James water blast, illustrated in Fig. 34, was developed in South Africa and now is used in many of the mines on the Rand. In a recent report by the Royal Commission on Mines on New Zealand, this type of dust allayer was recommended for adoption in the mines of that country. The blast is intended especially for allaying dust and absorbing noxious powder fumes immediately after blasting. As shown in Fig. 34, the appliance is simple, consisting of a short piece of 6-in. pipe about 10 ft. long closed at each end by a flange which has been tapped and threaded so that the 6-in. pipe can be installed in any part of the 2-in. compressed-air mains. When

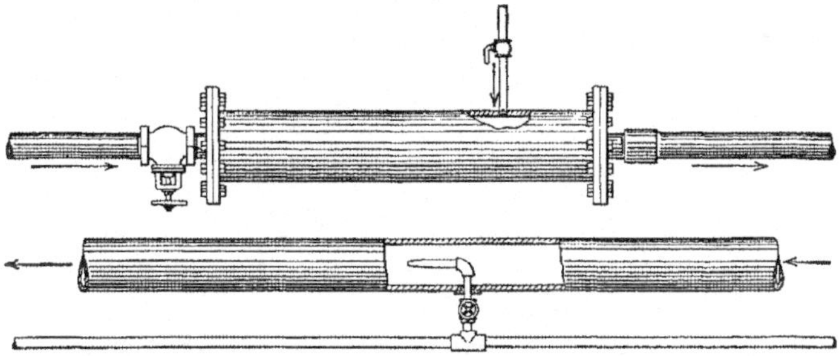

FIG. 34.—WATER BLAST AND DRAFT INDUCER FOR ALLAYING DUST IN DRIFTS.

placed in position the side of the 6-in. pipe is tapped and threaded for a small pipe by which connection is made with a supply of water under low pressure. This may be obtained conveniently by providing a tank or cistern for water storage in some part of the mine above the level in which the water blast is used. Valves in the 2-in. service pipe and in the small water supply pipe are used to control the flow of air and water. While drilling is going on the air valve is left wide open, but the water valve is closed. When ready to blast the miners close the air valve and open the water valve; the 6-in. cylinder is quickly filled with water. As soon as the blast has been fired the air valve is opened suddenly causing the water contained in the 6-in. pipe to be ejected into the face of the drift in the form of a fine spray. The spray of water is effective for a distance of 30 or 40 ft. back from the face. To promote further ventilation an induction draft pipe as shown in the sketch may be used to deliver air at the face or to withdraw air from it. The water blast not only allays the dust made while blasting, but wets the broken material so thoroughly that little or no dust is raised by shoveling.

IV

SHAFT WORK

Methods in Use—Timbering—Use of Steel and Concrete—Shaft Stations and Skip Pockets

Shaft Sinking at the Pioneer Mine.—The vertical shaft of the Pioneer mine, Ely, Minn., is being sunk 200 ft. from the 1400-ft. level. Fig. 35 shows the method followed in this work. The sinking had to be done without interfering with the hoisting of ore. A station about 25×60 ft. was cut on the

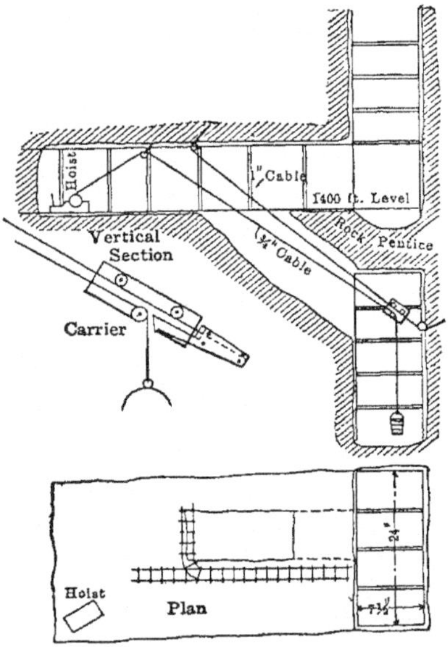

FIG. 35.—SHAFT SINKING UNDER ROCK PENTICE, PIONEER MINE.

1400-ft. level and an inclined winze started near the center of the station. This winze was extended on about a 45° slope until it intersected the line of the main shaft leaving a rock pentice in the bottom of the shaft just above where the new work was to begin. After the completion of the winze a 1-in. cable was anchored in the roof of the station and in the side of the shaft extension, as shown. This cable was used as a track upon which a carrier was operated. A small hoist was placed near the farther end of the station, and a 1-ton bucket was used for handling the broken rock. The bucket is drawn up by a 3/4-in. cable until the bail strikes the carrier on the stationary cable. The carrier then conveys the bucket to the station where it is emptied into a car and trammed by hand to the shaft and hoisted on the man and timber cage, thus not interfering with the hoisting of the ore. The shaft is 7 1/2×24 ft. and has four compartments.

Rapid Shaft Sinking in Butte (By C. J. Stone).—The following notes concern more particularly the equipment and the methods employed in sinking the shaft of the Butte-Alex Scott Copper Co. below the 1400-ft. level, rather than any general description of methods in the Butte district. During April, 1910, an effort was made to attain the greatest possible speed at shaft sinking, consistent with good work and safety to the miners, and as a result 106 ft. was sunk from the 1400-ft. level in 30 working days.

The shaft has but two compartments, each being 4 ft. square in the clear. The rock was all hoisted to the surface in straight-sided buckets 27 in. in diameter by 42 in. deep, swung from the bottom of a skeleton sinking cage of light construction. The sinking cage measures 16 ft. from its bottom to the top of the sinking shoes. A previous sinking campaign had developed serious trouble from the loaded bucket swinging and striking the wall plates of the shaft at times when rapid hoisting was necessary. To eliminate this the bucket is hung from two chains close to the bottom of the cage, only sufficient space being allowed to permit its being detached while at the bottom of the shaft. A ring is welded into the bucket at each side and a finger hook, such as is used on logging chains, is passed through the ring and locked in place by a slip ring. A screw-eye fastens the chain to the cage and furnishes the adjustment. With this device an adjustment can be secured on the chains that will permit only the least amount of swinging of the bucket in the shaft, and hoisting can be done at any speed desired and with perfect safety to the miners below. The chains may be quickly detached to remove the bucket.

The working crew consists of four machine miners and one pump man on each shift, and three eight-hour shifts constitute the day. One of the miners on each shift acts as a working boss and he is paid 75 cents extra per shift. Two 3 1/8-in. Ingersoll-Rand drills are used under an air pressure of 85 to 90 lb. at the compressor. The cut holes are drilled from 8 to 9 ft. deep and a wedge bit is used on the finishing drill. The side or back holes are 6 ft. deep. It requires from 16 to 19 holes to break the ground, which is for the most part a hard granite with the partings or cleavages running the long way of the shaft. The blasting is rarely perfectly satisfactory. Should the ground be particularly soft and the cleavages favorable, a blast will probably break to the bottom of the holes. Under ordinary circumstances, however, from 18 in. to 2 ft. will have to be fired again.

The practice in some large shafts is to blast the cut holes first and after mucking, blast their bottoms until the cut is entirely out, when the remainder of the holes are fired. Experience has shown that better results are obtained if the cut holes are fired with a battery, but the damage to the timbers when sinking in hard rock is so great that the method has not found favor in Butte and the old method of blasting with waterproof fuse maintains. Forty per cent. gelatin dynamite is used.

The water is handled with a No. 7 Cameron sinking pump. The air exhaust is passed through a check valve into the water or discharge column. This eliminates the roar of the exhaust in the shaft and makes it possible for either a Knowles or a Cameron sinker to lift water 200 ft. in place of 100 ft., which is the normal lift of a No. 7 pump. In the sinking of the Alex Scott shaft the flow of water varied from 20 to 30 gallons per minute and no time was lost during the month because of water in the shaft.

The hoisting was done as rapidly as possible. During the mucking hours the bucket was brought to the surface from the 1500-ft. level in from 30 to 45 seconds, according to the engineer. The hoisting engine is of the first motion type, built for high pressure; the cylinders are 12×36 in. and the drum is 5 ft. in diameter. It was built by the Nordberg Manufacturing Company.

The timbering is the usual shaft set. The sets are of 10×10-in. black larch and fir timber, placed 5 ft. between centers and lagged with 2×10- or 3×11-in. plank. Each set is thoroughly blocked and wedged and absolutely no cutting is allowed. The shaft must be broken sufficiently large to hang the sets free from the walls and the lagging must be placed loose to permit later swelling of the ground. For blasting timbers heavy channel irons are used, the channels being bolted tight to the bottom set before firing. Openings are cut in the channels for the nuts of the hanging bolts. The ends and the centers are protected in this way as well as the wall plates. A marked difference is noted in the physical condition of the timbers by the use of the channel irons in place of the ordinary 5×10 blasting timbers.

The bonus or premium system was employed as one means of securing rapid work. The ordinary speed of shaft sinking below the 1200-ft. level in Butte is from 65 to 85 ft. per month. As a basis for the bonus, therefore, 75 ft. were taken and the shaft miners and pump men were given each one dollar per foot for every foot that was accomplished above the base during the month. In this instance it amounted to $31 bonus to each man as a reward of merit. The bonus cost per foot amounted to $15, and the entire or actual labor cost for the 106 ft. accomplished, including the bonus, amounted to $36.54 per foot. Should only ordinary speed have been made and the bonus system not employed as an incentive for hard and faithful labor, the cost would have been $45.46 per foot. However, as the actual amount of sinking that otherwise might have been accomplished is an unknown factor, the latter figure is only an assumption on the base or average measurement.

Shaft Sinking at Stella Mine, New York.—The drawings in Figs. 36 and 37 show the methods of shaft sinking used by the St. Lawrence Pyrites Co., DeKalb Junction, N. Y. In each case it was desired to sink the shaft and at the same time continue hoisting ore. As there were no levels below the shaft sump, there was no opportunity for extending the shaft by means of a raise from lower levels. The Stella shaft is inclined at an angle of 18°; the Anna at 45°.

At the Anna shaft advantage was taken of an existing crosscut, and a stope in an upper branch to start the new section of the shaft, back of and above

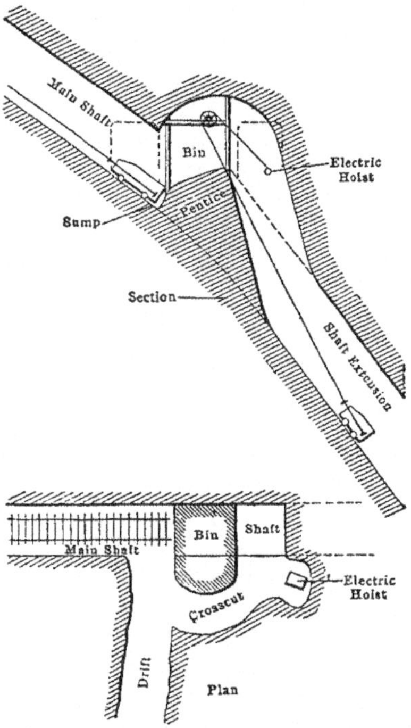

FIG. 36.—DETAILS OF ANNA SHAFT EXTENSION.

the sump. The hanging wall was taken out sufficiently to allow the construction of a small ore bin, as shown in Fig. 36, and at the same time provide space for the sheave. An electric hoist was installed in the crosscut a little to one side of the shaft. The ore bin is constructed just above the incline shaft so that the ore can be dumped directly into the skip and taken to the surface. A small self-dumping skip is used in the shaft extension and all the material from the shaft is dumped into the ore bin.

The method employed at the Stella shaft is to drift out about 20 or 25 ft. from one side of the main shaft and then sink on the incline at such an angle as to meet the line of the permanent shaft at a distance of 30 to 35 ft. below the working level. One reason for not cutting out the hanging wall, as was done at the Anna, was that the roof was barren. In the Stella, all of the work was done in the main orebody. When the line of the main shaft was reached, this auxiliary shaft was then deflected to as to be a continuation of the desired shaft. The ore from the auxiliary shaft is dumped into the bin and transferred by car to the main shaft. Another advantage of sinking the shafts by these methods is that the sump is maintained in perfect condition and the pumps are able to take care of all of the water so that the work below was comparatively dry.

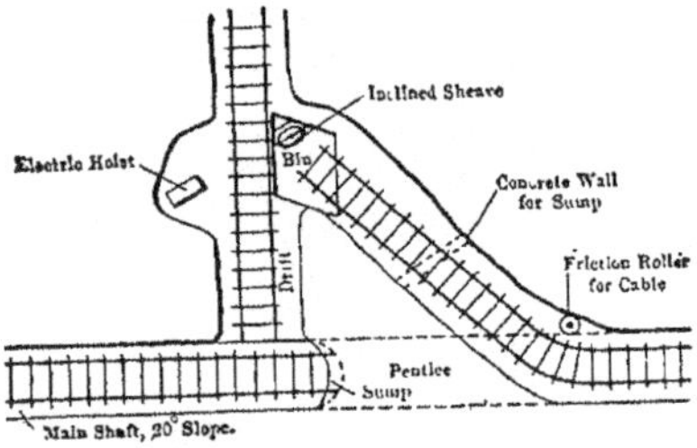

FIG. 37.—PLAN OF STELLA SHAFT.

Bucket Trolley for Shaft Sinking (By L. E. Ives).—The problem of maintaining a normal production of ore from a shaft in which sinking is being carried on, is one that constantly confronts the mine superintendent. In Fig. 38 is illustrated a method used in the Michigan copper country, at a number of important mines.

The sketch shows a section of an inclined shaft, but does not show details of timbering. *A* and *B* represent the two bottom plats from which ore is being mined and hoisted. Below the bottom plat *B*, and not indicated in the sketch, is a pentice, or bulkhead, which prevents the skip, in case of accident, from descending upon the miners who are engaged in sinking at the working face *M*. While affording protection to the miners, the pentice at the same time precludes the possibility of loading the muck from the working face, directly into the skip. Again, even were this possible, it is undesirable for the reason that the skip would be taken out of ore-hoisting service for too long periods.

The pentice is extended across both skip compartments, but terminates just short of the ladderway. Over the center of the ladderway, and at the proper spacing, holes are drilled in the hanging wall into which iron eye-pins *N* are inserted and wedged. From these eye-pins, by means of chains *G*, is supported an I-beam *F* in such a way that the flat side of the flange is parallel to the hanging wall and about 10 in. or a foot from it. Along this I-beam runs the trolley *K*. The latter is triangular in form, but the angles vary with the dip of the shaft in which it is used. It is so constructed that when the trolley and bucket are being hoisted, or in other words, when the trolley is running along the I-beam, the bucket is suspended vertically beneath the pulley, which is farthest up the shaft along the I-beam. Only one rope *E* is used and in lowering the bucket *L*, when the trolley is stopped by a projection, placed for the purpose at the lower end of the I-beam, the rope continues to lower the bucket to the working bottom, as indicated by the dotted lines. Here the rope is unhooked from the bucket, if desired, and the latter may be moved to any part of the

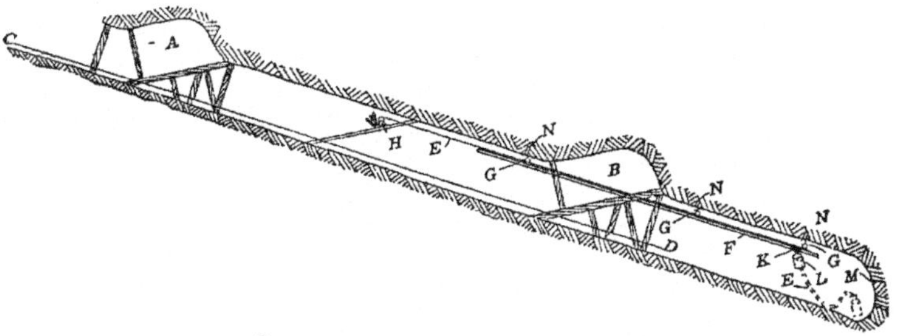

FIG. 38.—BUCKET TROLLEY FOR SHAFT SINKING.

bottom. In hoisting, when the bucket reaches the trolley, the former automatically clamps into the latter and with no delay whatever, trolley and bucket together continue to move upward, until the point for dumping is reached.

In dumping and loading into the skip two methods are in use. In either case the bucket is dumped into a chute. In one case, however, the contents run immediately into a tram car and this, when filled, is pushed around and dumped into the skip, just as ore would be. In the other case, and this is used preferably where only one compartment is being used for hoisting ore, the bucket is dumped into a bin which is built in the unused compartment. The mouth of this bin is so placed that when the latter is full and a skip is available, the gate is opened and the entire contents of the bin are dumped at once into the waiting skip. For hoisting the bucket and trolley an ordinary wire rope is used and a compressed-air hoist *H*, usually called a puffer, is placed on a platform, built at a point between the two deepest levels. The hoist is operated easily by a boy. *C D* indicates the top of the skip rail.

A Two-way Shaft.—We once observed in Colorado a unique method of prospecting a vein, which is not to be generally recommended, but in this case well served its purpose, and conformed to the cardinal principle of prospecting, namely, "Follow the ore." A vertical shaft had been started on vein outcropping at *A*. About 50 ft. down, at *B*, the vein was found to split. The chances seemed to be that it was going around a horse, the latter appearing to be of large size, and both branches of the vein looking equally good. In order to follow them both, the vertical shaft was converted into a two-way shaft as shown in Fig. 39. Hoisting was done regularly from both branches. Skids were laid in branches *B C* and *B D*, the buckets sliding down and being dragged up upon them. The upper end of the skids were extended by a movable switch, constructed of two pieces of timber, pivoted at the lower end. In hoisting from *B C*, the switch was thrown in the position *Z Y*. When it was desired to hoist from *B D*, the switch was thrown over to *X Z* causing the bucket to descend in the desired direction. Of course, it was necessary to place rollers for the cable at *X* and *Y*, which are shown in the sketch in an exaggerated form, in order to make the arrangement quite clear.

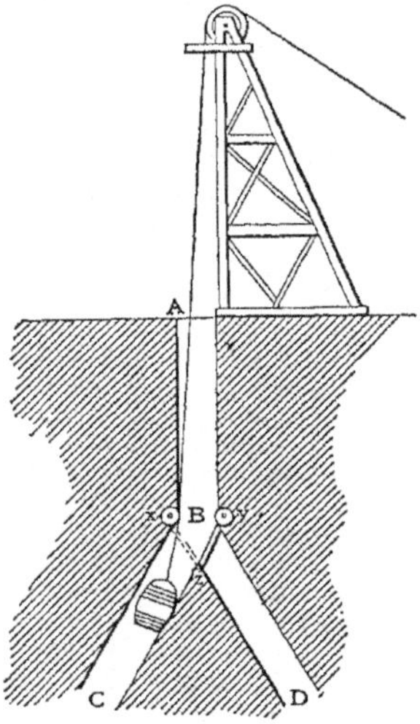

FIG. 39.—AN UNUSUAL TWO-WAY SHAFT.

Securing Loose Rock by Bolts.—In the Camp Bird mine the wall rock is quite hard and in most cases stands well, but in the driving of the main raise extending from the haulage level to the upper workings, through which supplies and men are hoisted, several large slabs of rock began to loosen after the timbering of the raise was well under way. Cracks began to open behind these slabs which would have come down in pieces weighing several tons had is been necessary to blast them. As this could not be done without considerable danger to the timber in the raise, the slabs were bolted in place in the following manner: A slab was securely spragged in place by temporary stulls running across the raise, so that is could not move or at least without giving ample warning. Single-jack holes were drilled through the slabs, which were themselves quite solid, and into the solid ground behind them for a distance of a foot or more. Then the depth of the hole was measured, and an old piece of drill steel, with its head well upset so as to form a strong head like that of a bolt, was cut off to the right length and the lower end split. A narrow steel wedge was then inserted in the split end, which was pushed into the hole, wedge-end first, until the wedge was against the bottom; then, by hammering, the split ends of the drill were spread until the bolt was securely anchored in the wall rock behind the slab of ground, thus holding the slab tightly in place. The sprags were then removed. These slabs had been held by bolts 7 or 8 years when my attention was called to them. Apparently they had not moved a fraction of an inch since they were bolted to the wall behind them. Several bolts were used in each slab placed according to the direction of the crack behind them. The method of securing the bolts in the wall rock is, of course, the same as that of anchoring bolts in concrete when for any reason new bolts have to be used in old foundations.

TIMBERING

Necessity of Strong Partitions in Shafts.—It is usual and proper, in a mine shaft, to separate the compartment reserved for the men from those through which the ore is hoisted to the surface. In some mines it appears to be considered that any sort of timbering is good enough to form the partition, and it would seem as if the temptation to use thin planking which can be nailed in place almost as quickly as it can be sawed, is almost too strong to be resisted. In one such case an accident of a serious nature was only avoided by a mere chance. A skip was being hauled to the surface, containing a number of large pieces of ore. It had been nearly filled, and by some means the top piece became dislodged and fell over the skip into the shaft. In falling it glanced against the timbering of the partition, then went down the shaft, bounding from side to side. As, however, it passed down, its momentum increased, until near the bottom of the shaft, the thin wooden partition was not strong enough to throw it off again, and it crashed through the timber into the compartment reserved for the men. By the merest chance, no one was there at the time. If there had been a workman in the way, he would most assuredly have been killed as the fall was a long one. It is just this sort of thing which a manager of a mine should foresee and prevent as far as possible by making his constructions amply strong, even at increased expense.

Corner Framing of Shaft Timbers (By W. H. Storms).—There is some diversity in the style of framing timbers for shafts. The difference is found chiefly at the corners, and while each method has its advocates, any of these several methods will answer all purposes under certain conditions, for all are in practical use. The drawings shown in Fig. 40 illustrate the several styles. Only the ends of the wall plates are shown, it being understood that the end plates must be framed in exact conformity with the wall plates. In each case the daps in which the posts must rest are either shown or are provided for. In Fig. 1, the dap is shown on both upper and lower sides; in Fig. 2 it shows only on the under side, the corresponding dap for the upper post being cut in the end plate, which is not shown; in Fig. 3 the edge of the dap is shown both top and bottom, and in Fig. 4 it is shown on both upper and lower sides of the plate.

The style of framing shown in Fig. 1, was in use at least 60 years ago, and is about the only style of framing shaft timbers illustrated in old works on mining, such, for instance, as Overman's "Metallurgy," published in 1850. There are mines where this peculiar style of timbering is still in use, but it has nothing in particular to recommend it, while the placing of the end plates, unless they are provided with separate hanging bolts, is accomplished with difficulty, which

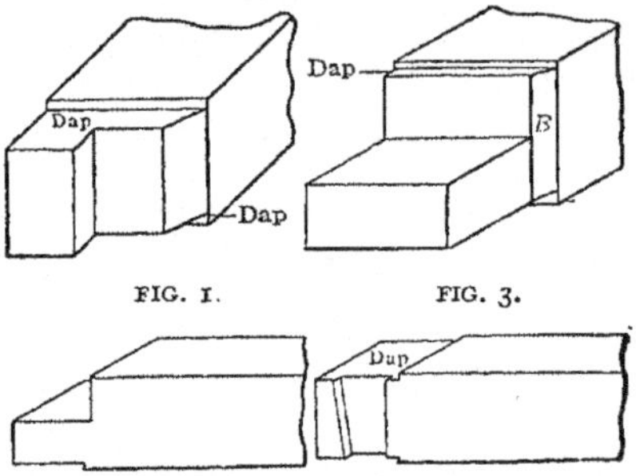

FIG. 1. FIG. 3.

FIG. 2. FIG. 4.

FIG. 40.—SHAFT TIMBER ENDS.

is not the case with any of the other methods here shown. Fig. 2 represents the end of a wall plate halved and provided with a dap on the lower side to accommodate the post of the set below, at that corner, and a corresponding dap must be cut on top of the end plate that will rest upon it. This is a simple and common way of framing timbers for vertical shafts. The method of framing shown in Fig. 3 is similar to that in Fig. 2, the only difference being in the bevel, shown at *B*. This bevel is for the purpose of reducing the tendency of the timbers to split under heavy side pressure, but ordinarily, where the side pressure is sufficiently great to cause timbers to split when framed as shown in Fig. 2, the bevel will make but little difference. Neither the style illustrated in Fig. 2 nor Fig. 3 is well adapted to shafts which depart more than 15° from the vertical.

Timbers for inclined shafts are now framed almost universally as shown in Fig. 4. This method is equally adapted to either inclined or vertical shafts. It will be noticed that the so-called dovetail mortise is beveled on one side only, the other side being normal (at a right angle) to the top of the timber. A modification of this is sometimes seen where both sides are cut with a bevel. This has no advantage and is also more troublesome to cut. All the marking in the laying out of shaft timbers can and should be done with the use of a templet. This should be made of a good sound piece of wood, for the back, to which should be secured steel plates of the proper size and shape to indicate the various cuts to be made. When a templet is used there are fewer mistakes

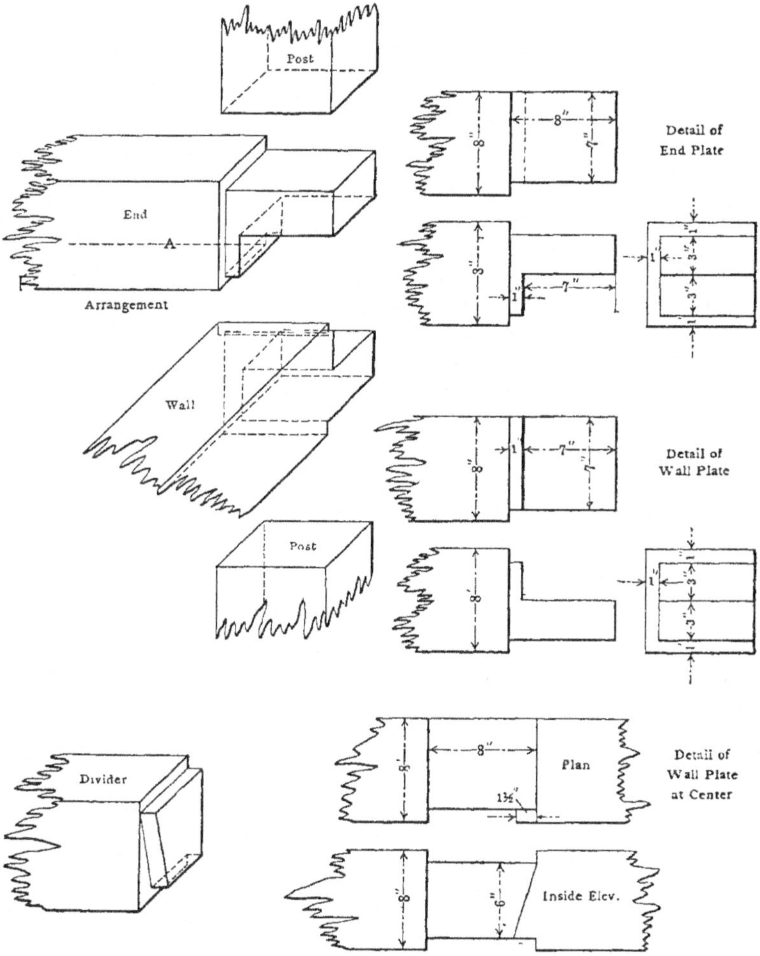

FIG. 41.—METHODS OF FRAMING TIMBER.

made in laying out the work. A man may be an excellent carpenter and yet know nothing whatever of laying out marks for framing shaft timbers. Then too, with the use of the templet all the timbers are framed exactly alike, all cuts corresponding in position and size, which is sometimes not the case when the work is laid out with a square, unless the man be a most careful workman.

Figure 41 shows the method of framing shaft sets from 8-in. timber which is probably in most common use. It combines the maximum strength with the minimum labor of framing. Angles other than right angles are to be avoided as far as possible. They are hard to fit. To frame this joint, fasten templet to side *A* of wall and end plates and frame top and bottom faces of tongue and shoulder, squaring from the templet. Also frame daps for posts at center of wall plates and mark top and bottom of dap for the divider. The sides may be framed and squared from these faces when finished. The wedge-shaped gain on divider holds the divider on the bottom set and avoids the deep cut in the wall plate necessary for a tongue.

The false set which protects the bottom timbers from the blasts is easier to handle if made of half-round logs without framing. The false wall plates should completely cover the shaft wall plates and be attached by bolts through the holes drilled for the hangers. The false end plates and divider need only reach from one false wall plate to the other and holes should be drilled in the shaft-end plates to accommodate bolts in the same manner as in the wall plates.

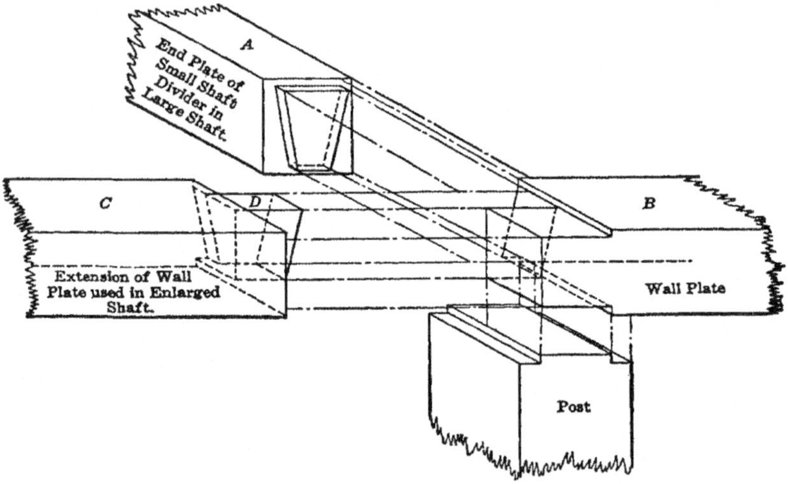

FIG. 42.—FRAMING FOR SHAFT TIMBERS TO ALLOW FOR ADDITIONAL COMPARTMENT.

Method of Extending Shaft Timbers (By D. A. McMillen).—In timbering shafts it is often necessary to devise some means of converting an end plate into a divider and extending the wall plate so as to add another compartment. In the ordinary procedure, when adding an extra shaft compartment, it is often cheapest to retimber entirely that portion of the shaft which is to be enlarged, as the ordinary wall plate serving for a two-compartment shaft will not do for one of three compartments. If the end plate of the two-compartment shaft is framed in the beginning, as shown in Fig. 42 at *A* and the wall plate on the side to be extended as *B*, it is comparatively easy to add an extension *C* to the wall plate and to fit these together, making the timbers *B* and *C* act as wall plates, and *A* as a divider instead of an end plate. The scheme thus simply resolves itself into a matter of cutting the wall plate *B*-*C* into two parts that can be afterward fitted together. A block to conform with the shape of *D* is usually fitted into the open space that is left before the timbers of the extra compartment are added. This system has been adopted in several places in the Globe district and has proved satisfactory.

Shaft Timbering at the Keystone Mine (By William H. Storms).—The Keystone mine at Amador City, Calif., is one of the largest mines on the famous Mother Lode. It has been extensively developed by a number of

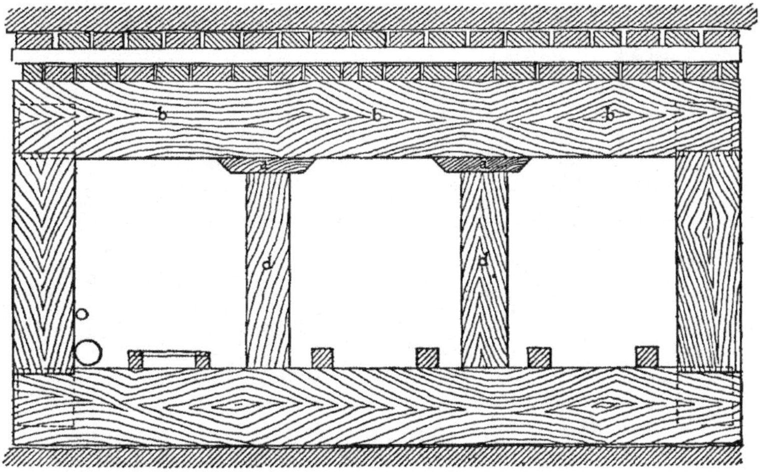

FIG. 43.—METHOD OF TIMBERING THE KEYSTONE SHAFT.

incline shafts of varying depth, the deepest being the Patton shaft near the north end of the property, which was started by W. H. Patton over 40 years ago, and now has a depth of 1573 ft. This shaft, beginning in the slate country rock, at a depth of about 400 ft., is approached by a fissure which occurs in the hanging-wall side of the shaft and continues in or near it to about the 800-ft. level. The ground is very heavy, and has for years been a source of annoyance and expense in keeping the shaft open. Time and again the timbers have been reinforced and renewed until, in some of the worst places the shaft is supported by a solid crib work of great timbers. One effect of the heavy downward pressure of the swelling ground on the hanging wall side of the shaft has been to force the caps down upon the ends of the dividers, causing the latter to cut deeply into the caps, thereby weakening the timbers and reducing the size of the shaft, hence necessitating frequent repairs. A few months ago a scheme was introduced in the retimbering of the Patton shaft which has been found to give complete satisfaction and which is evidently going to result in the

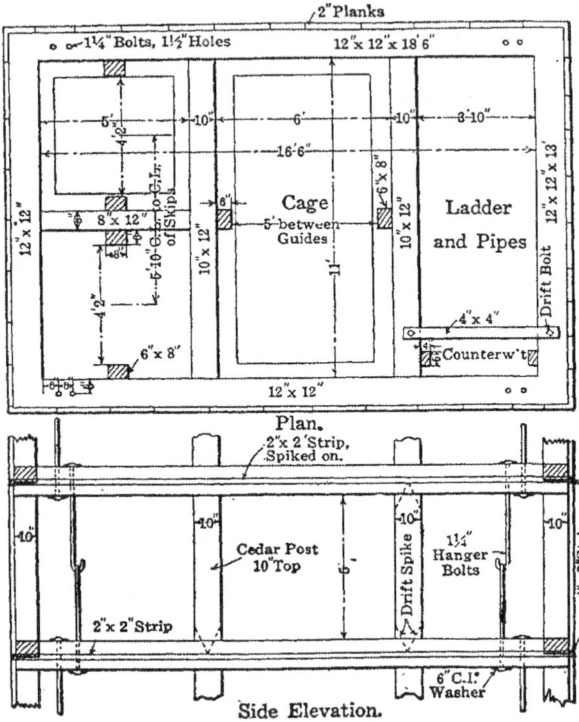

FIG. 44.—ROGERS SHAFT BELOW CONCRETE PORTION.

saving of much time and timber in repairing this shaft. The idea is illustrated in Fig. 43. It is simply the insertion of a heavy head block *a* between plate *b* and divider *d*, as shown. No mortise or dap is cut in the wall plate, and the divider is sawed off square. The plate timbers are 18 in. square, the dividers 12×18 in., and the head block is 18 in. wide, 24 in. long and 4 in. thick. It is cut as shown, and is slipped in between the wall plate and the top of the divider and held in place by driving in shingle wedges. The pressure soon exerts itself, however, and wedges then become superfluous.

Rogers Shaft at Iron River, Michigan (By H. L. Botsford).—The accompanying sketches, Figs. 44 and 45, illustrate the size and timbering of a new shaft which the Munro Iron Mining Co. is sinking at a property near Iron River, Mich. A concrete drop shaft has been sunk through the overburden by The Foundation Co. of New York. From this point on, the sinking will be continued by the mining company and the shaft will be rectangular in section,

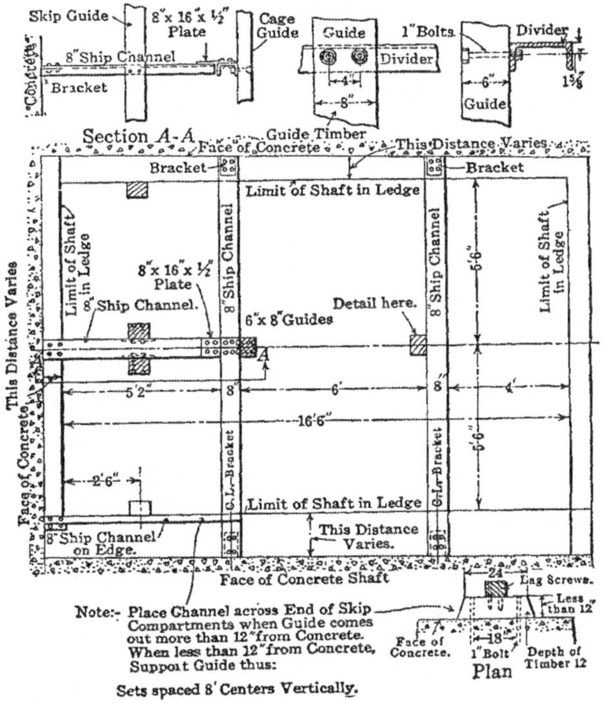

FIG. 45.—STEEL WORK IN CONCRETE PORTION OF ROGERS SHAFT.

11×16 ft. 6 in., inside dimensions, with two skip-compartments, a cage-way and a combined pipe- and ladder-way. One end of the latter compartment is arranged to accommodate the counter-weight for the cage.

Figure 44 shows a plan and side elevation of the shaft as it will be constructed below the concrete portion. The skips, which will be operated in balance, are of liberal width to facilitate their loading through the chute at the bottom of the shaft. In the past it has been found that where the skip is narrow, and blocky ore is hoisted, considerable trouble is experienced by the clogging of the ore in the loading chute; with the width of skip as designed, it is expected that this difficulty will be obviated. The cage compartment is 6×11 ft. and will permit the operation of a cage large enough to take a 55-cu. ft. saddle-back tram car, this being the size of the underground car which will be used in the mine. The size of the cage will also permit the hoisting of rock on it, if this is found to be necessary during the ore-shipping season. Another advantage of the large-sized cage is, that cars may be loaded with mine timbers at the timber yard on surface, trammed to the shaft, lowered into the mine, and taken to their destination underground without further handling of the timber.

The steelwork which will support the guides in the concreted portion of the shaft, and the bracket for attaching the steel members to the concrete walls is shown in Fig. 45. These sets will be spaced 8 ft. between centers. Bolts which fasten the brackets to the concrete will be split 2 1/2 in. on the bottom end and a wedge inserted before driving them in the holes drilled for them in the concrete wall; in addition they will be leaded in.

While the initial cost of such shaft construction is high, the advantages are many. Water difficulties are overcome to a large extent and stability and alignment are assured.

Combination Post and Set Timbering in Shafts (By Claude T. Rice).—The ground in the inclined shafts that follow the Calumet & Hecla conglomerate is so heavy that a crew of seven men is kept at work easing the timbers. The manner of supporting the roof when it becomes bad is novel, in that the main part of the top weight is carried on posts while the scalings from the roof are supported by lagging carried on regular shaft sets. When the shaft is first sunk, only the posts, or the "end timbers," as they are called locally, are put in, and the shaft is made 22×9 ft., but after a time the weight begins to come on the roof and the shaft pillars begin to flake away under the pressure, increasing the width to about 25 ft. by the time that the shaft sets are placed.

The posts are put in with foot and head blocks, built up of 6-in. pieces criss-crossed to make a head block 18 to 24 in. thick, and a foot block about 12 in. thick. Then when the weight comes on the posts the crushing of these thick head blocks gives the ground a chance to adjust itself to the new conditions before the posts are injured. The posts when first put in are about 6 1/2 ft. long, from 2 1/2 to 3 ft. in diameter. Posts of Georgia pine last about two years before they need to be reheaded and will stand three reheadings, each of which lengthens the post's life two years, so that the posts have a life of seven or eight years in these shafts. Some hemlock posts are used, which last about two years before they have to be replaced. No reheading of the hemlock posts is possible, as they rot too fast for it to pay. A few hard-wood posts have been tried, but lasted scarcely a year, rotting into a mush-like pulp, in an amazing manner.

The post is reheaded as soon as the head block has been compressed to the limit and before compression of the post itself has begun. Owing to the pieces in the head block being laid so that they take the weight across their fibers instead of along them, the compression is confined entirely to the block until the last, when the wood fiber has become so compacted that it is as firm as the fiber of the posts. This reheading consists of sawing and chiseling out a cut across the top of the post so as to allow the post to be knocked out. About 6 in. is lost in this way at each reheading. Then a new set of foot and head blocks is put in, and the post is good for another two years, being practically as strong as when first put in. When the second set of head blocks has been compressed to the limit, the post is again reheaded, but before the post has had to be reheaded, the roof will have begun to scale off partly under the weight and partly because of swelling, caused by oxidation of lime minerals in the hanging wall. Lagging has, therefore, to be put in over the skip compartments, and this lagging is carried on regular sets such as are commonly used in inclined shafts. Owing to the swelling of the foot wall, which gives especial trouble through raising and warping the tracks so that the skips will not stay on them, no sill is used under the posts of the shaft set; instead foot blocks of liberal proportions are used.

The lagging rests on 14×14-in. caps of pine. In a shaft having two hoisting compartments, the cap is in two pieces that butt against one another over the manway, which is placed in the center, as shown in Fig. 46. Above the cap proper is carried a false cap of round timber whenever there is sufficient room. This rests on blocks placed on the cap pieces directly over the posts. Under the cap and between the posts and the cap are squeezing pieces of 6×14-in. timber, about 24 in. long, with their ends beveled so that they will bend and have less tendency to cut into the caps. The squeezing pieces take most of the crushing in these sets, as by them the pressure is distributed over a larger area on the cap than on the post. In other words, there is a concentrated pressure on the underside of the squeezing pieces, and a distributed pressure on the top side. Consequently, the post cuts up into the underside of the squeezing piece, but the squeezing piece does not in turn cut into the cap. As the weight comes end-on on the fibers of the post, it sustains little injury when it cuts into the squeezing piece.

Studdles are put in to brace the timbers, always at the bottom and generally also at the top of the posts. In order to provide room for the circulation of air between the stull, or end-timber posts and the square-timber posts, as the shaft sets are called, a 3-in. block is put in between. This prevents decay starting on the posts where the two would otherwise be in contact.

In easing the timbers, the men work on top of the lagging that is carried on the square timbers, and throw the rock that comes from the easing of the roof off at the far sides of the shaft so it can be scraped down to the level and loaded on the sides. Some of it cannot be prevented from falling into the manway and is worked down to the level that way. By this arrangement the timbermen are able to work over the skipways without interfering with the hoisting; indeed it is to accomplish this that the square timbers are used.

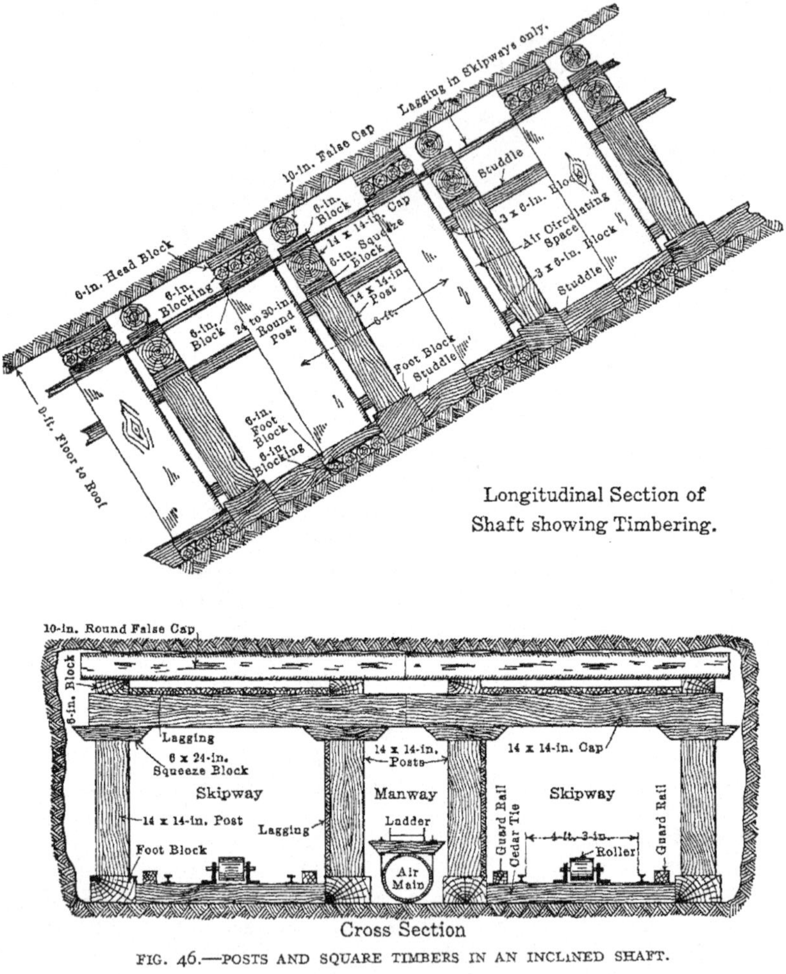

FIG. 46.—POSTS AND SQUARE TIMBERS IN AN INCLINED SHAFT.

The combination of posts to take the bulk of the top weight with sets to carry the lagging is a most admirable scheme for timbering inclines in heavy ground, for it throws most of the wear and tear on the round posts, which are the cheapest elements in the combination to replace. Moreover, by using the squeezing pieces the caps are saved from injury and the life of the square-timber sets is increased.

The manway is put in the center of the shaft, so that the men who replace the rope idlers can inspect the track between idlers without entering the skip compartments. The ladder is carried on pieces of timber clamped to the air main, the ladder being bolted to the crosspieces by staples. There is some objection to this arrangement, as if any repairs have to be made on the air-main, the ladder has to be removed. Still there is the advantage that the ladder is kept clear of the ground, so that rocks cannot accumulate under it which is one of the things that has to be guarded against in an inclined shaft.

Timbering Swelling Ground (By George C. McFarlane).—In swelling ground it is noticeable that the swelling is always at right angles to the foliation, and drifts paralleling the stratification may require double setting, while the crosscuts will stand without timber. I have also noted cases where drifts in the upper workings had stood for years without retimbering, while below the oxidized zone heavy sets were crushed and broken in 2 months. As a rule this swelling ground is not difficult to retimber. Ground that is heavy because the rock is loose and full of slips often comes down in large masses; when a couple

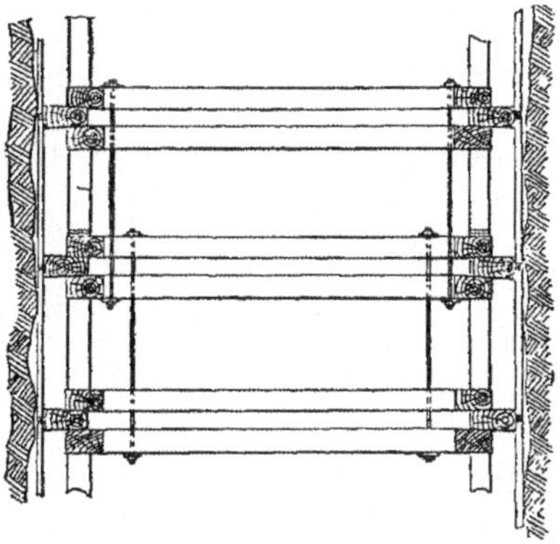

FIG. 47.—SHAFT TIMBERING IN SWELLING GROUND.

of sets break, the fall may bring down adjacent sets. On the other hand, in a drift in swelling ground the timbers may be crushing and binding and when a broken set is knocked out, only three or four wheelbarrow loads of rock will come down.

For replacing broken shaft sets in swelling ground I devised the form of timbering shown in Fig. 47. Two sets of 6×10-in. timbers with a 6×12-in. filler between the wall plates are used. One side of the 6×10-in. piece and both sides of the filler are sawed on a bevel of 1/2 in. as shown. In sawing, the bevel cut is made by placing a bar of flat iron across the bunks of the saw carriage in front of the head blocks and canting the stick back on the bar. In placing the sets, the posts must be put in tight, using two or more jacks to bring the plates and filler to a solid bearing at each post. In sinking a new shaft with this kind of timbering the set would be gripped together by hanger bolts until a set of dead logs were placed. As the walls of the shaft swell, the filler is of course forced in between the plates, causing the posts to sink into them.

If the plates and filler grip tighter on one side, the set will be crowded toward the slack side as it takes the squeeze. In an hour or so between shifts this can be remedied by jacking in a couple of extra posts on the slack side, and in this way the guides can be kept in good alinement. This form of set allows the walls of the shaft to swell 4 in. without breaking the timbers. Should the swelling force the filler flush with the plates, pieces of lagging can be removed, a few at a time, the wall cut back a few inches and the lagging reset by inserting blocks between it and the wall plates. The posts can then be removed and the filler jacked out with a couple of pipe jacks, the posts reset and the temporary blocking between the lagging and the wall plates removed. One set of posts will have to be chopped out to release the fillers of the first set; after that, many of the posts and part of the lagging can be recovered in shape to use again.

Placing Shaft Timber.—At the Iron Blossom mine, in the Tintic district, Utah, shaft sets are put together at the bottom of the shaft and then hoisted into position. When a set of timbers is to be put in, the framed pieces are lowered on the cage, temporary guides being used so as to allow the cage to drop below the point to which shaft timbering has advanced. The wall plates are laid upon a 5-ft. board placed across the bottom of the cage. The end plates and dividers are then dropped into place and the sets drawn tightly together. Wooden dowels may be used to secure the framed ends to the wall and the end plates. When the set is put together, the cage is hoisted to the proper point and the rigid set drawn up against the posts by hanging irons from the next set above. By thus making up the shaft set before it is put into position it is claimed that time is saved and more rigid sets are insured.

Supporting Guides or Runners in Shaft.—The scheme shown in Fig. 48 for the support of shaft guides is used at the Tobin and Dunn mines at Crystal Falls, Mich. The guide itself is 5×8 in., and is fastened to the shaft timbers by two 3/4-in. lag screws. In addition to this a 3×3-in. angle iron is used every 10 or 15 ft. in the shaft. This angle iron is bolted to the runner with two 1/2×6-in. bolts and fastened to the shaft timber with two 1/2-in. lag screws. The original method was to use only one lag screw in the center of a 5×6-in. runner. This was found to be entirely too weak, and the screws almost invariably broke at a point where the thread begins. With the present arrangement the runners do not work loose, except as the acid waters may corrode the bolts.

Holding Shaft Timbers with Wire Cables.—The Fremont shaft, at the Fremont Consolidated mine, near Amador City, Calif., has two compartments and dips at an angle of 52°. It is 650 ft. deep with a 50-ft. sump, and is true throughout its depth, being unquestionably the best inclined shaft on the Mother Lode. In sinking this shaft, some heavy ground that caved badly was encountered. It was impossible to get a bearing for the wall plates or caps, and the more the ground was trimmed away to secure a bearing for these timbers,

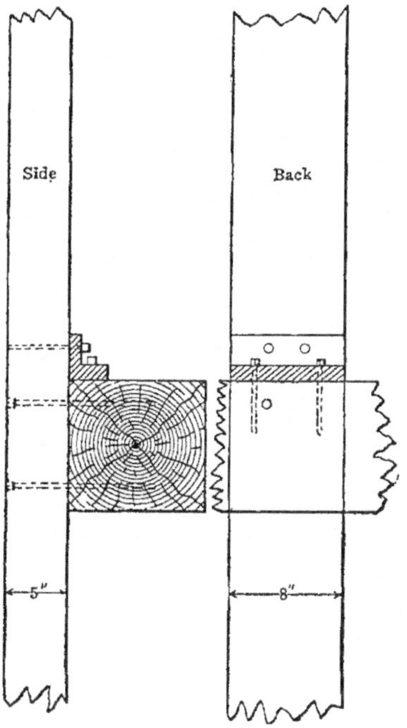

FIG. 48.—METHOD OF SUPPORTING SHAFT GUIDES.

the worse it caved, until a large cavern was formed above the shaft. In order to timber the shaft through this ground, the expedient of securing the timbers in place with old hoisting cable was tried and proved quite successful. The sets in the caving zone were tied with the cable to those above which had firm bearings in the wall rock. This hanging of the timbers was continued until firm ground that would give sufficient bearing for the timbers was again encountered. Stringers were then placed over the suspended shaft sets and upon them a cribbing built up in the opening; old timbers and waste were stowed in it until it was entirely and tightly filled. No trouble has since been experienced with the shaft at this point.

USES OF STEEL AND CONCRETE

Steel Shaft Sets on the Mesabi Range (By F. A. Kennedy).—The steel sets used by the Shenango Furnace Co. in its Whiteside mine near Buhl, Minn., are illustrated in Fig. 49. The inside measurements are 6 ft. × 18 ft., 8 in. The wall plates and end pieces are 5-in. H's weighing 18.7 lb. per foot, with 10-in. I-beams for dividers. The sets are spaced 4 ft. center to center and held together with eight 3 1/2-in. angle studdles. All angles are

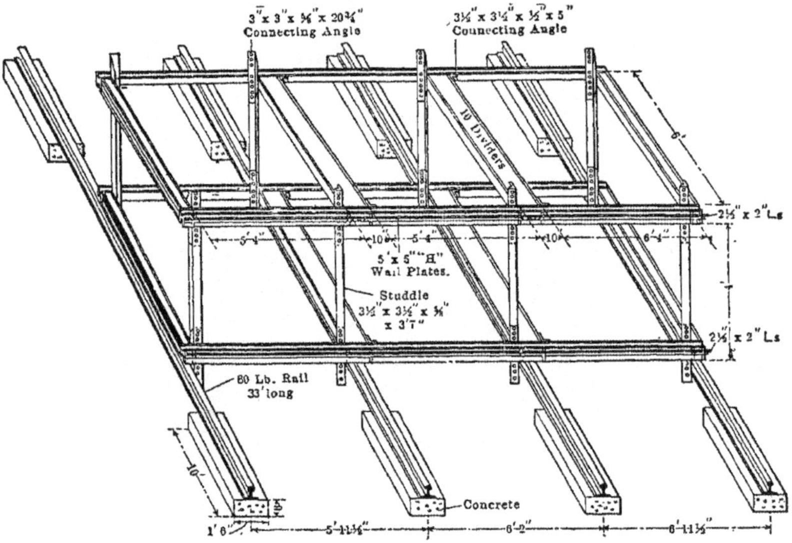

FIG. 49.—STEEL SHAFT SETS AT WHITESIDE MINE.

shop riveted to the end pieces and dividers so that little time is lost in putting a set in place. Machine bolts were used with nut locks for bolting the steel together. The shaft was wet and it took but 2 hours' labor at the most to put a set in place. Norway planks were used as slats and are held in place by means of small 2×1/2×2-in. angles shop riveted to the back of all wall plates and end pieces. Later on it is proposed to replace the slats by concrete. The sketch shows how the first bearing set was placed under the second set. The rails are 33 ft. long, resting on concrete foundations each 10 ft. in length. Another bearing set of 12-in. × 14-ft. I-beams was put in when the rock formation was encountered. Bearers like the last named were put in every 60 ft. down the shaft. This is found to be a satisfactory method of sinking.

Arrangement for Guiding a Drop Shaft.—The sinking of the W. F. 2 shaft at Obernkirchen, South Hanover, Germany, encountered just below the surface an 18-m. zone of watery sand and clay, necessitating the use of drop-shaft methods. The manner in which the shaft was guided in the true vertical direction is of interest, and is illustrated in Fig. 50.

The inside diameter of the finished shaft was required to be 4.5 m.; the concrete wall of the drop shaft was therefore molded to an inside diameter of 5.5 m., with walls 77 cm. thick. The outside of the wall was coated with cement plaster and then smeared thickly with brown soap, whereby the friction of the shaft against the guides was greatly reduced. It was only necessary to build

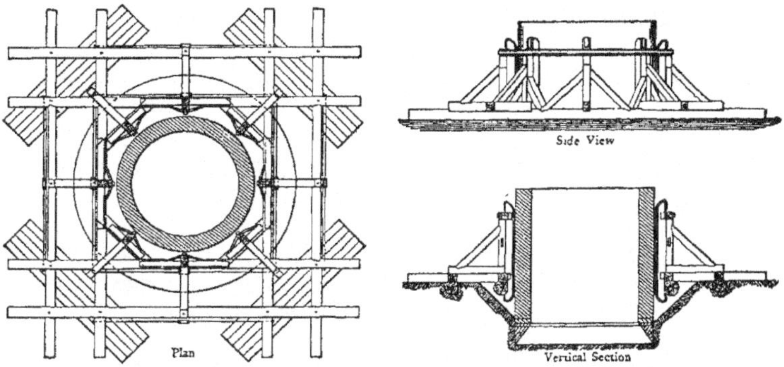

FIG. 50.—ARRANGEMENT OF GUIDES FOR DROP-SHAFT SINKING.

the walls 16 cm. above the top of the guides to maintain the weight required to give a steady downward motion. The sinking went on without incident to a depth of 14.5 m., when the wall refused to drop further, even though it was heavily weighted and the top built up to a height of 3.16 m. above the guides.

It was ascertained by boring that a further depth of only 7 m. was necessary to reach solid strata, and it seemed possible to gain this distance by forepoling. Before this plan could be put into operation, however, an inrush of material under the sinking shoe made it imperative to adopt a second, interior drop shaft, made of sheet iron. The bottom of the concrete shaft was firmly puddled with clay, and the iron drop shaft, of 4.95 m. inside diameter and 15.8 m. high, was lowered into place.

Sinking by this means went on rapidly until solid strata were encountered at a depth of 21.76 m. The shaft was continued into the rock to a further depth of 6 m., leaving 3 m. of the iron wall reaching up inside the concrete wall of the outer shaft. The space between these two walls was then filled with cement grouting (equal parts of quick-setting cement and sand) by boring holes through the iron plates and connecting the holes with pipes reaching to the surface. The upper 20 cm. of the space was filled with pitch-pine wedging.

The first 5 m. in the solid rock were lined with cast-iron tubbings and a masonry bearing ring, behind which all the spaces were thoroughly grouted. At a depth of 32.6 m. sinking was begun in the ordinary manner, with masonry lining.

Concrete in Inclined Shafts (By Sheldon Smillie).—When a well organized, well advised mining company commences to sink a shaft, the question of durability of equipment outweighs that of first cost, and for this reason concrete has in some places gradually superseded timber for supporting the walls through the overburden. Its long life in wet shafts and its freedom from the danger of fire especially commend it. For use in vertical shafts through water-bearing ground it has been employed for some time, but only recently has it been applied to metal mining and inclined shafts.

In the Lake Superior copper region all new shafts are equipped with concrete collars. Concrete runners under the rails have not yet gained the popularity they deserve, but are coming into use more and more. The collar or cribbing makes an absolutely water-tight joint with bedrock, keeping out all surface water, is free from rot and should a fire occur in either the shaft or in the rock-house above, large iron doors may be closed to make an effectual barrier.

Managers have installed concrete runners experimentally and in a critical mood have found absolutely nothing to complain of except that a fast-moving skip makes much noise. It was thought that with rails laid rigidly on the concrete there would be considerable breakage from crystallization, but this seems not to be the case. In my opinion, if it proved a source of trouble, both the breakage and the noise could be eliminated by inserting a strip of wood between the rail and the concrete. The strip could be replaced without much trouble when rotten, and would furnish little combustible material in event of fire. The only real objection ever made was that if a skip went off the track it would either break the runners or would become so tightly wedged that the runner would have to be broken to get the skip out and valuable time would be lost making forms and allowing new concrete to set. This might happen with small, poorly balanced skips, but in a period of 4 years at one of the largest mines of the district, where a hoisting speed of 4000 ft. per minute is usual, I do not recall a single derailment.

The accompanying sketches, Figs. 51 and 52, show the usual designs employed. In preparing for the collars the shaft is carried down to firm bedrock, the size of the outside dimensions of the concrete, the temporary lining making the outside form for the concrete. When sufficiently firm ground is reached, the size of the shaft is contracted to its regular dimensions, forming a ledge upon which the concrete is started. The inside form need only be built high enough to give the concrete at the bottom a good set when the lower frames may be moved above. The collar can be built to any height required by dump facilities and tracks.

The walls are proportioned according to the depth of the shaft, the concrete mixture and whether or not reinforcing is to be used. When well reinforced on the hanging walls with old rails, rope and the usual collection of old iron found around a mine, 12 in. is probably the minimum thickness. To make the cribbing water-tight, care must be taken that the ledge is clean, the concrete well tamped, and when left over night should be left rough and well wet the following morning. For this purpose a fire hose attached to the column-pipe will be found convenient. In filling in the concrete a chute is all that is necessary

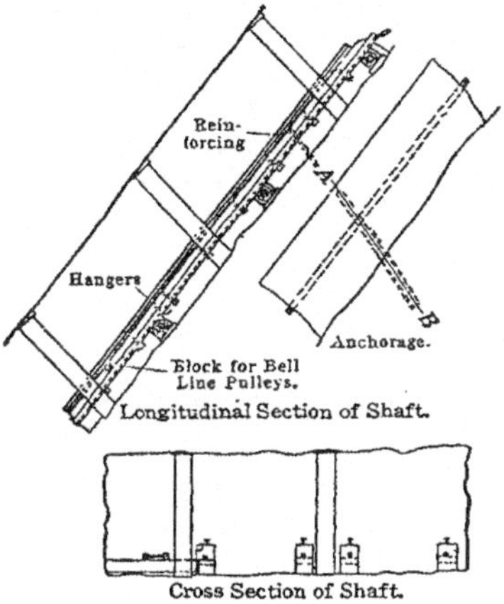

FIG. 51.—SECTIONS OF INCLINED SHAFT.

for shallow work, but if the fall is too great the sand and rock tend to separate and the concrete should be lowered into place. When first finished the ground water appears to seep through and below the cribbing, but this stops in time, due probably to clogging of the pores of the cement.

If considerable water is running when the cribbing is installed it may be necessary to drain in some special way and stop the holes later with plugs. In the cribbing the dividers and runners are built at the same time as the walls. The dividers between the compartments are concrete posts normal to the dip about 1×2 ft. in section and spaced at 6 ft. centers.

The dimensions of the runners vary with the type of skip and may be square or battered; 12×12 in. is a good section. One mine has runners 14 in. high with a 14-in. face and 16 in. wide at the base. The first rails are anchored by straps to the runners of the rockhouse and are held in place on the concrete at intervals of 3 ft. by a pair of cast-steel clips, cast to fit the flange of the rail.

Through the clips 3/4-in. bolts, heads up, pass through 3/4-in. galvanized pipes set in the concrete, terminating in No. 20 galvanized-iron boxes, 3×4×12 in., which are laid transversely in the runner. The box permits the insertion of a washer-plate and nuts. Some companies use bolts with threads at both ends, but this makes tightening difficult, especially if one nut goes on hard and the other easily, and the threads are apt to wear on the upper end with any motion of the rail. Square heads and nuts are preferable as offering greatest purchase for a wrench and least wear. For tight places bolts are made with

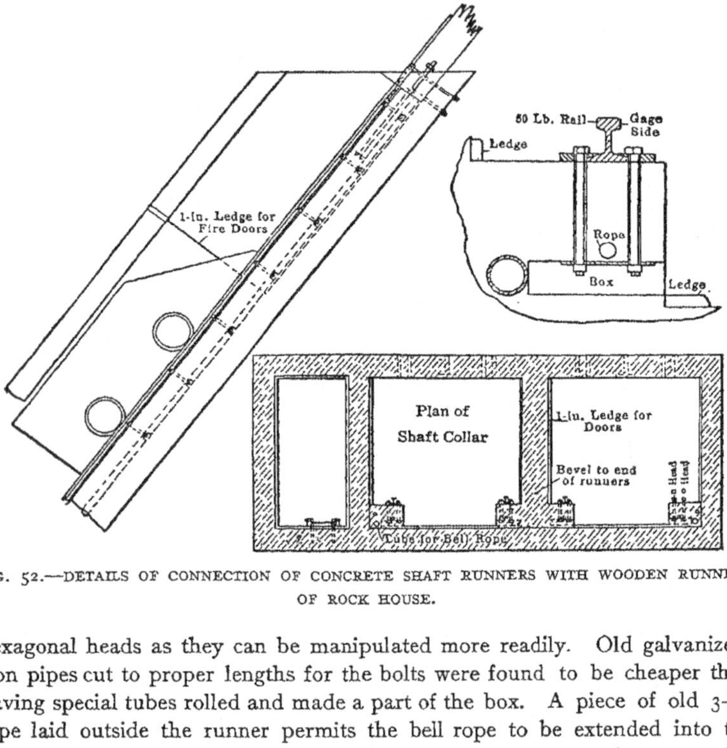

FIG. 52.—DETAILS OF CONNECTION OF CONCRETE SHAFT RUNNERS WITH WOODEN RUNNERS OF ROCK HOUSE.

hexagonal heads as they can be manipulated more readily. Old galvanized-iron pipes cut to proper lengths for the bolts were found to be cheaper than having special tubes rolled and made a part of the box. A piece of old 3-in. pipe laid outside the runner permits the bell rope to be extended into the shaft house, and a piece of old hoisting rope forms a good core for supporting the runners.

In finishing the top of the collar, provision must be made for attaching the runners of the permanent rock house. In the case illustrated in Fig. 51, the rock house was to have a channel section to which wooden runners would be bolted. Long 2-in. bolts were to be placed at the proper distance at the side of each runner; these would lie between angles riveted back to back on the back of the channel and the bolts would tighten down on a washer-plate at the upper end. If doors are to be hung over the shaft a 1-in. length should be made all around each compartment by thinning the walls by that amount, and the necessary holes for bolts and lock provided. An opening is left at the side of the ladderway to admit the air and water pipes from conduits or ditches.

The hanging wall of these mines needs little or no support, and the runners for the rails are built directly on the foot wall of the shaft, saving room that has hitherto been given to cross ties. They are similar to the runners in the collar, but usually a little higher to allow for the irregular foot wall. The adhesion to the foot wall is usually neglected, and an anchorage made of drill steels set at intervals in the foot wall to carry the entire weight of the runners.

In steeply inclined shafts the outer ends of the drill steels are supported by tie bolts from ring bolts set in the foot wall a few feet above. Tie bolts set at intervals in the concrete outside the rails support the rails by a long hanging bolt bent at the lower end and passing through the web. The guide sheaves for the ropes are supported on structural-steel cross pieces from runner to runner laid in the concrete. Pulleys for the bell rope are screwed into wedge-shaped rocks set at convenient intervals in the side of the runner.

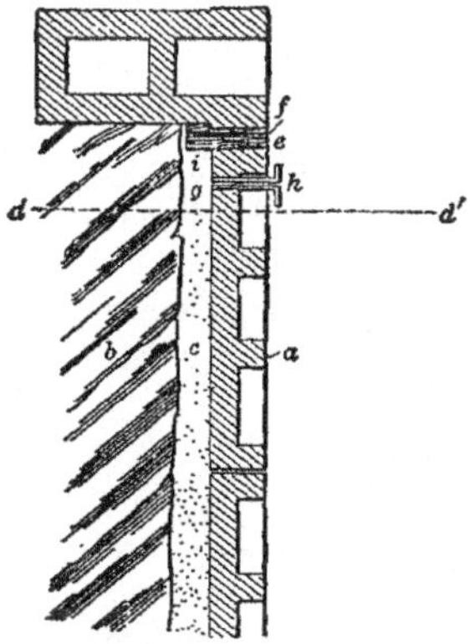

FIG. 53.—SHAFT TUBBING ARRANGED FOR INJECTION OF GROUTING.

Injection of Grouting Behind Shaft Tubbing.—An original method was employed at the Hildesia shaft at Diekholzen, Germany, for insuring a perfectly water-tight joint between the upper ring of a set of tubbing and the bearing ring above it. As the tubbing is erected from below, resting on a similar bearing ring at a lower point in the shaft, a small space, of variable size, always remains around the top of the uppermost ring, and this has to be carefully closed, generally with pine or poplar wedging.

In the case under discussion, illustrated in Fig. 53, after the upper ring had been put in position, the space between it and the rock wall was filled with cement grouting to the level $d-d'$, and the wooden wedging *e* was inserted in the usual manner. At four equidistant points around the shaft, 10-mm. holes *f* were bored through the wood. At four other equidistant points, 45-mm. holes *g* were bored through the web of the cast-iron lining, close under the upper flange; these holes were fitted with pipes and couplings.

By means of hose, one of the latter holes was connected to a pipe from a high-pressure pump, and cement grouting was forced into the space *i*. The three other holes of this set were closed as soon as cement began to come through them, with the escaping air. More cement was injected, until it began to escape through the holes in the wedging, and these were then tightly closed. Further additions of grouting, under a pressure of 80 to 90 atmospheres, were then applied, for the purpose of forcing the cement into every crevice of the rock wall, and also into the grain of the wedging. By exercising this unusual precaution, the chance of leakage, especially during the winter when the tubbing contracts, was entirely overcome.

Grouting in Quicksand.—A new method of grouting in quicksand is given in *Engineering News*. A pipe large enough to serve as a cement-injection tube is fitted at its lower end with an auger point and a helical-screw blade; just above this blade several holes are drilled in the wall of the pipe and fitted for discharging grout outward over the upper surface of the blade. This boring apparatus is twisted down, if necessary with the help of jetting, by water pumped through the pipe. When it reaches the desired depth, grout is pumped into the pipe, and at the same time the drill is turned backward so as to withdraw it. The grout flows out along the face of the blade, and becomes mingled with the layer of sand above by the rotation; the resulting mixture is passed by the turning of the screw blade to the space below, where it builds up in a cylindrical body corresponding to the volume swept through by the blade. When the drill is wholly withdrawn, it may be sunk again alongside the first location, and thus a large mass of contiguous, coherent cylinders of consolidated sand can be formed. The process is patented in Germany.

SHAFT STATIONS AND SKIP POCKETS

Shaft Station in Inclined Foot Wall Shaft (By Claude T. Rice).—At the mines with a flat-dipping vein, in the Michigan copper district, the skips are almost universally loaded by dumping the cars directly into them. Turntables are used for switching the cars, which are made to hold approximately 2 1/2 tons. Generally the shafts are sunk in the vein, and the spur tracks come up to the sides of the shaft; the main track continues through to the other side of the plat, along the hanging wall, and the turntables are placed in front of the shaft on the main track as shown in the small drawing in Fig. 54. The skips generally hold 7 tons, so that it takes three cars to make a load. Two cars stand ready to be dumped, while a third car is on the main track.

The arrangement of the station at the 31st level of Wolverine No. 4 shaft which is sunk in the foot wall about 25 ft. from the vein is shown in Fig. 54. On the south side of the shaft a recess is cut so that two of the cars can be dumped on the side, the third car being dumped from behind; on the north side only one car is dumped from the side turntable and the other two are dumped from behind, the empty car being switched over to the side-dump track to get it out of the way.

On the south side, in the back part of the station, it will be noticed that a square corner is cut. This was for the small hoist used in sinking the shaft

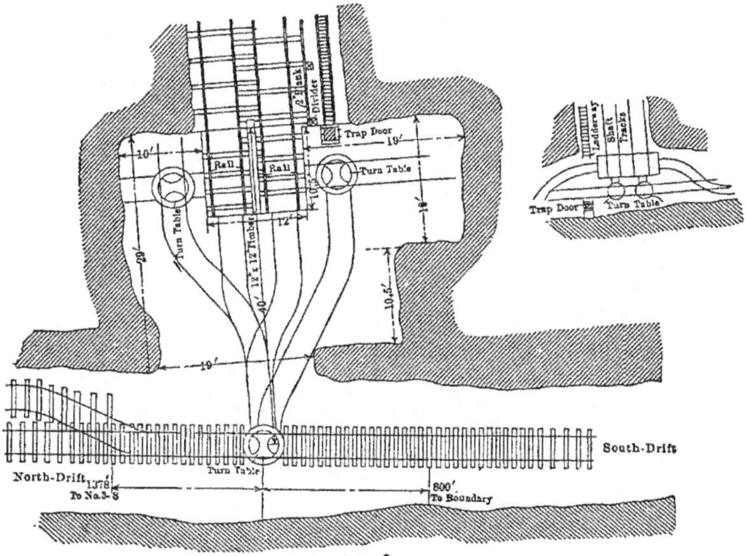

FIG. 2

FIG. 54.—SHAFT STATIONS IN INCLINED SHAFTS.

deeper from that level. When the station is cut in this way a brow of ground is left under the crosscut; to support this a reinforced concrete pillar is put in at the Mohawk mine before the brow has begun to give any trouble, and the floor of the station is laid with reinforced concrete, 6 in. thick, designed to stand a load of 400 lb. per square foot.

Large Underground Station in a Coeur d'Alene Mine.—One of the largest and most complete underground timber, boiler and hoist stations in the United States is newly completed at the Morning mine of the Federal Mining & Smelting Co., Mullan, Ida. Its construction involved several interesting problems. The station is situated at a point nearly 2 miles from the entry of the No. 6 tunnel, now the main haulage way of the Morning mine. In this tunnel electric haulage is used, ore being handled in trips of fifteen 5-ton cars. About 1000 tons of ore are produced each day, and practically the entire output must pass through this station. Ample space for the handling of the ore and timber trains was therefore a prime requisite in the laying out of the station.

The station proper is 100 ft. long, 36 ft. wide, and is 24 ft. high in the clear at the shaft, dropping to a height of 11 ft. at a point 200 ft. distant. There is a wide double-track approach. A boiler room, about 28×19 ft. in size, opens off from the farther end of the station. Adjoining the boiler room is the hoisting-engine room, 30×47 ft. in size.

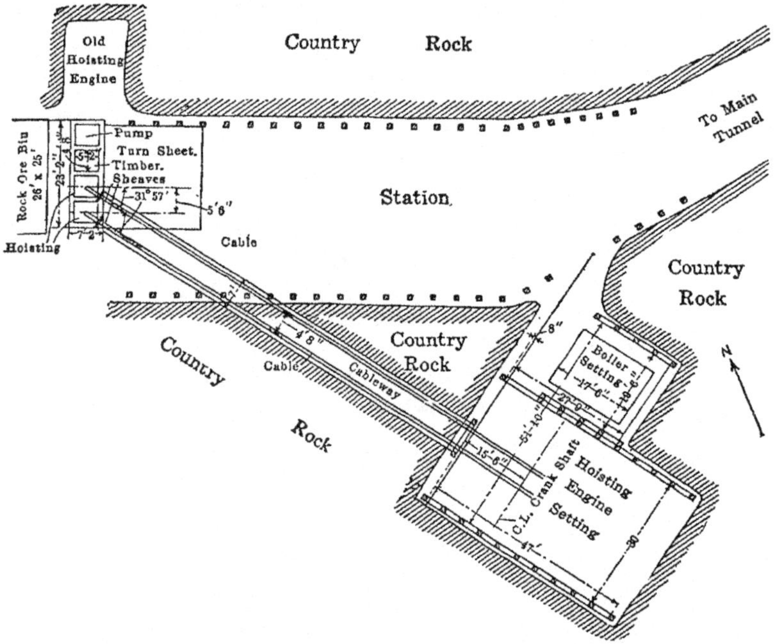

FIG. 55.—GENERAL PLAN OF TIMBER, BOILER AND HOIST STATIONS, MORNING MINE, MULLAN, IDA.

An interesting problem arose in connection with the placing of the engine. A shaft with compartments 4 ft. 8 in. × 5 ft. 2 in. in the clear had been decided upon, and this would throw the sheave wheels 5 ft. 6 in. apart. It was, however, deemed wise to use an engine similar to that in use at the Mace mines in order to facilitate repairs, etc. The reels on this engine are spaced 4 ft. 8 in. apart. For a while this promised to make trouble, until the expedient of setting the engine off at an inclination to the axis of the shaft was hit upon. The crankshaft of the engine is 111 1/2 ft. from the center of the shaft and inclined from its long axis at an angle of 31° 57′. This throws the sheaves at the proper distance apart.

The cableway from the engine to the sheaves is an inclined raise through solid rock so that no headframe structure is required. From the collar of the shaft to the center of the sheaves is 100 ft. An old hoist set in line with the long axis of the shaft will handle timber. (The sheave for this is only 45 ft. above the collar of the shaft.) The general layout of the station is shown in Fig. 55.

Five feet from the wall plate of the shaft is an ore bin 26 ft. long (parallel to the long axis of the shaft); it is 25 ft. wide and 52 ft. from toe to top, the bottom having a 45° slope. This bin was excavated out of solid rock and is armored on the front inside face with 60-lb. rails. Skips automatically dump ore into the bins from which it is drawn directly into the 5-ton cars of the electric trains.

In cutting out the station some interesting rock excavation was done. The face was advanced carrying its full height and width. To do this four 10-ft. bars set end to end and blocked tight with 3-in. planking were used across the face. The line of bars was arched slightly toward the face, from which it was braced with the wedge timber. This formed a "compression" truss and although many miners object to running two machines on a bar, on the score that the bar will not hold tight, three or four machines were continually operated on this series of bars, and no special trouble was experienced from fitchered holes. For this work 3 5/8-in. piston drills were used, and as many as 190 eleven-foot holes put into a round. The cuts and lifters were fired first, then the other holes. Electric battery firers were used in all cases. One round of holes usually broke enough rock to fill 400 of the 35-cu. ft. capacity cars. Only two settings of the bars were necessary for drilling the entire face: The first was on the muck pile and the second lower down after the face had been mucked clean.

Concrete Floors for Shaft Stations.—At the Mohawk and Wolverine mines, Michigan, the stations at the inclined shafts have reinforced-concrete floors which are absolutely fireproof and can be put in almost as cheaply as timber floors, taking into consideration the sets that would have to be used to carry such a floor. These concrete floors are made 6 in. thick and are designed to carry a load of 400 lb. per square foot. The reinforcement used by W. F. Hartman, the engineer who designed the floors, consists of a double layer of 4-in. triangular mesh reinforcement, threaded with 3/8-in. strands of old hoisting cable, running crossways to the strands of the triangular mesh, and at 6-in. centers. In order to protect the reinforcement from fire and corrosion it is placed about 1 in. from the bottom of the concrete.

Skip Pockets.—At the Bunker Hill mine, near Amador City, Calif., skip pockets are arranged to facilitate the handling of ore and waste. The usual custom is to have pockets beside each other, each discharging into a different shaft compartment. The objection to this is that it permits the handling of only one class of rock in each compartment from any level. It also means that trammers must switch their cars to the proper track when dumping at the skip pockets. The shaft at the Bunker Hill is inclined, having two hoisting compartments. To overcome the objections mentioned above, a waste and an ore

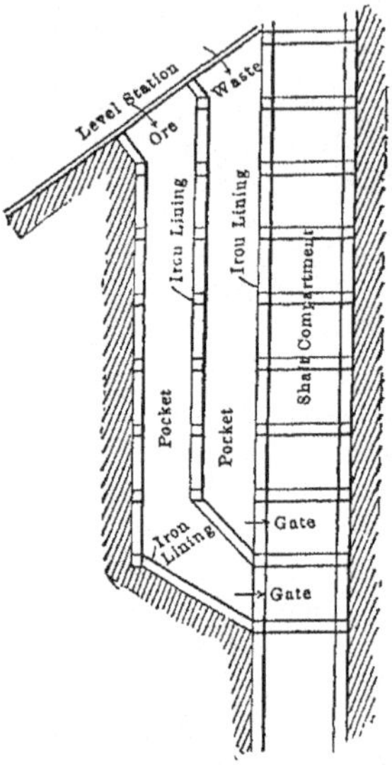

FIG. 56.—ARRANGEMENT OF SKIP POCKETS AT BUNKER HILL MINE.

pocket are arranged, one over the other, each pocket discharging into both shaft compartments. Below levels the shaft is widened out to three times its regular height and is carried so for nine sets. The compartment above the shaft is partitioned off as a waste pocket and the one above that as an ore pocket. The timbers between compartments are heavily lagged and lined with strips of iron off the guides in the shaft. By this arrangement each of the tracks at the station serves both pockets and no needless switching of cars is necessary. The accompanying drawing, Fig. 56, illustrates the idea of the skip pockets. Either ore or waste may be drawn at any level whenever desired. With such an arrangement only a single track is required at stations.

Skip Pocket and Station at Leonard Mine, Butte.—In Fig. 57 is shown the general idea of the arrangement and timbering of the skip pocket and station, on the 1800-ft. level, in the No. 2 shaft of the Leonard mine at Butte, Mont. The excavation for the skip pocket is started at a point five timber sets (25 ft.) from the shaft. It is carried straight down for two sets and then benched in three 5-ft. steps, the bottom being two sets wide. From the bottom of the pocket the excavation is carried down the width of the shaft for three sets.

A sheet-steel gate operated by a compressed-air cylinder controls the discharge of ore from the pocket into an apron, also of sheet steel. The lip of this apron, when lowered, extends over the edge of the skip so that the ore is run directly into the latter. The lip of the apron is raised and lowered by compressed air. To operate the gate and apron, a man stands on a platform on the second set of the compartment beside the shaft.

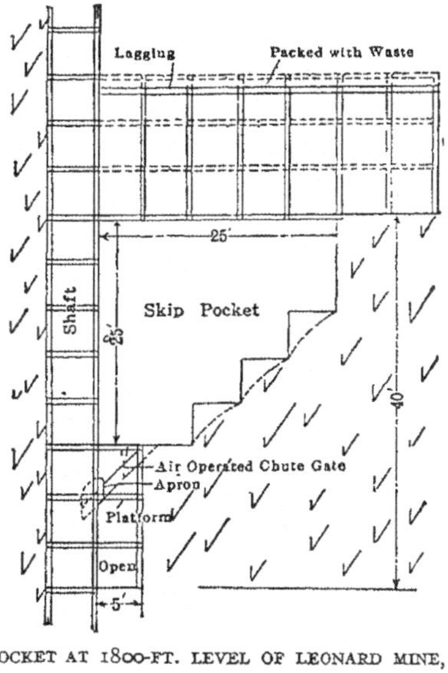

FIG. 57.—SKIP POCKET AT 1800-FT. LEVEL OF LEONARD MINE, BUTTE, MONT.

The level station is first timbered with square sets, as shown by the dotted lines in the sketch. Two 8×10-in. stringers are then run out under the top caps for three sets, one end of each stringer being blocked up from the shaft-wall plate, the other from a cap of the third set from the shaft. Support is thus given to the roof of the station, while the first two sets of square-set timber are removed and replaced by the permanent station sets. Stringers are then blocked up in a similar manner, to span the next two sets, timbers removed and replaced, etc.

The posts of the station sets are 13 ft. long, made of 14×14-in. timber, 12×14-in. material being used for the caps. The space, usually a couple of feet high, above the station timbering, is filled in with waste; 3-in. lagging is used over the top and sides of the station sets.

A 1-in. space is left between pieces of lagging to allow a free circulation of air and thus check rotting of the timbers. A station may be carried back as far as is necessary to provide ample room for handling cars. Beginning with the fourth set from the shaft, the bottom of the station should be raised 1/4 in. per set to give the necessary grade to the approach. The station at the Leonard mine is 21 1/2 ft. wide and provided with a double-track approach.

V

DRIVING ADITS AND DRIFTS

Practical Considerations and Methods—Timbering—Special Types for Heavy Ground.

PRACTICAL CONSIDERATIONS AND METHODS

Fast Drifting.—Speeds as high as 60 ft. per week are obtained in cross-cutting the slate formation in the Kennedy mine at Jackson, Calif. Three shifts are worked and two machines run on one bar. As soon as a round is fired, the bar is rigged horizontally across the face. Then, working on the muck, the two drillers put in the back, breast and cut holes. By the time these are in, the shovelers have cleaned up the muck pile enough to allow a lower setting of the bar from which the lifters are drilled. In this manner no time is lost by the machine men in waiting for the muckers to clear away the rock from the face, and each man can put in his full shift at the work to which he is assigned. Such small refinements of methods should be carefully observed in the operation of a large mine and may represent the difference between a profitable or losing operation.

Maintaining Grade in Driving.—In driving and tunnel work the miners unconsciously tend to increase the grade toward the face at a steeper angle than is desirable for drainage or for favoring the tramming of loaded cars. Too much grade is disadvantageous because the grade favoring loads is so great that the cars tend to run faster than considerations of safety should permit; greater effort is required to move empty cars up the grade; natural ventilation is interfered with, the results being especially noticeable at the face; and, in some cases, the unnecessary loss of ore in the backs above the drift may be undesirable. The grade of drifts in some of the older mines of Cornwall is as great as 5%, often 7%. At the present time a drift is rarely driven at a greater grade than 1%, which is twice the grade recommended by some authorities. To avoid driving at too great a grade, a template should be provided which the miners can use as often as they desire and without losing much time. Such a template may be made by cutting a board of convenient width and thickness so that it is exactly 100 in. long. The edges should be planed true and parallel. A line is then drawn from the upper corner of one end to a point 1 in. below the upper corner of the opposite end and the board sawed along this line. The board is then turned over, and a level-tube let into the edge, which is so adjusted that the bubble will be in the center of the tube when the edge of the board is in a horizontal plane. In use, the template is laid upon the floor of the drift with the narrower end toward the face. Then if the grade of the floor is 1%, the bubble will be in the center of the tube.

Alignment in Driving.—When driving on a vein that abruptly passes from a relatively uniform rock into a sheeted zone or one wherein many nearly vertical fissure planes exist in close proximity miners are quite liable to deflect the direction of the drift to one side. This is especially true when the vein crosses or enters such a zone at an obtuse included angle. Due to the way in which the breaking of the rock is influenced by the fissures or joints together with the tendency to advance slightly toward the direction of the joints, the drift in passing through such a zone is apt, when finished, to present a stepped in or jagged appearance. This tendency to deflect must be guarded against.

Placing Holes for Blasting (By P. B. McDonald).—The subject of placing holes for blasting, formerly left entirely to the judgment of the miner and foreman, has lately received attention from engineers and superintendents. Wrong judgment in placing drill holes is one of the most expensive mistakes in underground work; expensive because of the loss of explosives and time in the costly operation of reblasting.

The following is an incident common in underground practice: A miner drills a cut of say 12 holes and blasts them. If the ground is tough, the misplacing of one hole upon the breaking of which the effects of several others depend, may spoil the blast so that large ridges or corners are left. He then cleans or blows out the holes which failed to break and uses from 10 to 40 sticks of dynamite, costing 10 cents per stick, for blasting the same holes a second time. In almost every such instance one or two extra holes or a closer attention to the position of the holes drilled, would have made the difference necessary to produce a successful blast in the first trial. The time spent on reblasting is usually more than would have been required for the added care in placing the holes, and the greater powder cost and the time consumed waiting for the smoke to clear, are distinct losses. The nature of rock excavation is such that a 5% closer attention to details may mean a 25% gain in results.

The simplest form of blasting is slicing or breaking to an open face. In open-cut excavation, holes are drilled at a distance back from the face a little less than the depth of the hole, although this distance is shortened in tough igneous rock and where the blasted portion is held at the ends as in drawing back a narrow stope. A favorite arrangement of holes for use on bench work in open-cuts is shown in Fig. 1. (References are to the illustration numbered Fig. 58.) The horizontal holes are fired at the same time as the vertical holes, in this manner deepening the cut broken.

In drifting in soft rock, such as friable schist or crumbly slate, most of the standard arrangements of holes give good results and the misplacing of one or two does not usually matter. The arrangement shown in Fig. 2 is, of course, applicable only to soft rock where the blasting shatters the rock so thoroughly

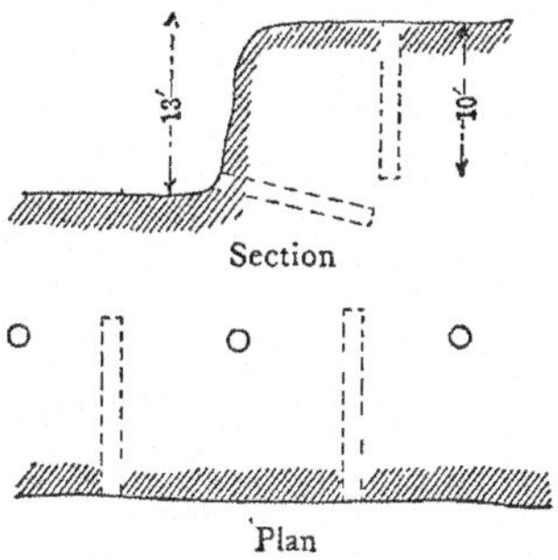

FIG. 1.

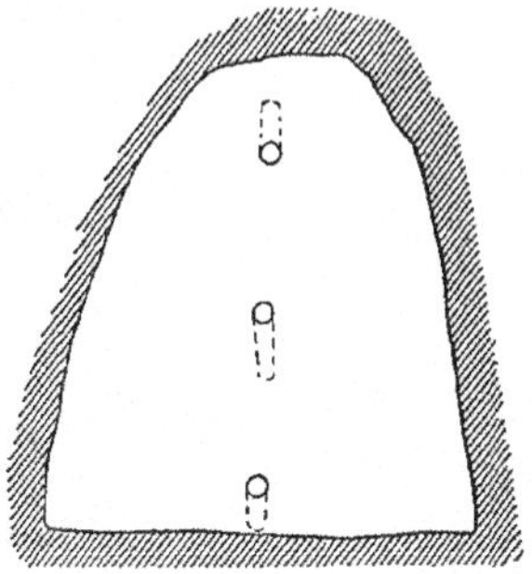

FIG. 2.

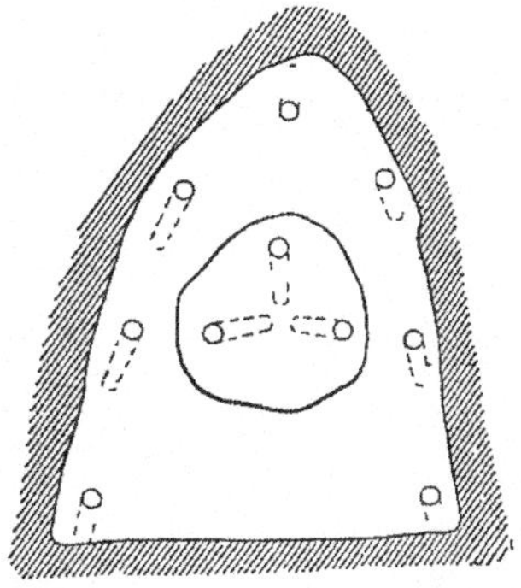

FIG. 3.

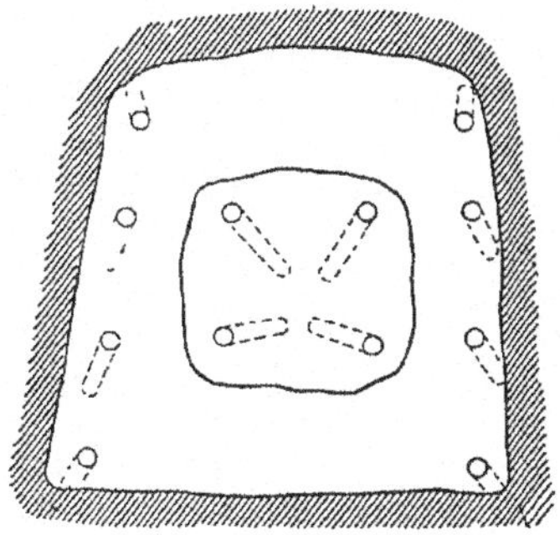

FIG. 4.

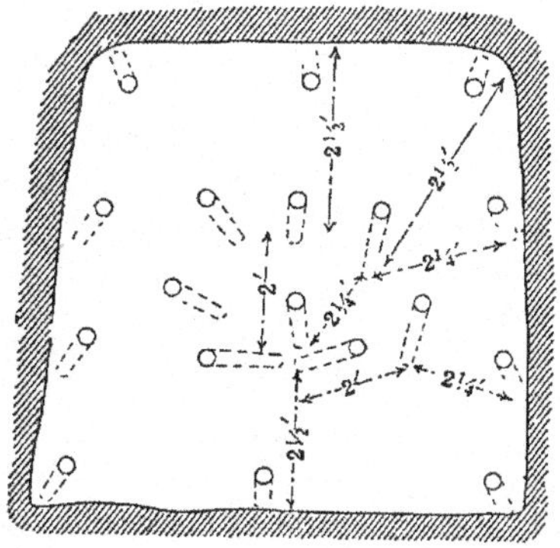

FIG. 5.

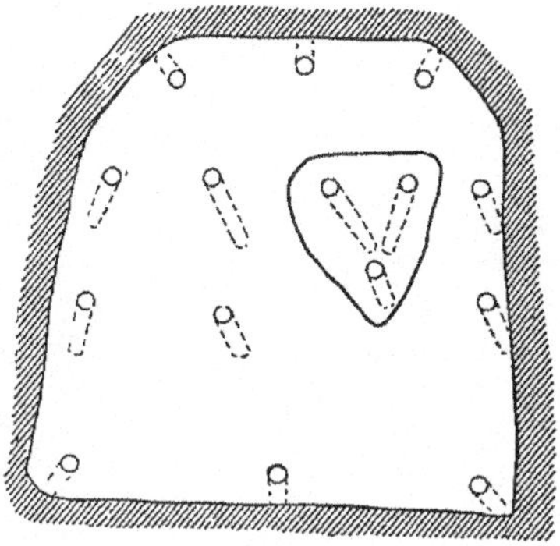

FIG. 6.

FIG. 58.—ARRANGEMENT OF DRILL HOLES IN OPENCUT AND TUNNEL WORK.

that it can be picked out. Fig. 3 shows a 10-hole cut with one back hole, frequently used in driving small drifts where it is desired to keep the back well arched. The arrangement shown in Fig. 4 is suited to larger square drifts. These holes would not break much hard rock because of the distance between the bottoms of the cutting-in and squaring-up holes.

It will be noted that Figs. 3 and 4, depicting standard cuts in soft rock, show two classes of holes, the cutting-in holes, and the squaring-up holes which determine the shape of the drift. In hard rock there are, in addition, what are conveniently called relief holes, because, situated and fired intermediately between the cutting-in and the squaring-up holes, they relieve the ground to be broken by the latter set. It is in the drilling of such a cut that the best judgment is required, for relief holes are a changeable feature and slight variations in the toughness of the ground and the size of the workings necessitate differently placed relief holes for successive cuts.

The most important rule to be observed in placing holes is that the determining factor is not the distance apart of the starting points of the holes but the distance between their bottoms. The upper few feet break out into the drifts comparatively easily, but the inner portion breaks with more difficulty. For instance, in Fig. 4 the tops of the holes are 2 1/2 ft. apart, while between the bottom of the cutting-in and the bottom of the squaring-up holes is 4 ft. of rock, so that in hard ground the cut would not break. Experience in any variety of rock will show what unit may be figured upon for the dividing distance of the bottom of the hole, usually 2 or 2 1/2 ft. The distance from the cutting-in to the relief holes is made less than from the relief to the squaring-up holes (figuring between bottoms in each case); possibly the former is 2 ft. and the latter 2 1/2, because ostensibly the squaring-up holes have a better chance to break out in the space enlarged by the relief holes than the relief holes have in the confined space made by the cutting-in holes.

In planning the arrangement of holes the first consideration is to get a cutting-in hole that will break well. The smaller number of holes used for this the better, because it is essential that cutting-in holes be fired simultaneously, and owing to irregularities in the rate of burning of fuse, this is difficult to accomplish when a large number of holes are to be fired; also, starting a cutting-in hole is often difficult because of the angle at which the drill point has to meet the face. The excavation made by three holes meeting at a point is almost as large as by five or six, so that it is usually better to use only three or four holes for cutting-in; and, if the ground requires them, to put the extra holes in as relief holes where they will break more ground.

After deciding upon the cutting-in, the squaring-up holes are placed along the sides, bottom and top, with the ground equally divided between their bottoms. Perhaps two relief holes, one on each side of the cutting-in holes will suffice; Fig. 5 shows two on either side and one above, helping the middle back hole, which is important because upon its breaking depends the successful blasting of the other two back holes. In the sketch the cutting-in holes are placed low; they might have been shifted higher and the upper relief holes placed underneath. Quite a common alternative is to shift the cutting-in holes to right or left, so that relief holes are required on but one side; thus in Fig. 6 it is seen that the cutting-in holes are to the right and high.

To sit in the office and figure the arrangement of holes for a cut is not satisfactory, unless the sketch is drawn to scale, because it is easy to draw the holes out of proportion without regard for practical considerations. The holes cannot be pointed at such an angle as is sometimes desirable because the crank end of the machine will strike the sides or back of the drift. It is generally good practice to drill as few dry holes as possible, and this throws out many of the cuts drawn by men who never helped drill a sticky back-hole; also it should be aimed to economize on movements of the arm and bar.

An inexperienced man examining the holes for a cut is apt to think that the deviation of the holes is at too small an angle, and that the holes are too straight. It is not an easy matter to tell definitely where the bottom of the holes will come, but by inserting long drills or tamping rods into the holes and noting the deflection or convergence of the protruding ends the relative location of their bottoms can be gaged.

Drifting with Stope Drills (By Horace Lunt).—In the Cripple Creek district the small hammer-type air drills have almost entirely replaced the

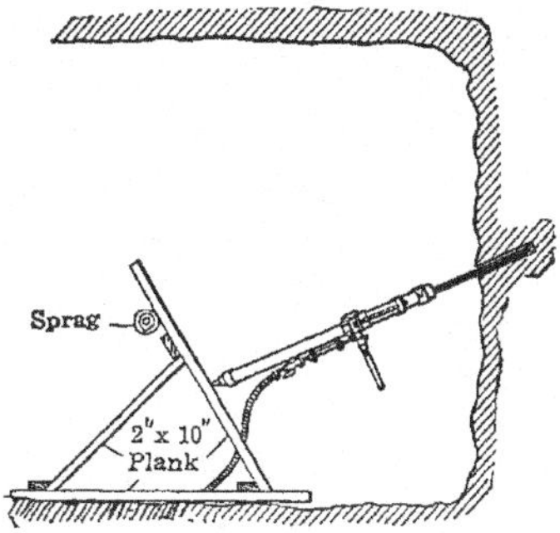

FIG. 59.—FOOT FRAME FOR STOPING DRILL.

piston drills in stoping. Much of the work is done by leasers, who are not, as a rule, able to maintain a large equipment of tools, and they have found that where the ground is not too hard they can successfully use stope drills for drifting. Sometimes the drill is supported by a sprag across the drift, but a better scheme is the simple A-frame illustrated in Fig. 59. This is constructed of 2-in. plank and is held in position by a sprag across the drift as shown. The sprag has to be moved once or twice during a round, but this does not take much longer than changing the height of the arm on the column of a piston machine. All the holes except the lifters are given an upward slope; they are put in nearly horizontal and cleaned with a blowpipe connected to the air supply.

Driving Inclined Raises with Stoping Drills (By Arthur O. Christensen).—The usual practice is to drive raises to connect levels for ventilation, or to prospect veins or the country above tunnels, as small as is possible to work in, to drive them perpendicular, and to timber with pairs of stulls 3 1/2 or 4 ft. apart. In the case of such work I have found two styles of raise easier, cheaper and safer to drive than the type mentioned. The first is a type of inclined raise which is described below. The second is a type of vertical raise described in the following article.

The inclined raise may be run at any desired angle in the vein, or may cut through the country rock. The advantages of it are that it requires no timber; it advances horizontally as well as perpendicularly; for short distances it prospects ground more cheaply, and is more rapidly driven than a drift or vertical raise. No mucking is required and a "buzzer" (hammer stoping drill) can be used. By making the raise incline about 40° the muck runs down into the chute at the bottom of itself. Only occasionally it is necessary to scrape down some of the dirt which packs on the bottom.

For raises steeper than 40° it is well to put a light stull or piece of lagging across the bottom of the raise every 4 or 5 ft., and not over a foot above the bottom. Dirt fills in behind these and they form steps making it easier to climb up, and aiding in rigging up a brace for the machine at the breast. These steps also serve to catch steel tools which are accidentally dropped by the miner.

The materials required to make a set-up are a few pieces of plank of varying lengths, wedges and spikes. With these it is always possible to brace the machine for any hole, although frequently moiling has to be done to get a suitable hitch for the plank on the roof. From Fig. 60 an idea may be gained of how to rig a plank against which to brace the machine and of a satisfactory round of holes.

By frequent use of the scraper, or by using hollow steel, flat or even down holes can be drilled, although such holes are seldom needed in driving a raise.

To clean the face, set up at the breast and be ready to drill, need not take over an hour, and if the tools do not have to be carried far it is often possible to get started in half that time. To put in the round of about eight holes, averaging 3 ft. deep, takes about 1 1/2 or 2 hours. In good ground with no mishaps 3-ft. holes with a small sized Waugh require, on the average, only 10 minutes each. The largest size drills, using larger steel, drive 4-ft holes in the same time. If a nipper brings the powder to the miner it is possible for him to shoot twice, thus making 4 ft. a shift. Otherwise 2 1/2 ft. is as much as can be expected for a daily average.

After advancing 30 or 40 ft. it is well to cut a station or pocket in the side of the raise in which to keep the machine tools, etc. To cut this pocket usually requires two rounds. Sometimes pockets are run 10 or 20 ft. in the form of inclined drifts from the raise with the purpose of prospecting the walls. Such drifts serve as very convenient stations in which to store planks, etc. It is often well to timber the pocket to protect the tools kept in it, and to keep them from rolling out or getting buried. Where it is not attempted to shoot twice a shift all this is easily done after the regular round is drilled. During two consecutive days rounds are shot out of the side, and on the third day the pocket thus formed is timbered. The timbering required is simple, and takes a man only a couple of hours to place.

By putting in a pocket every 30 or 40 ft. an inclined raise of this character can be carried up as far as 200 ft., although to climb up such a distance becomes difficult, and it requires time to keep the dirt from piling up on the bottom and filling the raise. The vertical distance made by such a raise is only 60 to 70% of the total distance covered, yet, by being able to shoot two rounds

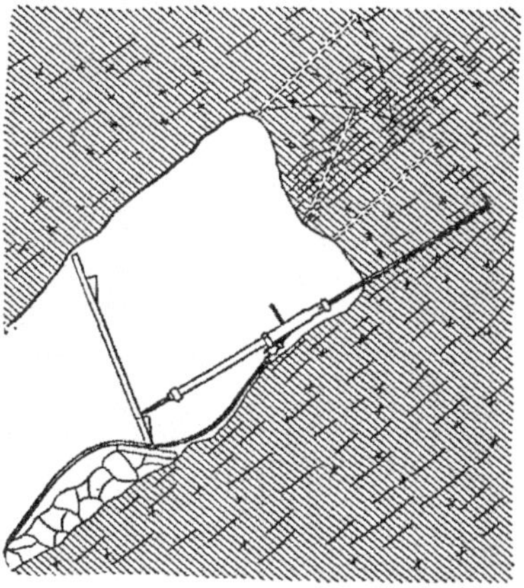

FIG. 60.—MACHINE SET UP IN INCLINED RAISE.

in the breast each shift, except on those days when cutting or timbering a pocket, it is possible to advance vertically faster than with a vertical raise. Then, too, ground is being prospected horizontally to the extent of 70 or 80%. of the length of the raise. For these reasons, for prospecting a vein short distances above a tunnel this form of raise seems cheapest and most speedy.

Driving Vertical Raises with Stoping Drills (By Arthur O. Christensen). —In driving a vertical raise, especially on a narrow vein, I find that greater speed can be made in the long run, greater safety maintained, more vein material broken in proportion to waste, and the raise can be run up indefinitely with but little increased labor as the raise progresses if the system described herein is used. Such a raise is made narrow, *i.e.*, 3 or not over 4 ft. wide, but

6 or 8 ft. long. This shape allows of easily putting in a round with the "buzzer" machine. The smallest-sized machine should be used.

Stulls are placed in pairs 4 or 5 ft. apart, one being put across the middle, and the other across the end of the raise. The stulls placed in the middle must be directly over one another, or at least in line, for they are to form the partition between the manway and chute. It is well to let the raise incline slightly toward the chute side so that the broken rock in falling will wear against the rock at the chute end of the raise rather than against the lagging between the compartments. Except for the last two pairs of stulls the middle ones are

FIG. 61.—STATION AND TIMBERING IN VERTICAL RAISE.

lagged up as the raise progresses and ladder and pipe line are carried along, as shown in Fig. 61. The condition of the raise after the round has been put in and the holes are ready to be loaded is here shown.

At the beginning of the shift it is first necessary to clean the muck from the platform at the top, and to pick down. This done, a few planks are placed across the chute from the middle stull of the next to the top set. This keeps any tools or steel from falling into the chute. I find it well to spread a piece of canvas over these planks to make them tight. Two long planks are then laid across the top platform, extending to the chute end of the raise. These are securely spiked down at the manway end. To prevent accident from the nails pulling out it is well to bind the ends of the planks down to the stull with baling wire. An opening is left between these planks on the chute end through which the miner can climb.

The machine, steel and tools are next brought to the top from where they are kept in a pocket or station cut in the side of the manway below. Short planks, blocks and such, necessary for rigging the machine, are kept on the top platform. They serve to prevent it from being broken by the rock shot down onto it.

This whole operation takes about 1 1/2 hours. A round of about 10 holes takes 2 or 2 1/2 hours, so that by dinner time the round is in and the tools and machine again put away. After dinner hitches must be cut or stulls put in. It is well to cut the hitches one day and the next day put in the stulls and move up the platform and ladder if needed. Every 30 or 40 ft. it is necessary to cut a pocket in which to keep the machine, tools, etc. To cut this pocket usually requires two rounds, which are shot at the same time as the regular round. An hour or an hour and a half is required to timber this. The pipe line must also be run up from the former pocket. To timber the raise, make these pockets and put in the pipe line and ladderway keeps one man pretty busy. Yet it makes a desirable job for a good miner and does not require him to unduly strain himself.

The scheme of placing the round of holes depends on the ground, the shape of the breast and the sort of vein being followed. In general it is well to put in two rows of holes, one on each side of the vein. These holes should be staggered rather than in pairs. Sometimes it is best to put from five to eight holes on one side of the vein, and then to bring down the vein with three or four holes placed back of it. In each case the round must be suited to the shape of the breast and character of the rock. By having the raise with its long axis with the vein a deep round can be drilled. If it were possible to keep up with the timbering a 4-ft. round could be drilled at each shift. When one man is doing it all, however, he must limit the depth of his round to the amount of timbering, etc., which must be done the next shift.

Not including cost of timber, tramming of dirt, sharpening steel and compressing air, the cost of driving a raise of this character, when run by a good hustling miner, should not be over $3 per foot, and under favorable conditions, considerably less. Doing every bit of work in connection with driving such a raise by myself I have driven 67 ft. a month, shooting each shift. During that time there was only one missed hole and no case of holes going out of turn. I attribute this success to the care taken in loading the holes and in spitting the fuses.

Staple for Temporary Staging (By B. M. Concklin).—In Fig. 62 is shown a staple for supporting temporary staging, that is used in the mines of the Oliver Iron Mining Co. on the Mesabi Range. One of these hooks is driven into each of the posts of two adjacent sets and on both sides of the drift. A pole is then laid in the two staples on each side of the drift and across these a platform of lagging is laid as shown in the illustration. The staples can be readily removed with a bar after the staging has served its purpose and has been taken away.

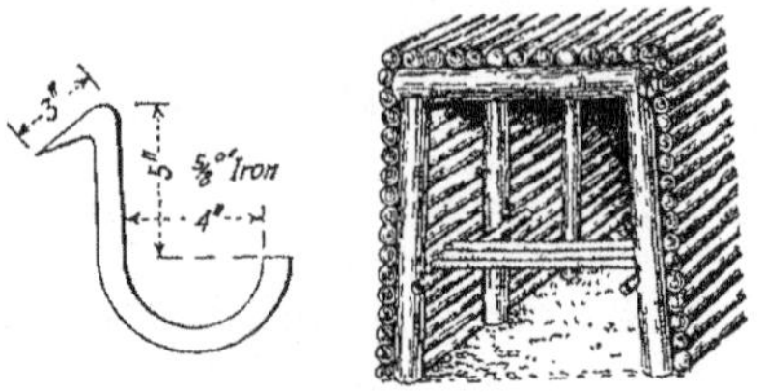

FIG. 62.—STAGING STAPLE AND MANNER OF USING IT.

TIMBERING

Special Types for Heavy Ground

Framing for Tunnel Sets.—One of the most satisfactory styles of framing tunnel-set timbers is shown in Fig. 63. This style is used in the Bunker Hill mine near Amador City, Calif., and also on the Comstock. This style of timbering is equally well adapted for round or squared timbers. The wide notches give a large bearing surface to take up side or vertical pressure, and make this method of timbering admirable for drifts in heavy and swelling ground.

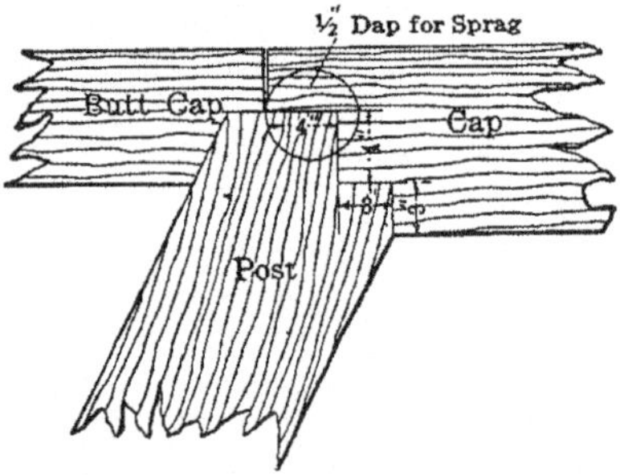

FIG. 63.—DRIFT SET FOR HEAVY GROUND.

Comparative Strength of Several Styles of Framed Timber Sets (By K. C. Parrish).—I would have some scruples about adopting the style of drift sets described in the preceding article because of the experience I had several years ago in holding heavy ground. Conditions in one place were such that the sets would be down in 6 weeks if not relieved; both caps and posts would be half moons. Sawed 10×10- or 12×12-in. caps and posts had to be relieved every morning in certain sections of the oreshoot. This meant that every stick was overloaded, hence the exceptional opportunity for trying out different styles of framing.

The desideratum is that a stick should be so framed that it will not be weakened by splitting, but will break first. Several methods of notched framing similar to that shown in Fig. 63 were tried, but it was found that any cuts in the post or cap weakened the timber, especially the post, which split most easily when cut with two notches. The simplest, cheapest and strongest method found was that shown in Fig. 64. For this practically no framing is required. Caps, posts and sills are sawed the proper length and shape and a 3-in. plank

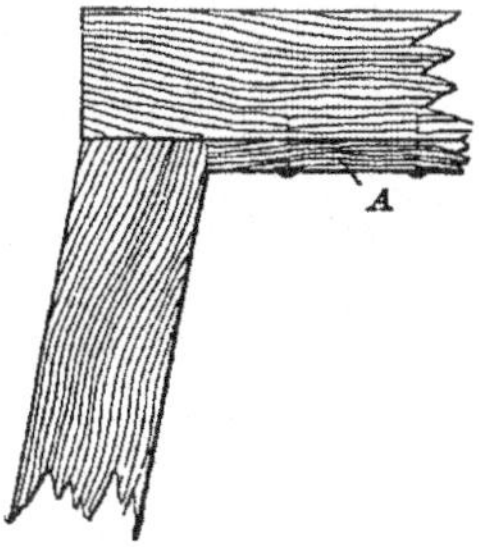

FIG. 64.—FRAMING FOR TUNNEL SET.

A is spiked to the lower side of the cap with six 30-dwt. spikes, thus obviating the necessity of cutting into any of the sticks. It was found that a 8-in. square timber framed in this fashion would apparently stand as much, if not more, than 10×10-in. material framed with notched caps and posts. A 1 1/2-in. plank was used with 8×8-in. timber, a 2-in. for 10×10-in timber, and a 2 1/2-in. one for 12×12-in. timber. These planks held the posts so that they would break at almost any point before they would split. The planks never buckled or slipped. This method was found to save much time in both framing and placing the sets and, above all, it gave much better service.

There are several possible reasons for the failure of the notched timber sets; one is the impracticability of getting a perfect fit, hence producing unequal strains, and causing excessive pressure at certain points, and thus a tendency to split. The post generally tends to split at the second cut and near the middle of the stick. This is probably due to the fact that the outer portion of the post is softer than the central portion and so, upon the compression of the post against the cap, the load is thrown on the inner cut, causing splitting there before the outer edge is sufficiently compressed to take its share of the load. In other words, it seems that even if a timber framed as in Fig. 63 is fitted perfectly, the difference in the texture of the wood in various parts of the same stick is such that the timber will be apt to split before breaking. To get the maximum efficiency of a stick, it should be under uniform compression, whether on the end or on the side, and separate fibers of the same stick should not be subjected to varying strains. Of course the strengths of separate fibers cannot be judged nor the compression regulated, hence the obvious advantage of distributing the load uniformly.

Reinforced Concrete in a Tunnel.—The Snake Creek drainage and working tunnel at Park City will be driven between 14,000 and 15,000 ft. from Snake Creek to secure deeper drainage for the mines of the district than is at present available, and to open the Bonanza Flat section in the southern part of the camp. About 2000 ft. from the portal some extremely bad ground was encountered, which could not be held with ordinary timbering. A system of reinforced concrete has been evolved which will be used for about 300 ft.

The section is egg shape with the small end down. The reinforcement consists of 50-lb. rails bent hot, in two pieces, these being set at 4-ft. intervals. At the top of the tunnel arch and above the rails is another rail, the whole being covered with triangular mesh screen weighing 109 lb. per 100 sq. ft. A shell of from 12 to 15 in. of concrete is put on the outside of the screen, and the space between the concrete and the roof filled with green quaking asp, to take up the first shock of settling ground. The tunnel has room for two tracks. It is 7 ft. high from the tracks to the center of the arch, and 9 ft. across, with a 4 1/2×9-ft. waterway under the track; 8×8 ties are placed at 4-ft. intervals. About 100 ft. of reinforced concrete has been completed, and this method of holding the ground promises to be successful. Where ordinary timber was used in certain places in the tunnel it was squeezed up to a 4-ft. opening, so that special measures were necessary. The part completed is holding well.

A Method of Mining in Heavy Ground (By W. L. Fleming).—The method here described is used extensively in the soft slates and serpentines of California. It is also useful in running through caved stopes and slide rock on mountain sides. Timbers are framed in the usual manner, but instead of being kept at a distance from the working face, each set is put in place as soon as room has been made for it. If blasting is necessary at all, the holes are drilled by hand and the light charge of powder required does little or no injury to the timber. Split lagging is mostly used though sawed plank is better if obtainable at a reasonable price. An assortment of widths from 3 in. to 8 in. is necessary. It is rare that face boards are needed to prevent running of the ground, but when occasion to use them arises, they are held in place by sprags from any available support, usually the nearest set. The ground will usually sustain itself until the drive is several feet under roof. Then the first two sets are placed and lagged as closely as is necessary, the lagging resting directly upon the collar set and upon a bridge on the breast set. Sets are placed 4 ft. center to center, and lagging 4 ft. 6 in. to 5 ft. in length is used. Sills are not used as frequently as they should be and the result is seen in the uneven appearance of many tunnels and the frequent repairs necessary.

In Fig. 65 is shown a set at a breast with the lagging *L*, resting on the bridge *B*, which is a piece of lagging or plank held from the timbers by the wedges *W*. In the spaces, *a a*, the lagging for the next set is inserted and driven forward as the breast is advanced. Referring to Fig. 66, which is a vertical section

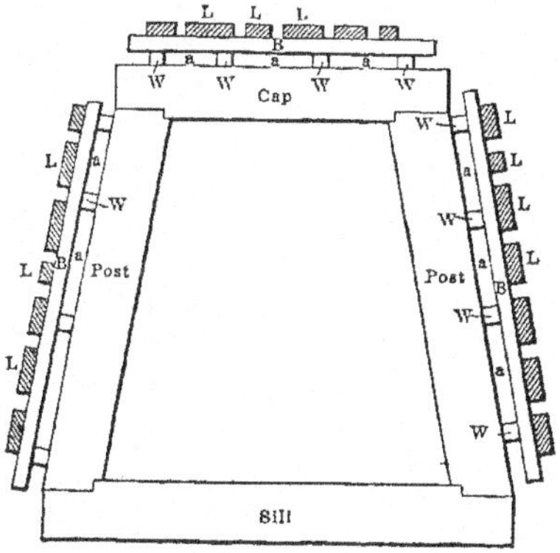

FIG. 65.—DRIFT TIMBERING IN RUNNING GROUND.

lengthways through the center of the tunnel, the operation is as follows. Suppose set *A* 's in place with the breast at *M N*. Lagging *L L*, is pushed under the bridges on posts and cap of set *A* and driven up to breast *M N*. The breast is worked forward by pick or drill and each piece of lagging is driven

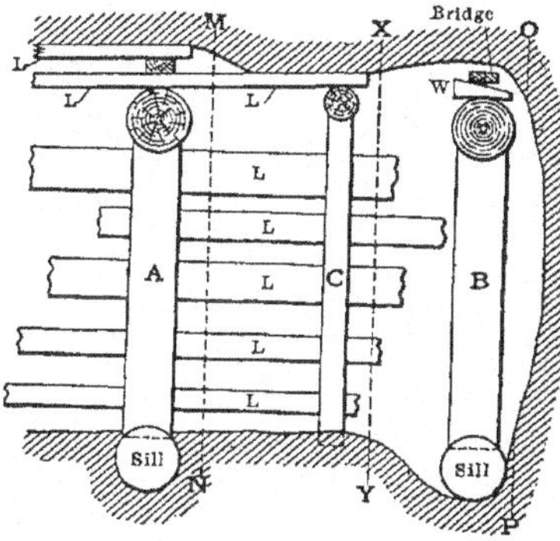

FIG. 66.—SECTION ALONG DRIFT SHOWING METHOD OF PLACING SETS.

forward as the ground in front of it is taken out. When the breast has reached *X Y*, a false set *C*, of light, roughly framed timber, is set in place to guide and support the lagging as it is driven forward. When the breast reaches *O P*, the set *B* is placed and the bridges put on. Then *C* is removed, allowing the driven

8

lagging to rest on the bridges of *B*. The wedges sustaining the bridges on *A* are next withdrawn, allowing the lagging of the previous set to press the bridge evenly against the forward lagging which rests directly on *A*. New lagging is now driven under bridges on *B* to the breast *O P*, and the operation repeated.

Driving in Loose Ground (By George J. Young).—In driving a drift through loose or running ground fore-poling is commonly resorted to. A modification of the method, in use on the Comstock Lode, is worthy of mention. This method is of particular utility where the drift must be supported by face boards. Ordinarily two settings of the face boards are required for each advance of one set of timbers, the first setting just beyond the position to be occupied by the false set, the second just beyond the regular set.

In the Comstock modification the top laths are driven one-half their length, the upper face boards being removed and the loose ground carried at such an angle as to prevent it from running into the drift. The laths are then supported

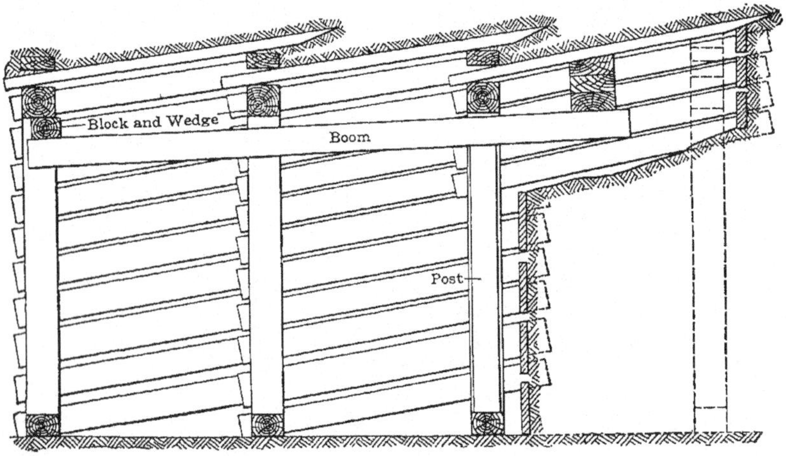

FIG. 67.—BOOM METHOD OF TIMBERING IN DRIFTING THROUGH HEAVY GROUND.

by a "boom," which is a wooden lever, an 8×8-in. timber, 9 to 10 ft. long. The boom rests upon the post which acts as a fulcrum and one end carries sufficient blocking to catch up the top laths, the other end being held down by blocking and wedges between the end and a neighboring cap. The boom is rigged along the center of the drift. The top and upper side laths are then driven forward to their full length and the upper face board placed in position. The remaining side laths are then driven successively and the face boards placed in position. On the completion of the advance the next regular set is placed. A single setting of the face boards thus secures an advance of one regular set. In square-setting old fillings the boom is also used and obviates the necessity of putting up temporary props in the working space. The method of rigging the boom is shown in Fig. 67.

False Set for Spiling Ground (By James Humes).—The tunnel being driven by the Austin-Manhattan Consolidated Mining Co., at Austin, Nev., is 7 ft. wide at the bottom, 6 ft. at the top and 8 ft. in height. The timber used is mostly 10×10-in. Oregon pine. The posts are cut 8 ft. long and placed in hitches. The caps are 4 1/2 ft. in the clear.

When I assumed charge of the underground operations of this company the old style of false setting for spiling ground, which was sure though slow, was in use. The cost was high on account of the large amount of timber consumed (which costs here $40 per thousand), for the same braces that would answer for one set of breast boarding would not do for the next and the tunnel was always littered with discarded breast-board braces. It was only rarely that we could get in one set of timbers in 24 hours, and as all costs were charged up against this work of retimbering they amounted to more than $40 per ft. for the spiling ground.

This excessive cost set us thinking, and the result was the production of a false set that increased the footage driven in the worst ground about 10 ft

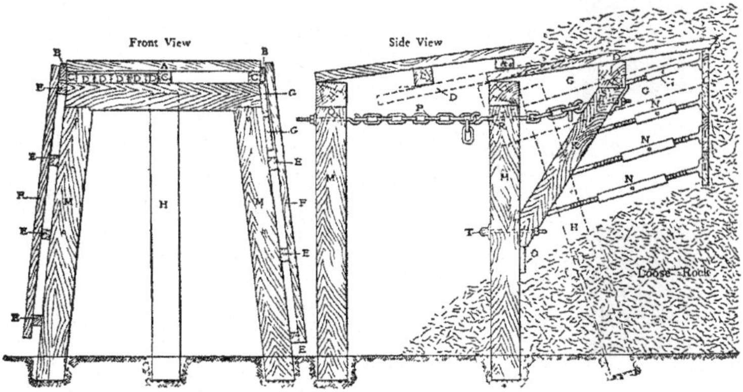

FIG. 68.—A FALSE SET FOR DRIVING THROUGH LOOSE AND HEAVY GROUND.

per week over our previous work. When we are sure we have reached ground that is to be spiled we stand the last set loosely, holding the posts in an upright position by spiking a 2×6-in. plank from the last set of timbers in place to each of the posts of the set about to be erected (*M* in Fig. 68). These pieces, and a temporary block placed on the center of the cap and against the ground, hold the timbers in position until the side bridging *F* is put in place and blocked.

We are now ready to place the top bridging *A*, which consists of a 3×6-in. piece supported on 4×4×6-in. blocks *C*, one on each end and one in the center of the cap. These blocks are generally removed as the spiling is placed in position. After the spiling is driven home the space between the bridging and the spiling is filled with wedges which are taken out when the further end of the spiling is let down on the bridging of the next set.

While driving the spiling ahead, a 6×6-in. tailing block *L* is used. This tailing block should be sufficiently long to engage at least half of the spiling; oftener it should be long enough to reach across all of them and should be moved up as the spiling is advanced. When driving the back spiling there should also be at least two side spilings advancing on each side; these should be kept as far up as the back spiling. While driving the back and side spiling, one man uses a light, sharp-pointed bar with which he removes the rock from before the spiling while his partner is forcing them up with a 12-lb. hammer.

When the spiling is advanced 3 or 3 1/2 ft. ahead of the last main set, the loose rock is pulled down to make room for the swing posts *K*, but none of it is shoveled out of the way into the cars until the swing set is pulled up tight against the spiling. Leaving this loose rock as shown in the drawing helps to sustain the mass hanging overhead. When the false cap is placed on the swing posts, the posts and cap are drawn up tight against the spiling by screwing up on bolt *Q*, the foot of the posts in the meantime being held in position by the bolt *T* and also supported by the block *O*. The fact that this set can be pulled up tight in this way is one of its advantages, for it obviates the necessity of driving wedges over the cap and against the spiling, the jarring of which sometimes results disastrously.

Everything is now ready to force the spiling in to its full length, and when this is done we are ready to go on with the breast boarding. The boards are held in place by the iron braces *N* which have right and left screws or threads. The center pieces or nuts are made of 2-in. pipes plugged at the ends and tapped for the extension pieces, and have holes drilled in them to admit a small bar.

The end or foot piece of the brace next to the breast board is a part of the rod bent at the angle shown. Each end of this foot piece is turned over into a slight point so that when it is backed up tight it presses into the plank and prevents slipping. The other end of the brace is rounded and fits into a hole about an inch deep bored in the swing posts.

These braces have not failed in a single instance, neither has any part of the swing set. It has often happened, in the old method of timbering, that downward pressure on the breast boards would push the braces below a horizontal position, and then, of course, they would fall out and leave the breast boards unsupported; but these iron braces can be placed at such an angle above the horizontal that the downward pressure on the breast boards cannot drive them out of place.

The leaning post *H* shown in the center of the tunnel is used for bracing the center of the breast boards when the extension braces are taken out to admit of the main posts being placed in position. It is also customary to place a plank upright against the breast boards and overlapping all of them, and it is against this plank that the braces are placed. The hole in the swing posts for the eyebolt *T* is made large enough to admit of the above adjustment. When everything is in proper position and the chains are pulled up tight, this eyebolt is also tightened.

The miners like this method, for when they reach a point where there is a cave of considerable height, they can put this swing set in position under the protection of the last main set. They can then place plank or heavier material from the last set over the false set and thus protect their heads until they have the next main set in position.

The side spiling is worked forward as fast as the breast boards are put in place and overlapping them; this and the support the side spiling receives from the swing posts is sufficient to keep them in place until the main set is put up. There are only two side spilings shown in place in the drawing, but it is frequently necessary to spile the sides to the bottom of the tunnel.

Drift Timbering for Heavy Ground.—On the Mother Lode of California especially where mining is conducted in the black Mariposa slates, great difficulty is experienced in keeping the drifts open. It is not uncommon to see huge drift timbers crushed and splintered within two weeks' time. Occasionally drift sets require retimbering so often that by the time a stope is worked out the drift below, originally 8 ft. high in the clear, will be barely high enough to allow the passage of an ore car. It is usually considered economical to run the drifts in the orebody. Above the drifts in the Kennedy mine at Jackson, a system of carriers on stringers is used to support the stope filling above. The usual method of timbering drifts in such heavy ground is illustrated in Fig. 69. The cap of the tunnel set is blocked against the walls and blocks set above the posts carry the stringers, locally termed sills. These in turn support the carrier timbers which extend across the drift and are also blocked against the walls by head boards. The flooring of the stope is laid on these carriers. In most cases the horizontal swell or pressure of the walls is sufficient to hold the carriers so firmly that they will support the stope filling during any necessary replacement of drift sets. In this manner the drifts are kept open for their full height and plenty of head room is assured in the main passageways of the mine.

Finger-pin Timbering in Swelling Ground.—The finger-pin method of protecting drift sets in swelling ground was devised on the Comstock Lode, and has been in use there several years. In it, as can be seen in Fig. 70, the lagging is held back against the swelling ground by means of finger pins, as they are called, pieces of fir 2×2 in. in section, about 6 in. long and sharpened at their small end to a diameter of about an inch. Two of these pins are put in at each end of a plank, with the larger end against the post as the pressure on that end is less per square inch than that on the small end of the pin. The pin therefore is forced through the lagging as the ground swells, and the lagging is impaled upon the pins. In this way the posts are protected. As soon as the ground has swelled so that the lagging is almost forced against the posts, the ground is eased off, pieces of 2-in. plank are put in to cover the holes, and again the planks are impaled on the pins.

The drawback to the simple sprag-pin drift set is that if the head timberman and shift-bosses fail to keep close watch of the sets, before it is discovered the

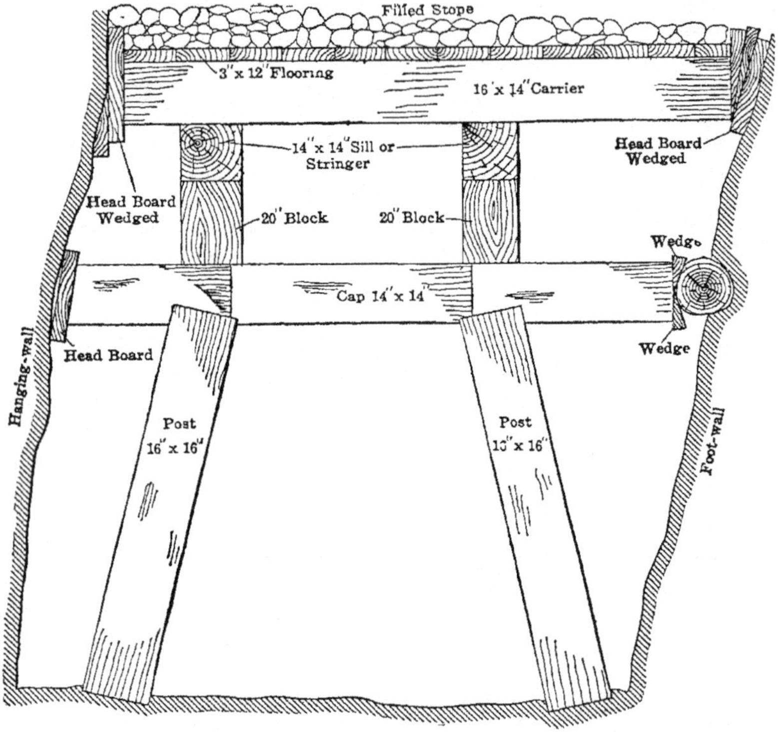

FIG. 69.—DRIFT TIMBERS AT KENNEDY MINE, CALIF.

ground is liable to have moved in to the post and broken the timbers. To guard against this the finger-pin method was combined at the Goldfield Consolidated mines with the old method of false sets. Then, in case the sets are forgotten only the timbers of the false set will be broken.

It is impossible to use these finger pins in supporting the top lagging as owing to unequal weight jack-knifing is sure to occur. Consequently, the procedure is to carry the roof lagging on false caps resting on blocks over the tops of the posts, but where old mine rails are obtainable, they probably make the best lagging for the roof in swelling ground. The rails are used with the bottoms against the caps. Then the ground as it swells will force itself down between the rails. This is the method used at the Copper Queen mines in supporting the roof of drifts in badly swelling ground. Rails are also used for side lagging in the worst ground.

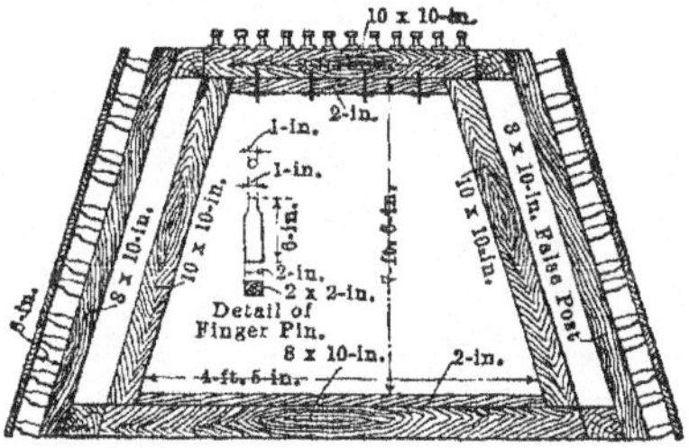

FIG. 70.—BRIDGE SETS WITH LAGGING OVER FINGER PINS.

In case the ground in the bottom of a drift swells, the best thing to do is to keep the bottom covered with several inches of standing water or sand. This keeps the air away from the ground and prevents further oxidation of the decomposing minerals, generally the feldspar. The oxidation of certain minerals in the rock is, in general, the cause of swelling ground. In holding swelling ground, therefore, the plan is to ease the timbers as little as possible each time so as to try to get a considerable layer of thoroughly oxidized ground exposed to the air so that further oxidation behind the outer layer will be prevented.

Joint for Drift Timbers.—The joint for framing the legs into the cap of a three- or four-piece drift set, shown in Fig. 71, is designed to give equal areas on

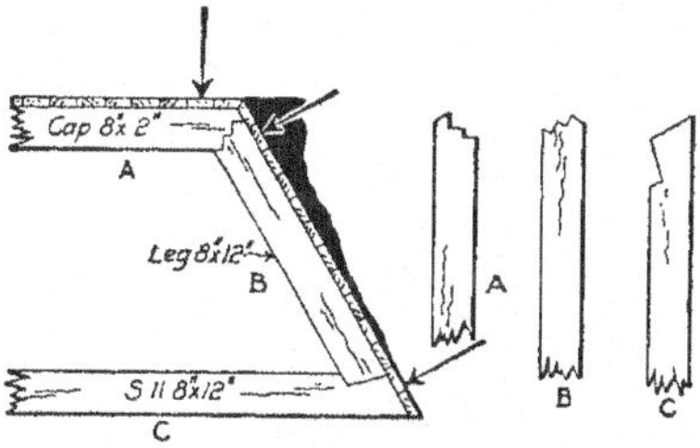

FIG. 71.—AN IMPRACTICABLE JOINT.

the ends of both cap and legs to resist both downward and side pressures. This joint would meet the requirements if it could be cut perfectly. It is in practice almost impossible to cut this joint so that all areas of the cut in the cap will bear evenly on all corresponding areas in the leg cut. For this reason the joint is not favored by experienced timbermen. A simpler and more effective joint can be made by bringing the ends of the legs to bear under the end o the cap in which no notch at all is cut, then to the under side of the cap a piece of timber is spiked, the ends of which bear against the upper ends of the legs immediately below the cap. This piece of timber resists the side pressure exerted against the legs, the cap, in so far as side pressure is involved, serving only to carry this piece of timber. The joint shown in Fig. 71 is liable to result in splitting of the timbers in those places where the wavy shade lines are drawn.

VI

STOPING

Variations of Practice—Timbering—Ore Chutes.

VARIATIONS OF PRACTICE

Stoping with the Slicing System.—The slicing system of stoping is being carried on at a number of the Utah mines where the ground is too heavy to be held by square sets. At the Daly West, Park City, and the United States mines, Bingham Cañon, this method is being successfully employed. The practice is simply to raise to the upper limits of the ore and mine over the whole orebody for a thickness of one timber set, using such timbering as is necessary to hold the back. The floor is then covered with cheap planking, lagging or any available lumber which will serve as a floor. Auger holes are bored in the timbers holding the roof, loaded and the timbers shot down, whereupon the back caves upon the flooring which has been laid. Another "slice" is then worked out in the same manner from the raise, one set below the portion previously extracted. To insure greater safety for the men the ore of the successive slices is best mined retreating; *i.e.*, a crosscut is driven from the raise to the other wall of the orebody, drifts run to the stope limits and the working face advanced toward the raise. A safe retreat through the crosscut, which is in solid ground and should be well timbered, is thus always available for the miners. By this method of stoping all the timber used to hold the roof as well as the different floors is, of course, lost. However, as the stopes need not be held open for any length of time, cheap lumber may be used for this purpose. Lumber in Bingham Cañon costs about $18 per thousand for local Uintah pine, and up to $22 per thousand for Oregon pine. In one of the large mines at Bingham, where ore is mined at about $1.90 per ton, using square sets in overhand stopes, the timber cost approximates 17 to 34 cents per ton of ore stoped. Under the slicing system the timber cost is probably not more than 5 cents per ton higher, and by it ground is worked which could hardly be held by ordinary square setting.

Stoping at Goldfield Consolidated.—On the 900-ft. level of the Clermont mine of the Goldfield Consolidated the Clermont stope is opened for about 150 ft. on the strike of the vein and shows a width averaging 40 ft. On account of its high grade, all of the ore must be recovered but it is desirable to use the minimum amount of timber. Under the present stope practice, introduced by J. Donnelly, mine superintendent, 8×8-in. material is used for the square sets and no trouble has been experienced from the timbers swinging. Square sets and waste filling are used.

The orebody is opened in sections eight sets long, and at each eighth set a manway is maintained. Every fourth set is used as a chute. The stope is filled with waste obtained from hanging-wall crosscuts, filling being kept up to the second floor below the back of the stope. Sill sets are 8 ft. in the clear and are framed of 10×10-in. material, but the stope is carried up with 8×8-in. timbers, the sets being 7 ft. high. The end set of each section of stope is left open so that there will be no danger of filling crowding in and wrecking the manways. Where necessary, cribbing is resorted to for confining the filling. By working the orebody in such short sections the danger from timbers swinging is eliminated and it is possible to work successive sections instead of alternate ones as must often be done.

A Modified System of Back Stoping (By J. E. Wilson).—The general scheme of a simple, safe and economical method of stoping where the cost of

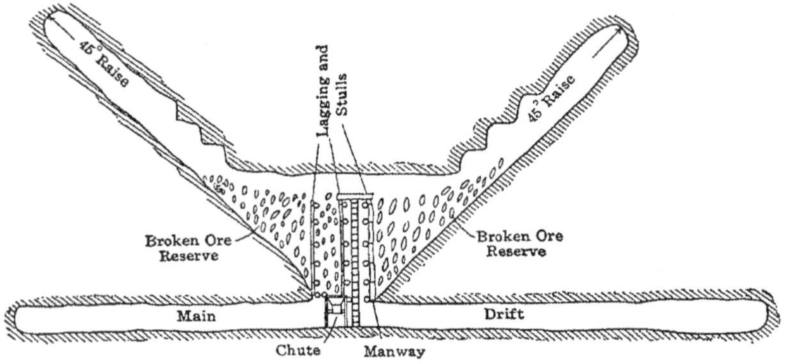

FIG. 72.—SCHEME OF BACK STOPING EMPLOYED AT THE DOLORES MINE.

labor and timber is a serious consideration is shown in Fig. 72. The method is a modification of back stoping. For the introduction of the modified stoping it is only necessary to drive one or two raises at an angle of 45°, or less, depending on the dampness of the ore to be mined. Where the ore is dry the raises may be driven at a flatter angle. Stoping can be started as soon as the raises are advanced about 20 ft. Work should begin above the chute, care being taken to cover the latter so as not to destroy timbers while the first few rounds are being blasted. The first bench or step mined will be slightly wider than the chute and manway combined and the length will increase as the stope advances, thus gaining stoping back for every foot raised. The broken-ore reserve will start from the first set of lagging put in, as only the overflow will go into the chute. The filling of ore serves for miners to stand upon while drilling the back.

The most advantageous method of breaking ground, according to my experience, is the bench or step system, which is illustrated in the sketch. This system eliminates the common and serious trouble of the cut or "relief" hole failing to discharge and thus impairing the result of the entire round. This failure of a round to break the rock properly means much added expense, is quite annoying, and worst of all, exceedingly dangerous as the following shift may accidentally pick or drill into the missed hole; this has often occurred with disastrous results.

The main features in this method of mining are the elimination of shovelers, as all ore broken will run into the chute by gravity, and of expensive scaffolding. No timber is needed, except that for the manway and chute, thus reducing expenses to a minimum. When the stope is mined to the level above the process of drawing, the reserve ore can be started either from the top or lower set of lagging, as the case may be. I would recommend, though, to draw from the top, as all boulders can be broken before entering the chute. I am now using this method in the Dolores mine, in Chihuahua, with satisfactory results.

Eliminating Shoveling in Square-set Stopes.—It is generally agreed that the shovel is the most uneconomic implement used about a mine. The ideal

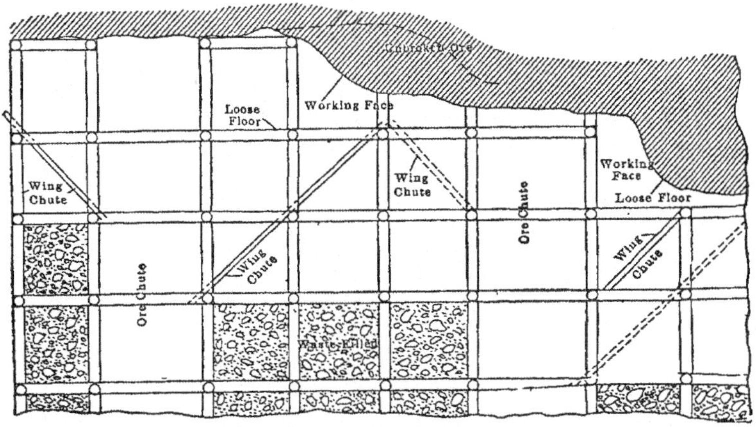

FIG. 73.—SECTION OF STOPE SHOWING DIVERTING WING CHUTES.

method of mining is that which necessitates the least possible handling or shoveling of the ore. An excessive amount of shoveling, and carrying of ore to chutes in small wheelbarrows, is done in most mines where square-set stoping is practised. At the Argonaut mine, Jackson, Amador county, Calif., this is eliminated in the stopes to a large extent by building plank wings between the timbers, which serve to divert the ore to chutes as it is broken. Every fifth set is usually maintained as an ore chute so that the wings need not span more than two timber sets (see Fig. 73). This makes it possible to keep the waste filling up to at least the second floor below that on which the men are working, an important consideration in the heavy ground encountered on the Mother Lode.

The wings are merely temporary sloping floors laid from one cap to the next above or below on either side along the stope, and extending to the ore chute. The men work on a few plank laid horizontally over the caps of the topmost timber set, but instead of breaking ore down on a tight floor below and then shoveling and wheeling it to chutes it is broken on the wings which divert it to chutes without more labor. Varying and unusual conditions in the stopes make it necessary, of course, to do some shoveling but the men are required to reduce this to a minimum.

In easily mined ground where little shooting is required and sorting in the stopes is not advantageous this system works admirably. The saving effected by eliminating much shoveling and handling of rock in the stopes doubtless makes it possible to mine profitably ore which, under the old system of mining, had to be left. This scheme was introduced at the Argonaut by R. S. Rainsford, superintendent.

Recovering Ore from Pillars.—The accompanying sketch, Fig. 74, illustrates in plan and section a method of robbing pillars at a mine in the Joplin

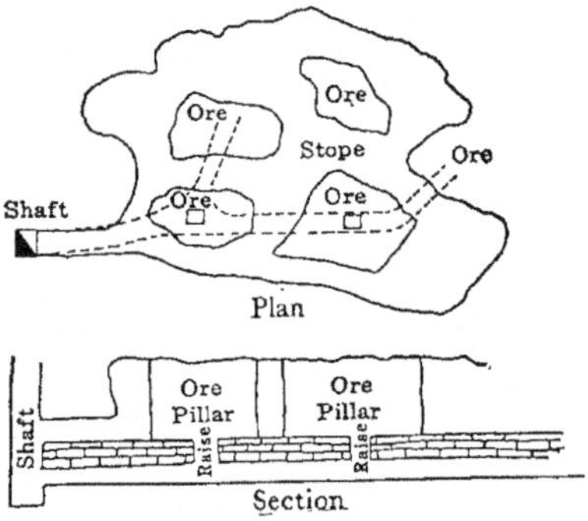

FIG. 74.—ROBBING ORE PILLARS.

district. The ore occurred in a pocket, where the roof was heavy and a large amount of timber had to be used to enable even half the ore to be extracted. The pillars themselves contained a high percentage of ore. Beneath the body was a solid, compact limestone stratum. The shaft was sunk a few feet into this, and a subdrift extended beneath the ore pillars with a 7- or 8-ft. roof. A raise was put up in the center of each pillar and the ore shot down into the drift below, and trammed to the shaft. This method gave a safe place in which to work and at the same time allowed nearly all the ore to be recovered.

Scaffolding for Drills in Wide Stopes.—In stoping in a wide vein it is often the case, when an overhand stoping system is in use, that some provision must be made for mounting the machine drills. If a shrinkage system of stoping is being used, it may also happen that so much ore has been drawn from the stope that the height of broken ore in the stope is not sufficient to enable the miner to reach the back. In some of the Lake Superior copper mines and in certain mines of the Transvaal, long drill columns are used some of which have a length of 28 or 32 ft. It is not always convenient to keep such long columns on hand or the slope of the stope may be such as to make the use of such columns impossible.

The accompanying drawing, Fig. 75, illustrates a type of scaffolding that is much used in the wide stopes of some of the mines of Western Australia. This scaffolding is built up of vertical posts *E* well footed and wedged in place at the top. At the required height steel dogs *A* are driven into holes cut in the backs

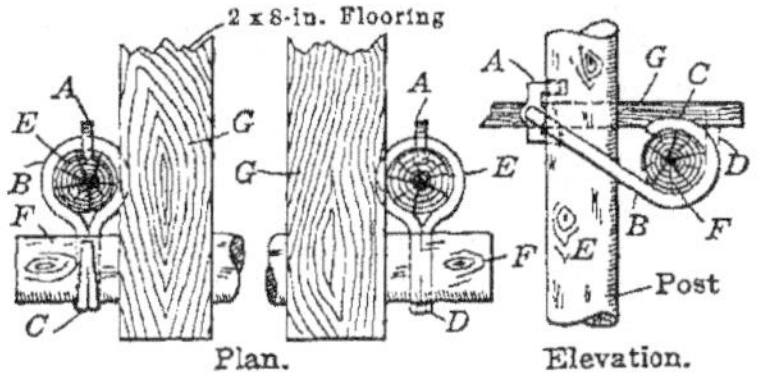

FIG. 75.—STAGE FOR DRILLING OR SAMPLING IN STOPES.

of the posts. Through a hole in the backs of these dogs, pass the ring-hooks *B*, made of 7/8-in. round iron. The hooks for the end posts are curved back as shown at *C*, while those between the end posts are curved as shown at *D*. The dog holds the ring part of the hook from slipping, but most of the holding power of the hook is due to binding against the vertical post. Horizontal members *F* are carried in the hooks, and form the support for *G* the plank flooring. Such a staging can be used for a variety of purposes about a mine besides that of supplying a stage for drilling. In sampling high backs a scaffold can be rigged with four posts set at the corners of a square with four horizontal members forming the sides of a square, that will give easy and safe access to the face to be sampled.

Placing Holes in Breast Stoping (By Harvey S. Brown).—In stopes 30 ft. or more in width, breast stoping usually leaves a stronger and safer back than back stoping by "upper" holes. In breast stoping, the manner of pointing the holes has a definite influence on the condition of the back. Holes drilled approximately at right angles to the breast, *i.e.*, parallel with the general direction of the advance, shake the back the least and are, therefore, the most desirable. Holes drilled parallel to the breast cause loosening of the back, necessitating more barring down, with attendant delays. The work can easily be arranged so as to permit pointing all the holes at right angles to the breast.

A Method of Blasting in Stopes.—The method of drilling and blasting used in many shrinkage stopes of California is worthy of far wider application than it receives. It can be employed in any stope where the muck from one round of holes is not depended upon to furnish a foundation for the bar or column for the next round. It may be used in either wide or narrow stopes. Only the rows of holes, *a b c* in Fig. 76, are blasted at one time, the remainder being left until the next shift. Thus the machine is always well behind the shots and need only be removed when all the holes possible from the set-up have been drilled. The troublesome and useless labor of carrying the drill and rig over timbers and muck in search of a safe place to leave it, every time a round is blasted, is avoided.

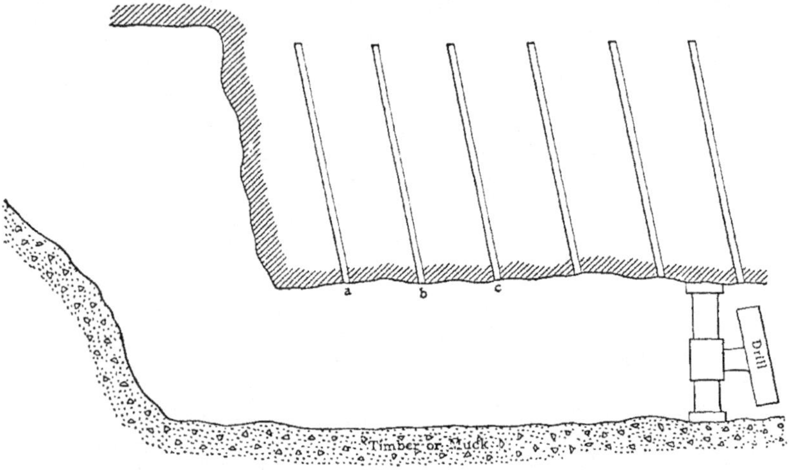

FIG. 76.—ARRANGEMENT OF HOLES FOR BLASTING WITHOUT REMOVING DRILL.

Obtaining Cheap Stope Filling.—A unique and satisfactory method of obtaining cheap waste for stope-filling purposes is being used on one of the lower levels of the Tramway mine, at Butte, Mont. It is on somewhat the same principle as the glory-hole method. Three crosscuts, converging like fingers on an open hand, were driven into the granite country rock for distances of 100 ft. or more, and chutes, for drawing the waste out into mine cars, were spaced along the crosscuts at regular intervals. The first floors over the crosscuts were then stoped out, after which the ground was allowed to cave, and as fast as the rock came down, it was drawn out into cars. Filling to the amount of 30,000 mine cars has been taken out in this fashion at a low cost.

"Sand Filling" Stopes in the Transvaal.—The "sand filling" process of charging the worked-out stopes is being put into operation at several mines on the Rand, where are prospects that in the immediate future it will be extensively adopted. The mill tailings supply exactly the class of material required. They are handy and enable the filling to be done in a cheap and efficient manner. An ordinary pack of large material, waste rock, is built next to the level, backed by smaller material against which the water-carried tailings are deposited until they completely fill the stope. The water drains off and is pumped to the surface to be used over and over again. Under favorable conditions, it is estimated that the tailings can be deposited in the mine for less than 2d. per ton. In a short time the filling becomes solidified and is capable of steadying the subsidence of the hanging wall.

The Use of Cyanide Tailings for Stope Fillings.—In West Australia fatalities from the vitiation of the air of mines by the fumes arising from the tailings from cyanide treatment used for filling stoped ground have been reported. In such cases it appears that wet, fresh tailings have been run directly into the mine without any previous exposure to the air by heaping on the surface. The West Australian Royal Commission, in dealing with this subject, recommended that tailings should not be used for filling: (1) In the wet state; (2) when they contain more than 0.01 % cent. of their weight of cyanides calculated as cyanide of potassium; (3) in any part of a mine where there is not a current of air passing freely. The Australian method of stope filling with cyanide tailings is a dry-filling system, whereas a water-borne system by which the dry sands are sluiced into the empty stope and the water drained off, leaving the sand in a compact mass, is used on the Rand, in the Robinson mine. The Transvaal Mines Department arrived at the conclusion that a solution containing prussic acid loses the latter rapidly by evaporation into the air. In the case of tailings being used for filling stopes, it is considered likely that the drainage from the sand containing cyanide will come into contact with acid water from the mines, and that some prussic acid will be formed. The Transvaal Mining Regulations Commission is therefore of the opinion that the percentage of cyanide in tailings used should be low and good ventilation should be required. A recommendation is made that the regulation of the West Australian commission in this matter be adopted, pending the results of further investigations to be conducted in the Transvaal.

Mining Dangerous Ground on the Mesabi Range (By B. M. Concklin).—Some of the simplest and most useful practices in mining are occasionally overlooked for no other reason than their simplicity. It may, therefore, be interesting to note a method of mining practised on the iron ranges of the Lake Superior region, which is simple in itself and also affords protection to the miners. In this case the ore is capped by tough taconite and paint rock, both of which are more or less disintegrated, while the bottom of the orebody is solid taconite. The orebody is from 18 to 24 ft. thick. The square-set method is usually used in the extraction of this ore, the orebody being too thick to be adaptable to the slicing method operated from one level and too thin to develop more than one level.

In the extraction of ore at the start of operations the back refuses to cave entirely, but arches for considerable distances, making it extremely dangerous to the miners working along the semi-caved area. To protect the workmen and also to extend and weaken the arch which holds the ground from caving properly, square-setting is started two or three sets back from the semi-caved ground. The work is then carried on just as if a new pillar were to be worked out. The slice or drift is first driven the full width of the pillar; then square-setting is started toward the open room. When the room is reached, square-setting is carried on from the slice drift or crosscut to the open room until the

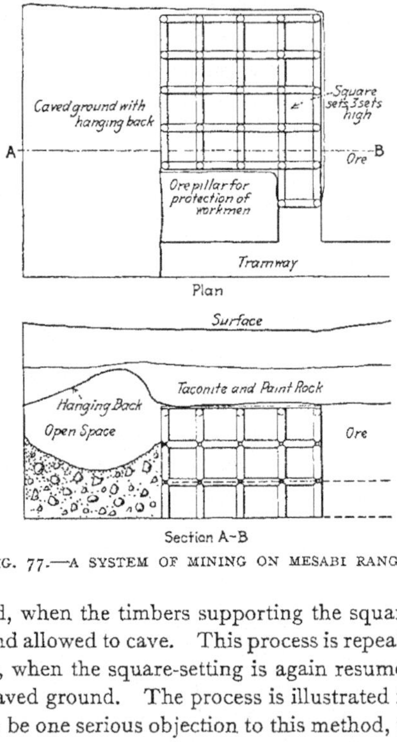

FIG. 77.—A SYSTEM OF MINING ON MESABI RANGE.

tramway is reached, when the timbers supporting the square set room are shot down and the ground allowed to cave. This process is repeated until the ground caves satisfactorily, when the square-setting is again resumed, this time immediately along the caved ground. The process is illustrated in Fig. 77.

There seems to be one serious objection to this method, the probable loss of ore in the last pillar between the square-set room and the tramway, due to the fact that the extension and resultant weakening of the arch has a tendency to crush this pillar as the timbers and pillar take weight. This disadvantage, and the additional cost of mining due to driving the crosscut through solid ore across the pillar, is offset by the greater protection afforded the miners.

Method of Rigging Ladders to Reach Stope Backs.—In the large underhand stopes in the Tennessee Copper Co.'s mines, where the miners practically never see the back, which in an open stope is frequently from 70 to 80 ft. above them, it is evidently necessary to keep the roof well trimmed of all heavy, or "balk ground." To insure this, a crew of men is continually kept at work, looking after the condition of the roof. This work is extremely dangerous and ready resource is required to enable the men to gain access to the back. Fig. 78 shows the method of rigging ladders to reach the roof over the benches of an underhand stope, open to its full height and for a width of from 50 to 150 ft. The ladders are securely lashed together, and, as shown, stayed by ropes secured to the drill steels set into the rock face. A small stoping drill is frequently slung from the ladder and used to put holes in the roof where much balk ground must be slabbed down. Shooting the roof is, however, a dangerous practice, as shattered rock is apt to be left to fall later, when the face of the stope has advanced and the back is inaccessible.

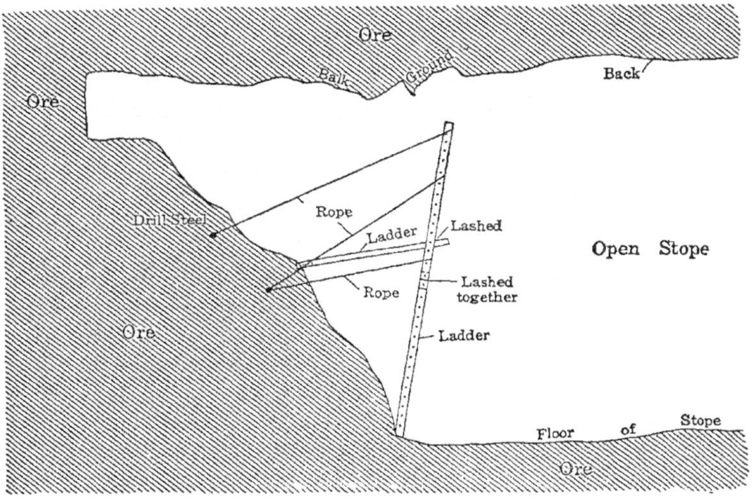

FIG. 78.—LADDER SCAFFOLD FOR STOPES.

Staging for High Set-ups in Stopes.—In stoping the flat-dipping veins of the Lake Superior copper mines, machine-drill posts 10 and 12 ft. long are sometimes used. When top holes are being drilled some staging must be used upon which the drill runners may stand. This staging is often made in the following way: At a point a little farther from the face than the drill post, a piece of drill steel or better a piece of 2-in. pipe, approximately as long as the drift is high, is wedged between the hanging and foot wall in a vertical position. A second upright is placed in a similar position to one side of the other, so that a line between the two is approximately parallel to the breast of ore. A chain is then bound around each in a series of half hitches and a clove hitch at a height above the floor a little greater than the height of the platform on which the operator desires to stand. A piece of drill steel or pipe, a few feet greater in length than the distance of the upright from the breast, is then slipped horizontally between two of the half hitches in the chain, until the end rests on a projection or in a hitch in the breast. One end of this horizontal pipe is supported by the hitch, the other by the chain. If properly hitched the chain will not slip down the upright pipes. One such horizontal pipe is supported by each upright. The platform is made by placing two horizontal planks so as to rest across the two horizontal pieces of pipe; one in front, the other in back of the drill posts. On these planks the operators stand while drilling the upper holes. The advantage of this staging is that it is secure, light, quickly erected, and can be adjusted to different heights by slipping the chains up and down on the uprights.

Chain Ladders in Waste Chute.—In the Utica mine, at Angel's Camp, Calif., waste is dumped into the stopes through raises from above. When an entrance to the stope is desired, chain ladders are used in the waste chutes. The ladders are built by connecting two chains with round iron rods at proper intervals. The connections are made by simply passing the ends of the rods through links of the chains and bending the rods back so as to have either end of each rod linked to a chain. Such ladders are practically indestructible even when used in chutes through which waste is constantly being passed.

TIMBERING

Notes on Placing and Cutting Stulls.—The hitch bottom should be level and not less than 2 in. in depth, and the back edge should be at right angles to the line of the stull or as nearly so as possible. For a heading, it is sufficient to merely smooth off the face of the rock so as to enable a good fit to be secured, no bottom or sides being necessary. A vertical bevel of from 10 to 20° will hold the stull firmly in place.

To measure and cut a stull, it is well to have a slidestaff and try-square, but a good fit can be made with a tape alone. Referring to Fig. 79, the tape is held at *a* and *a–b* is measured. Point *c* is next established vertically above *b* and at a distance from it equal to the thickness of the proposed timber, and *a–c* is measured. These are all the measurements required if the hitch and heading are square. The wall above *c* must be cleared of projections that would interfere with dropping the stull into place along the arc *c–x*. The bevel of the butt of the stull (the angle *b–a–d*) may be estimated by the eye. The butt of the stull is sawed first along the line *d–a* making the plane *d–p–a–o* and the long corner is lopped off with an axe making the plane *s–a–o–p*, and the edge *o–p* square with the line of timber. Then the sides are trimmed as shown making *o–p* equal to the width of the hitch. With the tape held at *o* or *p*, using the measurements *a–b* and *a–c*, the points *b* and *c* are marked on bottom and top of the timber allowing slightly for the bend in the tape caused by the partial wrapping around the stull.

It is frequently necessary to set a stull where neither hitch nor heading can be made square with the line of the timber. On the heading a circle is marked in some manner to represent approximately the outline of the stull. Referring to Fig. 80, point *a* is the back center of hitch bottom, and *o* and *p* the corners; *b*, *c*, *n*, and *m* are respectively the bottom, top, and sides of the circle marked on the heading; *a–b*, *b–o*, *b–p*, *n–o*, and *m–p* are measured. The face *s–o–p–a*, is first made by a cut with the axe and a mark made on this to represent point *a*.

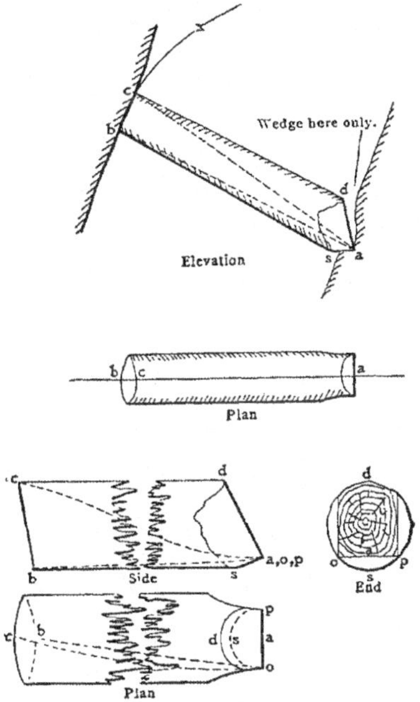

FIG. 79.—METHOD OF MEASURING THE STULL.

A–b is then measured and *b* marked; *o* and *p* are located equidistant from *a* by the measurements *b–p* and *b–o* from point *b*, care being taken that *o–p* is equal to width of hitch. The face *d–o–p* is sawed and the sides trimmed; *o–n* and *p–m* are laid off and the head is sawed through the points *b*, *m*, and *n*. Additional measurements to *c* may be made as a check. A stull of this kind should be braced laterally by a collar brace to the next timber.

Wedging must be done at the butt. The head should fit snugly against the rock at all points but especially at the bottom. Wedges driven at the head are a sign of poor work.

The saddleback system of timbering consists of the use of two ordinary stulls with their heads meeting upon the opposite sides of a 2-in. board so as to form an arch. It is used occasionally where the drive or stope is too wide to allow the use of a single stull.

Framing of Round Timbers (By Percy E. Barbour).—The application of square-timber framing methods to round timbers is illustrated in Fig. 81. In the upper drawing is shown in detail a square-set framed joint, assembled. However, when it is wished to produce a similar joint with the use of round timbers, the problem becomes a little more complicated. The lower drawing shows all the details necessary to prepare round timbers for this sort of joint. It will be noticed that, whereas with the square timbers there is but one bearing

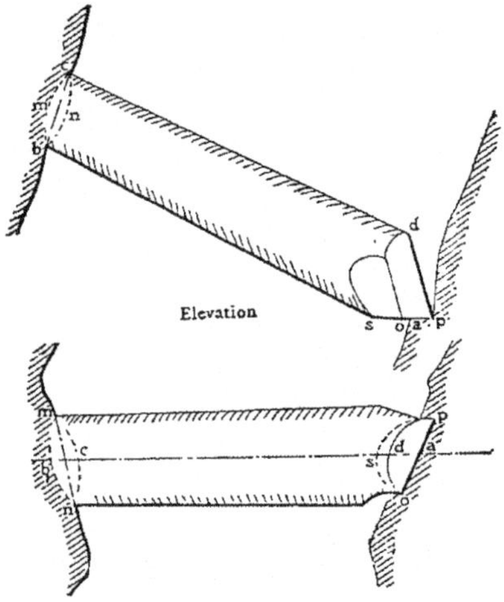

FIG. 80.—THE STULL IN PLACE.

surface at the end of a given piece, as at *A*; in the case of the round timber framing there are two bearing surfaces, as at *B* and *C*.

Leaning Stope Sets.—In the Argonaut mine at Jackson, Calif., and to a less extent in some of the other Mother Lode mines, leaning sets replace the usual square sets in stopes up to a width of 16 ft., which is the full length of the ordinary stull. The advantage of the leaning over the square set is in the fact that posts can always be set directly above each other. In the Argonaut the veins dip at such an angle that it is almost impossible to get in square sets so as to have posts rest on posts in the short space of time that the ground will hold. Simple stull timbering without posts would not hold the walls, which are blocky and in many cases must be lagged.

The so-called leaning sets are really stull timbering with posts and girts added. Or, from a different viewpoint, square sets of variable width, placed with the posts parallel to the walls of the orebody instead of vertical. The standard sets are framed of 8-ft. posts and caps and 4-ft. sprags or girts. Round stull timber is generally used. The greatest amount of pressure is from the swelling of the walls, and to take up this the posts are usually given a horn from 4 to 8 in. square. The sprags are not framed.

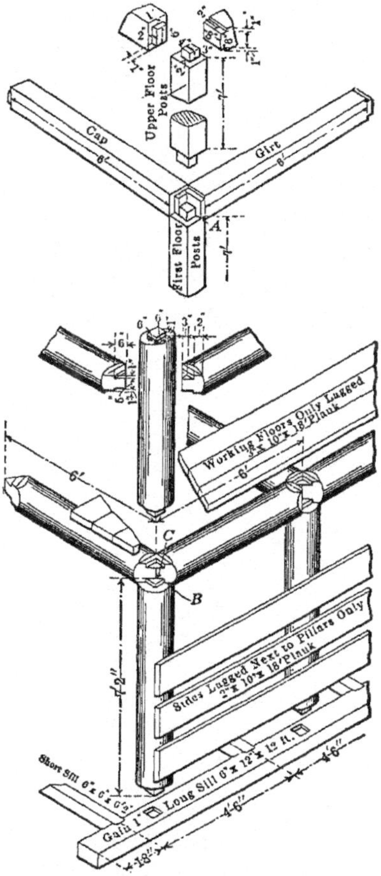

FIG. 81.—DETAILS OF SQUARE SET WITH ROUND TIMBERS.

The usual method of timbering drifts below leaning sets is shown in Fig. 82. In general, two stringers, one on either wall, are blocked up from the drift set and separated by a stull. They are wedged into position and the swell of the walls soon holds them so firmly that they will support the filled stope above, even after the drift sets below are removed.

Battery Method of Stull Timbering (By Claude T. Rice).—It is generally considered that stulls over 12 ft. in length have little supporting power in stopes. This is true where the ordinary method of stull or post timbering is used, but by standing the stulls in groups, or batteries, as they are called in the Lake Superior

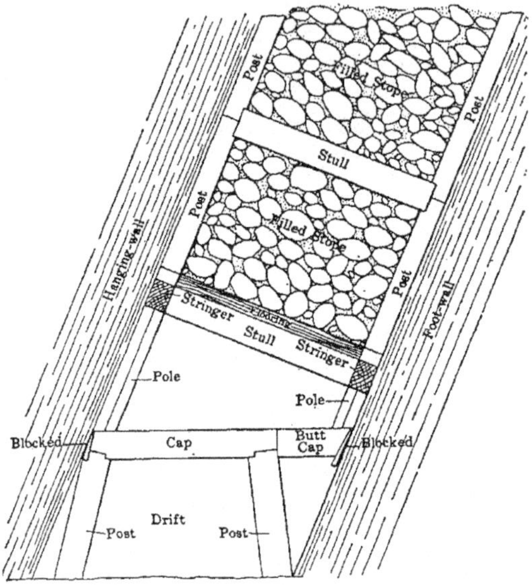

FIG. 82.—LEANING STOPE SETS USED ON MOTHER LODE.

copper district, stulls can be made to support heavy roofs in stopes as wide as 40 ft. Indeed, this form of timbering has replaced the method of square-set timbering formerly used in the Calumet & Hecla mines in which the caps were parallel with and the posts at right angles to the lode.

This grouping of the stulls gives them much greater strength than the same area of timber would have if placed in one stull, while by grouping them, poorer grades of timber, and smaller and therefore less expensive sizes can be used. In the Calumet & Hecla workings where this form of timbering originated, and where it alone is used at present, the battery consists of three posts or stulls. Two are placed in the line of the strike and above these, a somewhat smaller post that supports the wall plate or cap that carries the stringers over which the lagging for the roof is laced. Round timber flattened at the ends where it is to rest upon the posts of the battery is used for the wall plates. In case the hanging wall is good so that it does not require top lacing between the batteries, a block is put in between the front leg and the top blocking so that it can be knocked out and the wall piece inserted later if necessary. The timbers of the battery are stood directly on the foot wall, but the top blocking over them is interlaced so as to tie the group together at the top when the battery takes weight. The posts of the batteries at the Calumet & Hecla mines range in diameter from 16 to 30 in., according to the width of the stope and the condition of the hanging wall. These timbers are, therefore, quite heavy, but they are easily swung into place by means of a small air hoist placed out of harm's way at the bottom of the stope.

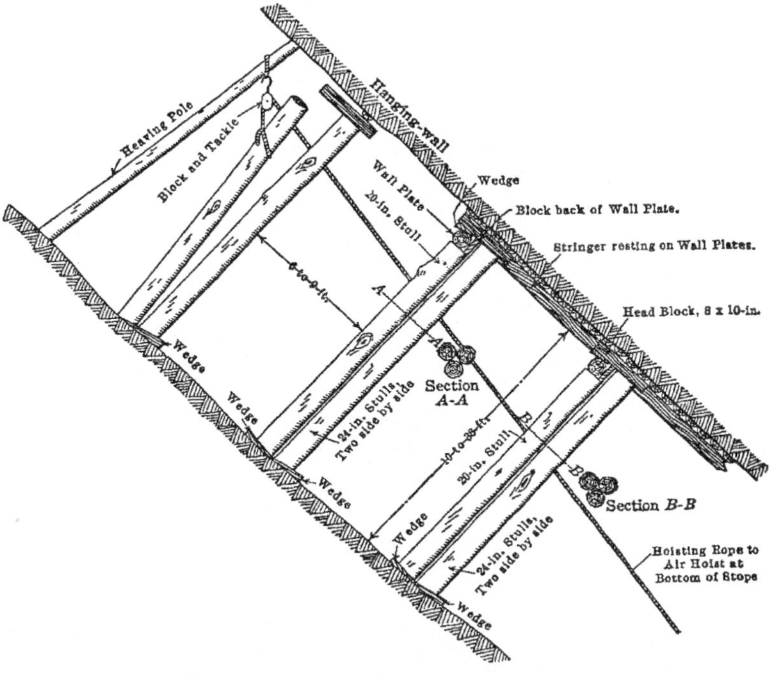

FIG. 83.—BATTERY STULLS IN CALUMET & HECLA STOPES.

There is a division of function in the different posts of the battery. The two main legs which are put in somewhat larger than the front leg carry most of the weight of the hanging wall and they also resist the bending tendency in the posts of the battery, while the main duty of the front leg is to carry the wall piece that supports the scalings that come from the roof later in the life of the stope. As the top post rests on the other two posts of the battery, the lower two help take care of bending strains that come on the front leg. Consequently, two shorter pieces of stull can be put in, butt to butt with a tying piece of soft 3-in. fir between them, to serve as the front leg. This spliced stull, unless the hanging is especially heavy, serves almost as well as a stull all in one piece would, for the pressure on it is mainly longitudinal. In this way short pieces of stull that have been recovered from batteries in old stopes before they caved can be used in the narrower stopes.

In the conglomerate stopes of the Calumet & Hecla company these batteries are put in at intervals of 6 ft. in the direction of strike of the lode and from 6 to 9 ft. or more apart on the dip, according to the heaviness of the hanging wall.

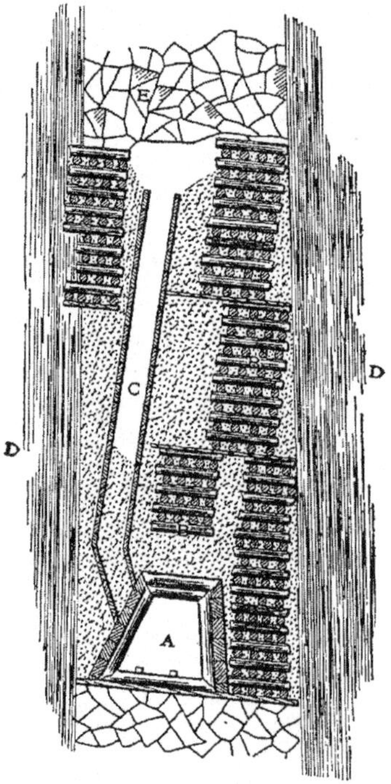

FIG. 84.—END VIEW OF STOPE, WALLAROO MINES.

After the timbering of the stope is about halfway to the level above, the size of the stulls used in the batteries is reduced for it is in the bottom half of the stope that the most weight comes. In the Hecla workings where the stopes have a width of from 14 to 16 ft. on an average a gang of six timbermen can stand and lace two batteries in a nine-hour shift, while in the Red Jacket workings where the stopes vary in width between 16 and 25 ft. and have an average width of 20 ft., and the hanging wall is not generally as good as in the South Hecla stopes, a timber gang can only stand and lace on an average three batteries in two shifts.

The resistance of this battery system against bending strains is amazing for in some parts of the Tamarack workings batteries are being used in stopes 42 ft. wide. In such cases foot blocking as well as head blocking is used, and in order to take care of bending stresses the three stulls of the battery are either wound with wire rope at their middle or else braced halfway down by struts from the other batteries. In this way the batteries are made to stand in such stopes for 3 or 4 months before the weight of the hanging wall becomes too great for them.

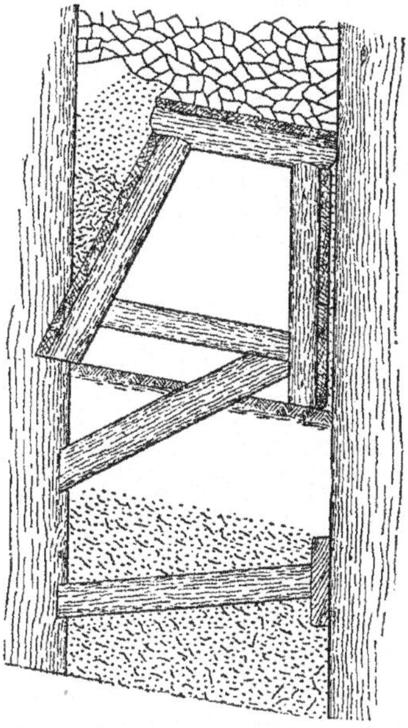

FIG. 85.—TIMBERING NARROW STOPES IN TREACHEROUS GROUND.

The experience of the Calumet & Hecla company would seem to show that, for timbering stopes where the dip is flat and the roof heavy, the battery system is stronger and cheaper than the use of square sets and rock-filling. Fig. 83 illustrates the manner of placing the batteries of stulls.

Timbering Wide Stopes.—H. L. Hancock, general manager of the Wallaroo & Moonta Mining & Smelting Co., South Australia, recently visited the principal mining centers of the United States in search of information which his company might profitably utilize in the mining and smelting of copper ore. In his report Mr. Hancock stated that, in his opinion, he observed no methods of mining at any mine visited which could be substituted to advantage at the Wallaroo mines.

The "stye" and filling system used at the Wallaroo mines is illustrated in Figs. 84 and 85. This method is old, and is used in securing treacherous ground in wide stoping widths. In some of the old mines at Clunes and Edgerton, where the stopes were 100 ft. wide in places, firewood was used, as shown in Fig. 84, in building what are called "horses," when made of timber only, and "pigstyes" when an outer frame of timbers is filled with waste. The latter kind of support is, of necessity, much used where mining timber is scarce. Fig. 85 shows the system used in timbering narrow stopes in treacherous ground, when approaching a level. The endeavor is made to dispense with "horses" or "styes" as much as possible where it is practicable to use filling.

Placing Sills beneath Square Sets Already in Place.—The correct practice in square-set timbering is to place the sills upon the floor of the stopes before the sets are put in. Yet, through false ideas of economy, the placing of sills is frequently neglected with the results, when it becomes desirable to stope the ore up to the floor of one level from that next below, the operation can only be accomplished by catching up the square-set posts from below at the expense of much time and labor.

On the Mother Lode of California there are few mines where sills are placed on the floors of the levels, it being claimed that the sills will rot before a stope is worked through from one level to that next above. Owing to the large sectional area of most stopes in these mines and to the heavy swelling ground commonly encountered, this statement, generally speaking, is without doubt true. These conditions might be met by cutting stopes of smaller sectional area, *i.e.*, if the stopes were worked in sections extending from foot to hanging wall but only for 30 or 40 ft. along the vein.

Generally, however, the stope is opened on the level of the gangway to the full width of the vein and for the entire length of the oreshoot, before any considerable upward stoping is undertaken. This practice in most cases eventually results in giving the mine management more or less trouble later on, which is rendered worse by failure to provide sills on the floor of the level, and by the failure to fill the stope completely as work progresses upward. Close filling of the stope is often neglected as the waste rock, in most cases, has to be broken from the walls, and this entails considerable extra expense.

In many cases sills may be inserted beneath the posts of the sets long after they have been in place, and the ore from the back of the stope next below removed with safety, and usually with little loss. It is a great advantage in connecting levels to have the timber sets in exact alignment, both longitudinally and transversely of the vein, on each level; having the posts stand immediately over each other makes the connection much less difficult and expensive. The proper place for the sills on each level can be easily established by the mine surveyor, and the lines once given on the level there is no difficulty in the timbermen keeping the sets in line, as all the members of the sets are of standard length (or should be), and consequently the sets of one level conform to the position of those both above and below.

When sills have been omitted at the time the stope was started and the placing of the sets commenced, and it becomes necessary to place them later, this may be accomplished by spragging the posts of the sill floor as tightly as possible, both longitudinally and transversely of the stope, and sawing off the foot of each post at the proper height and slipping the sills beneath. As a matter of course this cannot be done in a stope that has been even partly filled.

When the stope is still open and it is desired to place the sills as suggested above it may be accomplished in the following manner: The sills should be laid so as to butt against each other at the ends, or they may be framed so that the ends will overlap, by cutting out the upper half of one and the lower half of the

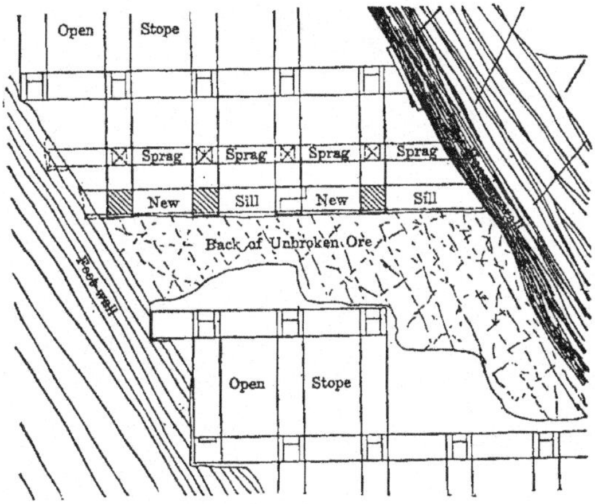

FIG. 86.—TIMBERING ARRANGEMENT FOR REMOVING BACK.

other, which will facilitate in no small degree the connection of the stopes. This must be done with great care and should not be left to inexperienced hands. The work must be done in small sections, and begun only after all the sills have been placed in position in the stope. The sills should be as long as it is possible to handle them—not less than two sets long, and three sets would be better. They should be placed across the vein, from foot to hanging wall, and the timber should be of good size, as they may be called upon to sustain a greatly increased pressure when the stopes are connected.

An idea of the method of placing sills in stopes where they have been omitted may be obtained from Fig. 86. The same method may be successfully applied should it be deemed advisable to replace old and rotten sills with new ones.

ORE CHUTES

Centennial-Eureka Chute Pocket and Gate.—Hard silicious ore will quickly cut out the bottom of almost any sort of chute gate or inclined ore pass. The ore at the Centennial-Eureka mine, at Eureka, Utah, is of such a character that it will quickly cut through even a double lagged, inclined, bottom of an ore chute, so a special type of chute in which a bed of rock forms the bottom has been devised.

An ore pocket is formed by building up from the level three sets of square set timbers into which ore from the stopes is delivered. Single 2-in. lagging is used to line the bottom set of the pocket or chute, and this set is filled with waste rock. On the side from which the ore is to be delivered to cars, the posts are notched, the cap being dropped 7 in. and placed with the faces 45° from the horizontal. A plank lip is then nailed to the cap; this extends only a short way into the pocket, but far enough into the drift to deliver ore over the edge of a car. An ordinary gate of planks sliding between wooden guide grooves is used to control the discharge of ore from the pocket. The lip on some of the Centennial-Eureka chutes is 35 in. wide and auxiliary posts are placed under the cap at either side of the lip. The top two sets of the pockets are lined with double 2-in. lagging.

In this construction the waste filling the bottom set of the chute or pocket forms a bed upon which the ore drops and over which it slides in its passage to the discharge gate. The wear from the movement of the ore is all taken up at this point and all trouble with the bottom of the chutes cutting through is eliminated as the waste forms the bottom. Owing to the large cross section of the pocket the movement of ore is slow (if the pocket is not entirely drawn at any time) so the lagging in the upper sets is not subjected to excessive wear, and in fact, seldom has to be renewed. Such a chute pocket is about as satisfactory and as near fool-proof as any to be found, and it has the additional advantage that it can be quickly built from the material used for ordinary mine timbering, and hence usually in stock. As stated, this ore pocket and gate are particularly useful for handling hard, silicious ores.

Steel Ore Chute for Use in High-grade Stopes.—When high-grade ore is being mined it is always advisable to exert every possible care to see that the fines, which often run high in gold and silver, are not lost. To this end in the square-set stopes of the Centennial-Eureka mine, at Eureka, Utah, every other floor is tightly boarded over so that the fines cannot drop through. All ore is handled to chutes on the tight floors. Steel ore chutes or passes are used between floors and to deliver ore to the haulage levels. Stope sets are 7 ft. 4 in., center to center.

The ore is broken down on 8×8-in. shooting timbers and dropped one set to a tight floor, there sorted and shoveled, or wheeled, to the steel chutes into which it is dumped. These chutes are built in sections, that is, they are carried from one tight floor to the stope floor immediately below on which ore is broken, and terminate about 3 ft. above the second floor, below which is another one that is tightly boarded. At each tight floor a temporary wooden hopper mouth is built to the chute so that ore from above will drop into the lower continuation, and so that ore from that stoped floor can be easily shoveled or dumped into it. The chutes being in sections, can be easily moved to another portion of the mine when one stope is finished. The steel ore passes are 14×14 1/2 in. inside measure, the sides being 3/16-in. sheet steel, bolted at the edges to vertically placed 1 1/2×1 1/2×1/4-in. angles. By placing the angles on the outside corners they are not subjected to any wear.

Such chutes are tight and durable, and their use in conjunction with tightly boarded stope floors insures the delivery of all ores from the stopes. Such refinements of practice (the additional costs thereby entailed) are unwarranted in handling large quantities of low-grade rock, but with such ore as is mined at the Centennial-Eureka, the loss that would result from careless handling of the ore through cribbed or loosely lagged chutes would probably be much greater than is the cost of extra installation.

Bulkheaded Ore Chutes.—Ore chutes are carried up through stopes, in many of the mines on the Mother Lode of California, by simply lagging around

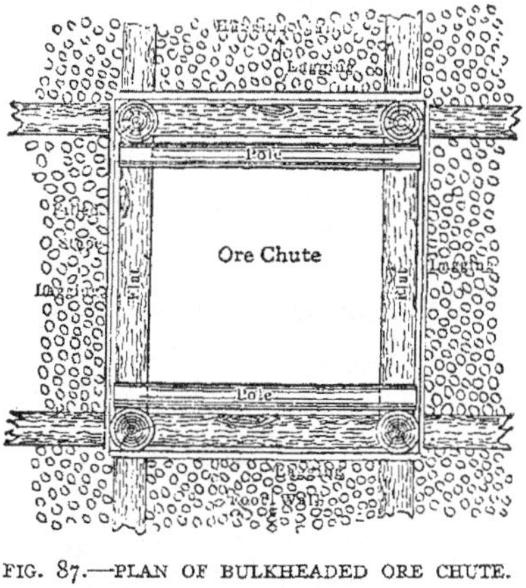

FIG. 87.—PLAN OF BULKHEADED ORE CHUTE.

timber sets. No lining is put inside of the square-set timbers in the chute. This construction serves well enough unless a rock happens to carry away one of the exposed horizontal timbers in the passage way, in which case the light lagging is quickly torn out, letting in waste and often resulting in losing the chute. The lagging may not hold the waste anyhow if the walls are heavy or timber sets swing.

A type of ore chute used in the South Eureka, near Sutter Creek, Calif., is far better for heavy ground and large stopes where the pressure from waste filling is liable to be considerable. Fig. 87 shows how the chutes are built. Stull timbers flattened off on two sides like ties, "flats," are first fitted in (across the vein) between the posts. Across the ends of these flats are then laid poles, or round timbers, with their ends extending to the outside of the flats. The poles are set inside of and tangent to the posts. Alternate flats and poles are then laid as the stope progresses. The lagging is not put on until filling commences, as ore can be shoveled between the bulkheading into the chute.

The advantage of this type of chute is in its durability and strengthening effect on the stope timbering. The lining is, of course, strong and wears in most cases as long as the stope lasts. Also broken timbers can be readily replaced without danger of the chute caving. Then too the flats reinforce the posts of the chute set and prevent adjoining sets from swinging. Thus instead of being as usual a weak place in the stope timbering, such a chute acts as a reinforcement.

Lining for Ore Chutes.—The majority of mining companies usually line the bins and ore chutes with sheet steel. It is seldom that the steel sheets are worn uniformly and as soon as a hole occurs it is necessary to renew the entire sheet, which often means a waste of material and extra expense. At Mineville, N. Y., Witherbee, Sherman & Co. found it more economical to use bar steel. At present they use $3/4 \times 6$-in. bars. By using bars it is less trouble to do repair work and it is necessary to remove only the worn-out part and put in a new piece. All the steel, even short bars with one end worn thin, may thus be used. Railroad rails of heavy weight are also good for lining the front of ore bins at a point directly opposite the place where the cars dump. They are occasionally suspended by the upper end and are free to move at the lower end. They form an excellent buffer for heavy ore as it comes from the mine car or skip.

Safeguarding Ore Chutes.—Where winzes, ore chutes or ore passes open into a floor of a drift they are a constant source of danger to the miners and trammers who have to pass through the drift. In a large mine there are usually many such openings in the floors of the various drifts and unless they are protected in some way, accidents, such as men falling into the openings, are sure to occur.

It is an excellent plan to cut all such passes so that they will open to one side of the drift or into a niche cut into one wall. In such cases there need be no opening in the floor of the drift itself, but it is not always possible to offset the winzes in such a manner. Wherever there is an opening in the floor of a drift through which miners pass, some provision should be made so that a man cannot fall into it. Preferably the opening should be closed by a gate or else by a rough grizzly of logs or timbers. Where the run-of-mine ore is coarse, it may not be possible to use log grizzlies. In such a case some warning should be given the passer-by in order to make him aware of the fact that he is approaching the collar of a winze.

At the Ray Consolidated mine in Arizona a device, similar to that used on railroads to warn trainmen that they are approaching a bridge, is employed at all places where there are openings in the floor of a drift into which a man might fall. The device consists of a beam set in the roof of the drift about 4 ft. from the edge of the opening; one beam is used on each side of the opening. From this beam a number of heavy cords, knotted at the end, hang low enough to strike the face and shoulders of anyone who passes below it. He is thus made aware of the fact that he is approaching a winze or chute and must be careful about passing that place.

Gate for Ore-bin Chutes (By Algernon Del Mar).—The accompanying illustrations show what in my opinion is one of the best gates for ore-bin chutes.

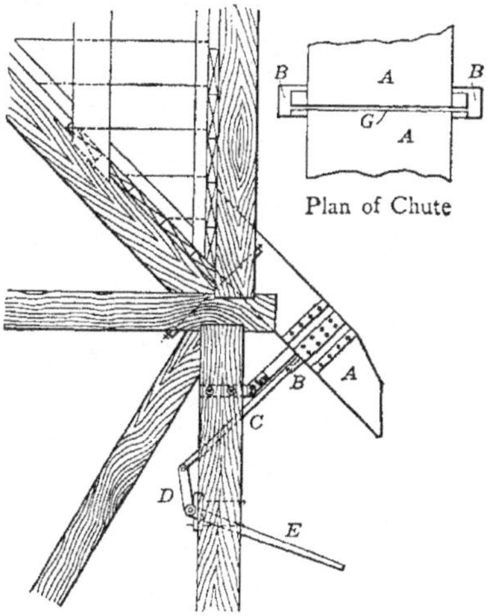

FIG. 88.—GATE FOR ORE BIN CHUTES.

It does not choke readily and can be used for coarse or wet as well as fine or dry ore. A gate of this type was furnished by the A. Leschen & Sons Rope Co. for the automatic loading station of their aerial tramway that I erected in Modoc county, Calif. The sliding gate acts from below by a lever and slides up and down on wheels in a recess on the side of the ore chamber, but in such a way that the ore cannot get into these recesses. A boy can operate it and can load cars as fast as they can be run under the lip of the chute. When loading must be done expeditiously, for example where a train of cars must be loaded, I believe they could be loaded by the use of this gate while in motion. The amount of ore going through the chute is that sliding over the top when the gate is let down and nothing can choke the return of the gate to the position where it cuts off the stream of ore. Referring to Fig. 88, *A* is a steel chute, *B* a sliding gate actuated by the levers *C*, *D*, and *E*. By lifting the lever *E* the gate is lowered allowing the ore to run over the top of the gate. As the top of the gate can be lowered until it is flush with the bottom of the chute there is a clear passage. When it is desired to shut off the stream of ore the lever *E* is depressed throwing the gate up. *AA* is the ore chute, *G* is the slot in the bottom thereof through which the gate moves in a vertical direction, *B*, *B* are the recesses or chambers in which the wheels which are attached to the sides of the gate slide up and down, so that the movement of the gate takes little effort.

Ore Crushing Plant Underground.—A crushing and screening plant is being installed underground in the Round Mountain mine, at Round Mountain, Nev. It will be placed on the 450-ft. level, and will be similar to a plant already installed in the "Stringer Section." As mined, the ore contains many large boulders of rock, which itself is barren, but may envelop pieces of ore. In order to save this ore it is necessary to crush the boulders and sort the crushed material. The plant will comprise a 12×20-in. jaw crusher, and an 18-in. belt conveyor to carry the crushed material to a trommel 10 ft. long by 40 in. diameter, which will be set above loading pockets cut out of the rock above the level. The fine material will be trammed to the shaft, hoisted and sent to the mill. The waste, or coarse material, will be hauled by mules, through a tunnel, to daylight. According to the last annual report of the Round Mountain Mining Co., the crusher installation will effect great economies in handling and sorting the material mined. A better grade of ore will be produced, as under these new conditions about 66%. of the gross tonnage mined will be rejected as waste whereas during the last year 54% was rejected. It will also increase the total recovery of precious metals as it will be possible to send to the mill the greater part of the ore contained in the boulders, which heretofore has been rejected with the waste.

Underground Grizzlies.—Much trouble has been experienced in the mines in the Ducktown district of Tennessee from the ore in the stopes breaking into such large pieces that it could not be handled or would not run in chutes without blocking. In the large underhand stopes, when the ore is milled down to the haulage level in benches from 5 to 7 ft. high, this difficulty, although annoying, is not serious, as the miners with blocking drills can easily drill the large lumps and shoot them into such pieces as can be handily loaded into the tram cars. However, in using a system of mining in which the ore is drawn from the stopes in chutes this difficulty at times becomes so serious as to preclude the possibility of drawing the ore from the stopes through chutes. At the Tennessee Copper Co.'s mines several types of chute gate have been tried in connection with their shrinkage or back stopes, but none has given any satisfaction, all being liable to choking from the large masses of ore which they have to deliver.

Recently, however, J. V. Bohn, superintendent of mines for the Tennessee Copper Co., has introduced a type of underground grizzly in the Burra Burra mine, which has proven so successful that in the future it will be used in connection with all the shrinkage stopes where large pieces of ore have to be dealt with, *i.e.*, on the hanging-wall side of the orebodies where the larger lumps tend to gravitate. These grizzlies, as shown in Fig. 89, are very simple and require practically no repairs for maintenance. An opening is made into the side of the haulage way or drift and a platform cribbed up with 10×10 timbers on either side, leaving a space into which a car may be run from the drift. Across the top of the opening left for the car, old track iron is laid at 15-in. spacing,

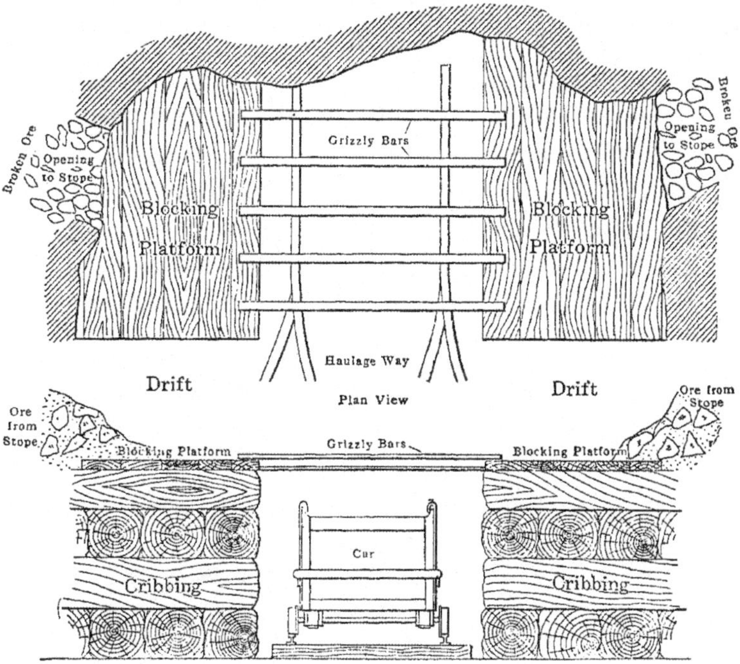

FIG. 89.—ARRANGEMENT AND CONSTRUCTION OF UNDERGROUND GRIZZLY.

the iron being seated in the timbers on either side. At either side of the platform openings are made to the stope above, these being made large enough so that ore of any size liable to be encountered will be readily delivered upon the platform.

To fill the car the trammer simply runs it under the grizzly and with a pick, draws ore from the mouth of the chute over to the grizzly bars, the fines, or materials of such size as can be handled by the crusher, passing through the bars, while large lumps are retained on the platform. These can be readily blocked while still on the platform, with no danger of destroying the grizzly, there being no gates or movable parts to the device.

A Movable Picking Floor.—A picking floor used by Witherbee, Sherman & Co. for ore that will not pass through a 6-in. grizzly consists of a movable chute 5 ft. wide and 12 ft. long, built of heavy sheet iron, with sides 1 ft. high. One end is supported or hinged just below the chute from the ore bin. The platform is practically level, and a ton or more of lump ore is discharged upon it at a time, and two men pick out the barren rock. When the platform is full of ore, the outer end is lowered by an air-hoist and the ore slides directly into the car for shipment. The second-class ore is tossed to a car on an adjoining track, while the waste rock is disposed of by means of a bucket suspended from a carrier which operates on an I-beam.

A Modified "Chinaman."—A modification of the chute known as the "chinaman," has recently been put into use in the shrinkage stopes of the Cop-

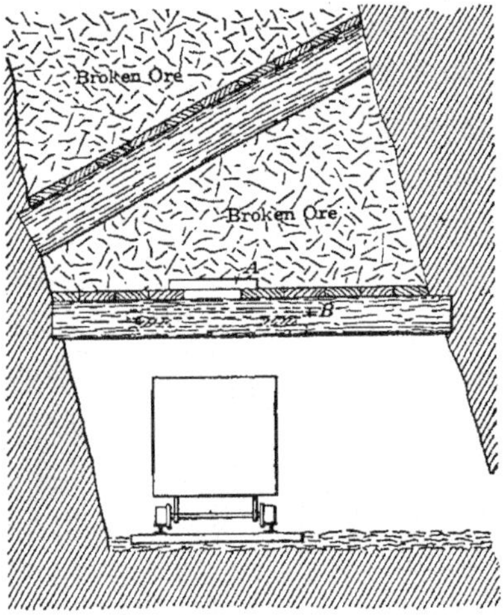

FIG. 90.—THE IMPROVED "CHINAMAN."

per King mine, at Clifton Ariz., by L. W. Armstrong, superintendent. It was found that the "loose, sliding" boards covering the opening in the center of the platform would not slide easily, or at all, when covered with ore from the stope, making it extremely onerous to open the chute and to close it when the car was filled. The difficulty was overcome by nailing a cleat of 2-in. stuff to the inside of each of the stulls supporting the platform and fitting in 2×6's from stull to stull on either side of the opening above; these 2×6's support loose boards *B*, as shown in Fig. 90, which can be easily moved by hand or with a short, sharp-pointed bar. Two of the loose boards *A* on the platform are removed, leaving a permanent opening. It was found expedient to raise the platform about 3 ft. above the top of the car as it enabled the trammer to get at the chute door more easily, and also avoided the inconvenient reduction in the height of the level at the "chinaman."

Ore Chute Construction.—Ore chutes of standard design are used in the mines of the Goldfield Consolidated company, at Goldfield, Nev., to deliver ore from stopes to main haulage tunnels. Light, steel, arc gates, with a long lever handle, are provided for the chutes. Fig. 91 gives the dimensions and details of the standard chutes. Posts of the drift or tunnel sets are placed at 5-ft. centers, allowing an opening 4 ft. 2 in. wide (10× 10-in. timbers are used) or 8-in. clearance on either side of the ore cars. The drifts are 8 ft. high in the clear. The bottom of the chute is inclined at an angle of about 35° from the horizontal and passes through the next set to one side of the drift. It is sup-

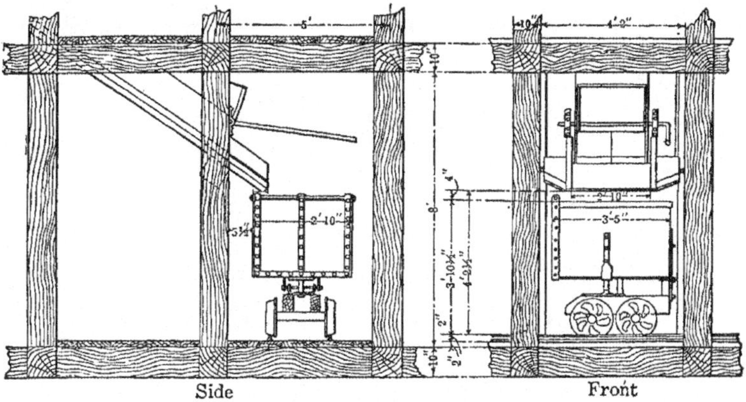

FIG. 91.—STANDARD ORE CHUTE IN GOLDFIELD CONSOLIDATED MINE.

ported by a piece of 8×8 in. timber set with its upper face parallel to the inclination of the chute bottom and let at its ends into the posts at either side of the chute. The lip of the chute extends 5 1/2 in. over the edge of the tram-car. A double lining of 2-in. plank is used on the bottom of the chute. The sides are made of one thickness of plank. The chute is carried the full width of the set from the stope to the lip, which is tapered down to a width of 2 ft. 10 in. The body of the cars used is 3 ft. 5 in. long. So as to allow the gates to be readily closed against the stream of ore, the gate should be set at such an angle that as soon as the stream of ore is intercepted its force tends to close the gate.

A Standard Ore Chute (By S. S. Arentz).—The type of ore chute shown in Fig. 92 is the standard in use at the mines of the Nevada-Douglas Copper Co. As a rule it has proved to be satisfactory in most places in which it has been used where no special conditions require a chute of different construction. At the Nevada-Douglas mines the chute is always built of the same size of timbers, the different pieces being framed at the surface and sent underground ready to be put together. In the illustration the 3-in. chute bottom is shown supported by a round timber at the top of the set, but it is often supported directly by the ground as the post supporting the main drift set is not used. While no originality is claimed for this design, such chutes are often built without adherence to any

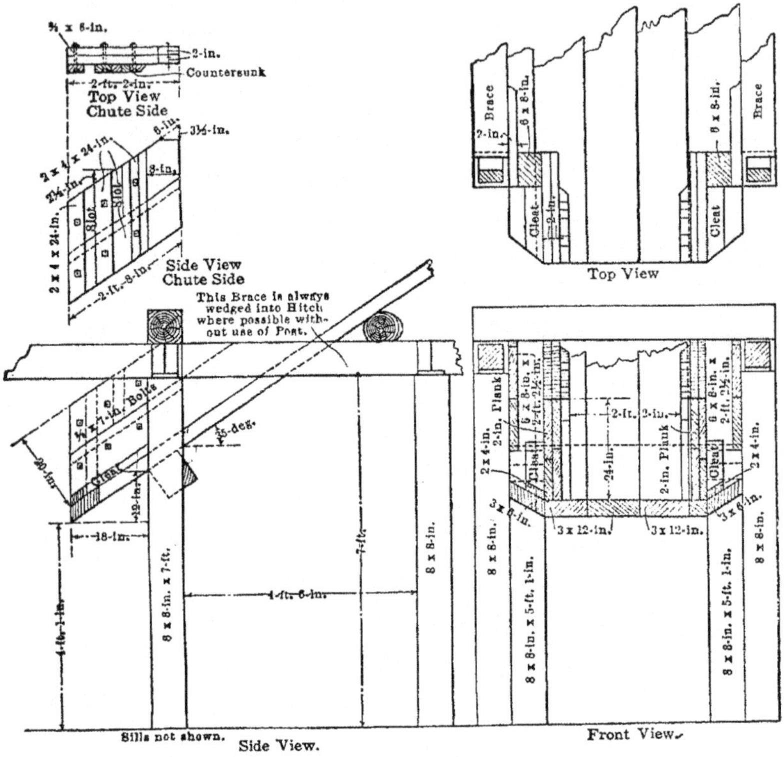

FIG. 92.—STANDARD ORE CHUTE USED AT NEVADA-DOUGLAS MINES.

standard of dimensions and those in various parts of the same mine vary in size and construction. There are many advantages in building chutes to standard dimensions other than those gained by being able to do all the framing at the surface.

VII

HEADFRAMES, CHUTES, POCKETS, ETC.

Gates for Chutes and Bins—Headframes, Tipples and Derricks—Ore Bins.

Gate for Ore Chute.—In both the Angels Quartz and Utica mines at Angels Camp, Calif., gates of very simple construction are used on ore chutes. The gate itself is a rectangular piece of sheet iron or steel, usually 1/4 or 3/8 in. thick, sliding between strips of iron nailed to the sides of the chute. A piece of round drill steel is fixed to the gate, being riveted and countersunk on the inside and having an iron ring shrunk about it on the outside of the gate, thus serving to hold it firmly. A long curved iron bar is pivoted on the side of the chute so that the short end engages the lug on the gate. Thus by moving the bar the gate is opened or shut. In the Utica a piece of drill steel is used for the bar, which may be slipped off at will.

Chute Gate at Mammoth Mine, Kennett, Calif.—At the mine of the Mammoth Copper Mining Co., near Kennett, Shasta county, Calif., an excellent

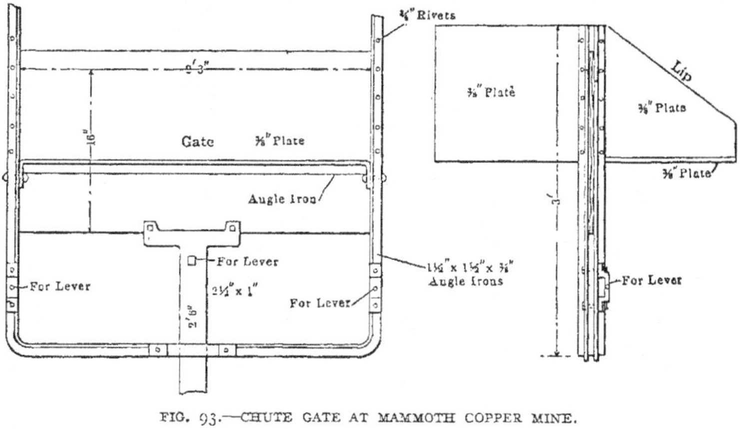

FIG. 93.—CHUTE GATE AT MAMMOTH COPPER MINE.

type of iron gate for an ore chute is in use on the large ore passes from the stopes where top-slice caving is being done. A large amount of ore must be handled quickly through these chutes, so that it requires a strong gate with a positive action. The details of the gate are shown in Fig. 93. The particular feature of the Mammoth chute gate is that it is closed by raising a door through the stream of ore passing from the chute instead of by lowering one, as in the ordinary types. Where ore is running rapidly through a chute, it is quite difficult to lower a gate into this quickly, whereas lifting the gate through the stream of ore presents no difficulty. The frame of the chute gate is made of two angle irons, bent as shown, between which the gate of 3/8-in. steel slides. One angle iron is cut away on the lower part of the frame. A bar of 2 1/2×1-in. iron is bolted to the lower side of the gate and slides through a guide at the lower part of the frame. The gate is operated by a lever connected to this bar and pivoted on the frame. Sheets of 3/8-in. steel cut as shown in the drawing, are riveted to the frame and form an extension of the sides of the chute and a projecting lip. The entire gates are riveted together and set up before being taken into the mine, so that they are ready to be set in place in the ore chutes. The all-steel construction of this chute gate renders it substantial, but at the same time rather expensive, so that its use is only warranted where large quantities of ore are handled.

Gate for Lump Ore Bin (By Guy C. Stoltz).—A gate commonly used on lump-ore bins at the iron mines, Mineville, N. Y., is shown in Fig. 94. The

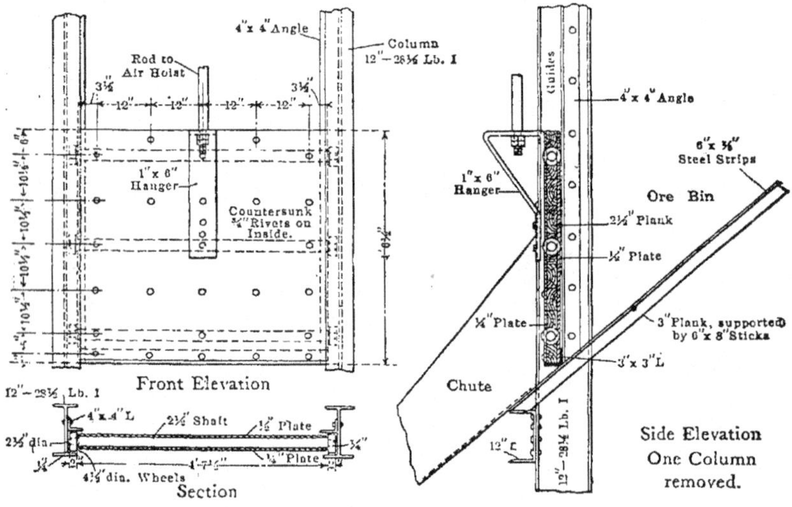

FIG. 94.—AIR-HOIST GATE FOR COARSE ORE.

gate is made of 2 1/2-in. plank, with an outside steel plate, 1/4 in. thick, 4 1/2 ft. square, and an inside plate of same area and 1/2 in. thick. Three axles, 2 1/2 in. square, turned to 2 1/4 in. diameter at the ends to receive rollers of 4 in. diameter and 2 in. face, are bolted between the plates. The 2 1/2-in. plank acts as a cushion, also gives weight and stiffness to the gate. The rollers run in the guides formed by riveting 4×4-in. angles to the two 12-in. I-beam columns.

On the bottom of the gate a heavy 3-in. angle iron is riveted to the plates to protect them from wear. The addition of two rollers to each side working at right angles to main rollers would improve the gate by lessening the friction due to side motion, for then any binding of the gate would be met by roll faces. The gate is fitted to a timber headframe by having, say, 10×10-in. posts take the

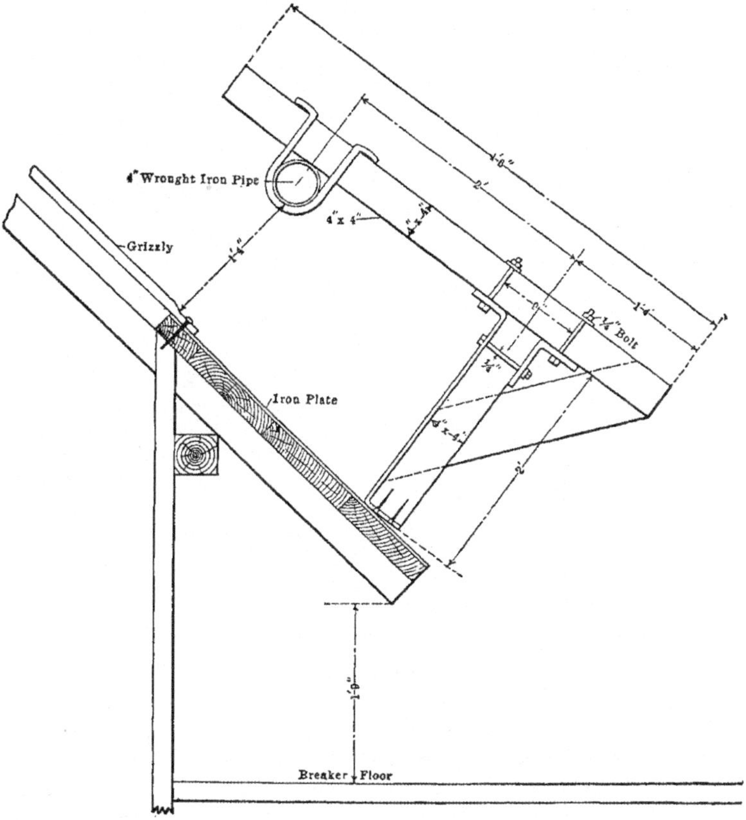

FIG. 95.—FINGER CHUTE FOR FILLING WHEELBARROWS.

place of I-beams and 4×4-in. hardwood strips bolted to posts to act as guides. This latter type is generally used at the underground storage pockets where it is essential to have a positive working gate which will close the instant the skip has been filled. S. L. LeFevre, assistant general manager for Witherbee, Sherman & Co., designed the gate.

A Finger Chute (By A. Livingstone Oke).—An adaptation of the well-known finger chute employed by me, while manager of a mine in Chihuahua, Mexico, is shown in Fig. 95. The ore coming into the breaker floor from an aërial tramway was being dumped direct on a grizzly, the fines going into the battery bin and the oversize accumulating on the floor above, whence it was shoveled into wheelbarrows and taken to the breaker. To avoid the shoveling, the finger chute was put in and two out of three peons were displaced. The fingers receive hard usage and should be built strongly. Their weight keeps the ore back, as, from the position of the fulcrum, they may be considered to have their center of gravity just where it is most effective, *i.e.*, in front of the sliding ore. They are easily controlled and saved all the hard work of shoveling. Other types of chute were tried, but failed to be of service.

Steel Arc Chute Gate.—A strong and durable arc chute gate of simple pattern is used on the flat-raise ore pocket in the Pittsburg-Silver Peak mine,

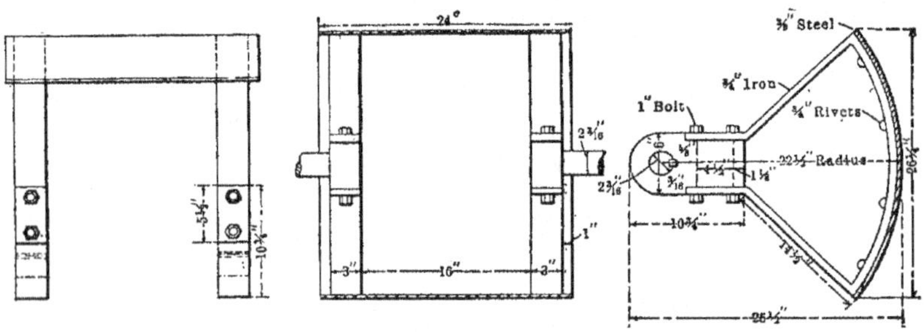

FIG. 96.—STEEL ARC CHUTE GATE AT PITTSBURG-SILVER PEAK MINE.

near Blair, Esmeralda county, Nev. The entire output of the mine, about 500 tons per day, is handled through these chutes; hence gates sufficiently strong to withstand the wear, and with a positive action, must be used. The type shown in Fig. 96 has given satisfaction. The frame of the gate is made of two pieces of 3/4×3-in. iron bent on an arc with a radius of 22 1/2 in. and turned back at either end, and bolted with 1-in. bolts, to the hub of the gate. These pieces of 3/4-in. iron are spaced 1 in. from the edge of the gate and fastened to the 3/8-in. sheet steel that forms the arc of the gate, with four 3/4-in. rivets. By using single pieces of heavy iron to fasten the arc to the hub and extending entirely across either end of the gate segment added stiffness is obtained. The hubs are 3 in. thick, 6 in. wide, 10 3/4 in. long and bored for a 2 3/16-in. axle. One of the chief advantages of this type of gate is in the few parts required for its construction, and hence the simplicity of setting it up. There are only five pieces to the gate and for putting them together, four bolts and eight rivets are required.

The components of the gate are a piece of 3/8-in. sheet steel, 34×26 1/4 in., to form the arc segment of the gate, two 3/4×3-in. iron bars, 33 1/4 in. long, for the frames or spokes, two cast-iron hubs of the pattern shown in the drawing, eight 3/4-in. rivets, and four 1-in. bolts.

Cananea Arc Type Gate.—The arc-type gate shown in Figs. 97 and 98 is used in the ore-bin chutes at the mines at Cananea, Mexico. It may also be

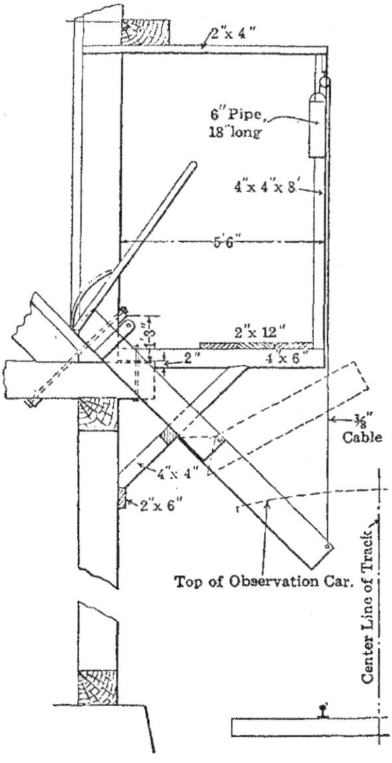

FIG. 97.—ELEVATION OF CANANEA BINS.

modified for use in underground chutes. The novel feature of the gate is the axle which is made of round iron bent as shown in the lower part of Fig. 98. The center of the rod, of which the axle is made, is flattened and bolted to the sheet forming the door. The position of the gate in the ore bin is shown in Fig. 97, in which illustration is also shown the manner of building the steel chute, below the gate, so that there is a hinged lower portion which is counterbalanced by a weight permitting the swinging of that portion of the chute up and out of the way of passing trains. The details of the steel chute are shown in the upper part of Fig. 98. A runway is built in front of the bins to give easy access to the handles operating the gates and so that the operator can take a position above the top of the car he is loading

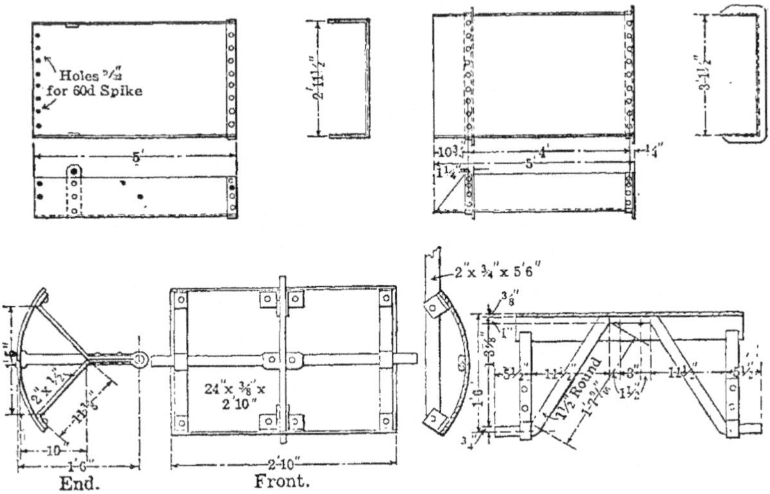

FIG. 98.—DETAILS OF METAL PART OF ARC-TYPE GATE FOR CHUTES.

SKIP LOADERS

Skip Loader at the Original Consolidated.—At the Original Consolidated mine, Butte, Mont., a novel skip-loading arrangement is being used in place of the ordinary ore pocket discharging directly into the shaft. The ground at this mine is rather heavy and it was not thought advisable to take away support from the shaft by cutting out the ground for skip-pockets. By using small apron chutes mounted on wheels, the skips were formerly loaded directly from cars. This method is, however, slow and requires too much labor shifting and dumping the cars, etc. To avoid this, the arrangement shown in Fig. 99 was devised and has already been installed on several levels of the mine. Above the ordinary station, 10 ft. high in the clear, an additional space, 6 1/2 ft. clear above the station proper, is cut and timbered with 12×12-in. material. The top of the station timbers forms a platform upon which a man can stand while operating the air gates on the chute. At the third station set from the shaft a two-compartment raise, inclined toward the station at 80° from the horizontal, is put up to the level above. This chute is carried 4 ft. 10 in. square overall and is timbered with 10×10-in. material framed in 7-ft. sets, with dividers of 5×10

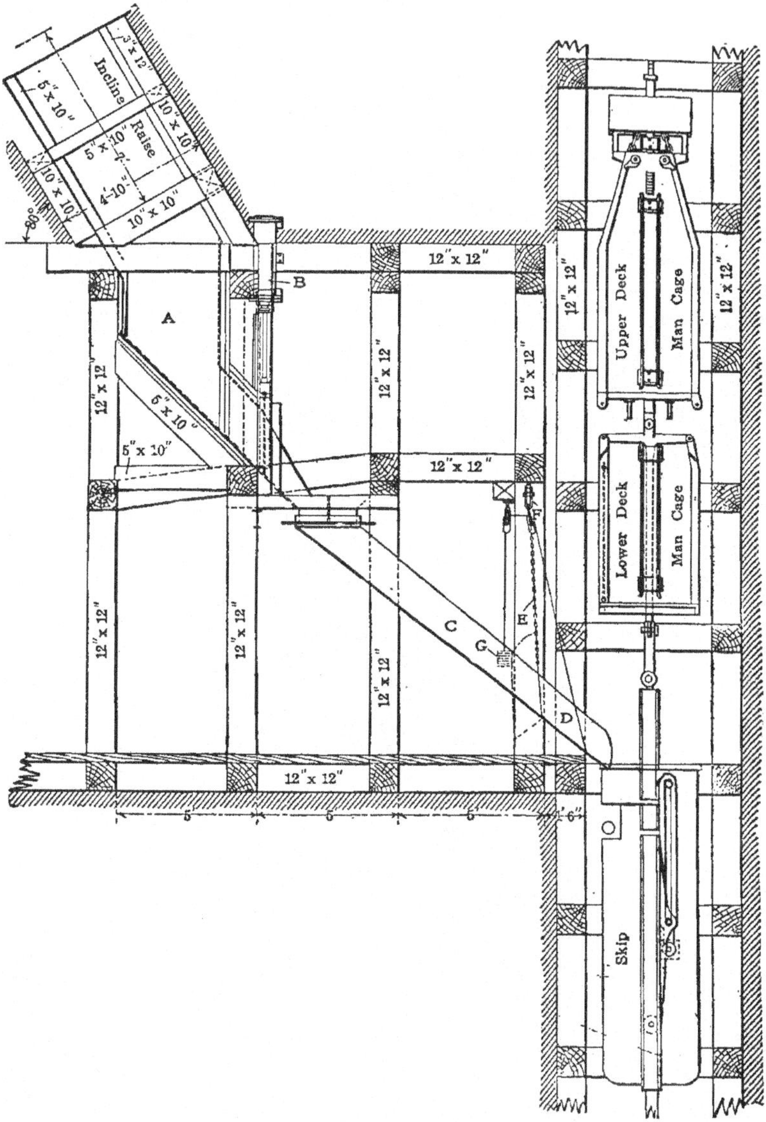

FIG. 99.—SKIP LOADING ARRANGEMENT FOR ORIGINAL CONSOLIDATED MINING CO., BUTTE, MONT.

material midway of each set. The chute compartments are lined with 5×10 material on the bottom, and 3×10 on the top and sides. The inclined raise terminates at its lower end in the hopper-bottom pocket *A*. The chutes are provided with steel gates, operated by compressed-air cylinders *B*. The dis-

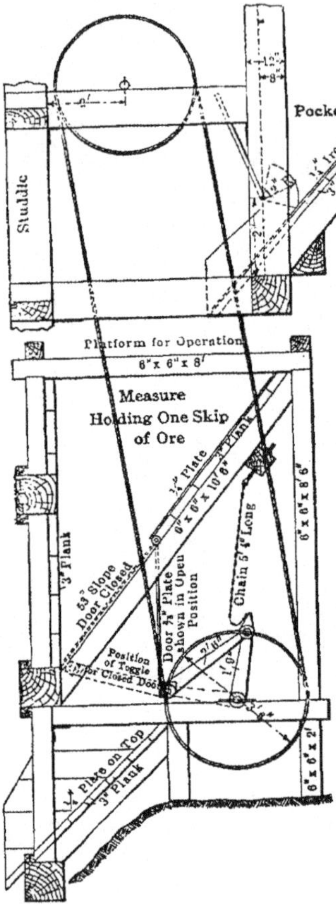

FIG. 100.—SKIP-LOADING ARRANGEMENT AT SCRANTON MINE, HIBBING, MINN.

charge is into a long sheet-steel, swinging spout *C*, the lip *D* of which, when turned down projects into the shaft far enough to deliver rock into the skip. The steel apron-chute is pivoted at its top end, so that the lower or discharge end may be swung to either shaft compartment. This chute is supported by a chain *E*, that is fastened to the pulley *F*. This pulley runs on a short track, thus enabling the spout to be easily swung. The lip of the spout is connected by a line, passing over two blocks, to the counterbalance *G*. This weight serves to keep the lip raised, so that the spout will swing clear of the shaft timbers. The counterbalance is lifted and the lip let down when a skip is to be loaded, the spout being swung out of the way when not in use.

When loading a skip, one man climbs up to the platform and operates the air gate on the raise (or pocket), while another swings the spout, lowers the lip and calls out when the skip is filled. By having the loading arrangement at a station instead of below in the shaft, time and labor are saved. The inclination of the raise carries it to the level above at a point far enough away from the shaft, so that the nuisance of having cars block the station is done away with. Having the approach to the shaft clear is an important advantage of this skip loader.

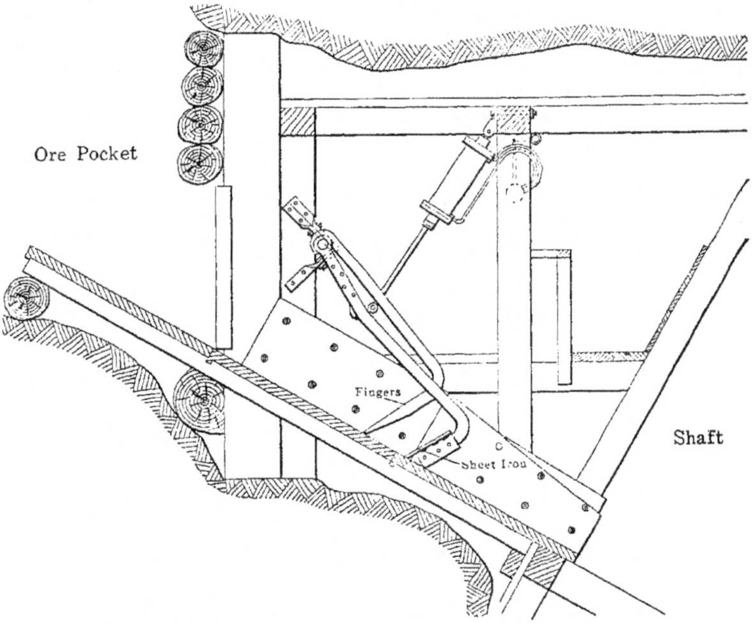

FIG. 101.—GATE FOR SKIP-LOADING CHUTE, GRANBY CONSOLIDATED MINES.

Measuring Pocket for Skips.—A skip pocket designed by C. F. Jackson for the Scranton mine at Hibbing, Minn., is shown in Fig. 100. The principal feature that commends this pocket is the fact that it opens in such a way that the shaft is clear at all times. A number of similar pockets are in use, but they open into the shaft and are more or less dangerous. In addition this pocket provides a safe place for the operator. He is on the platform above the pocket.

One man can both draw the ore from the chute and fill the skip from this pocket which holds just one skip load, 91 cu. ft. The pocket is opened by means of a rope and pulley. As the rope is moved it turns the lower pulley off center and

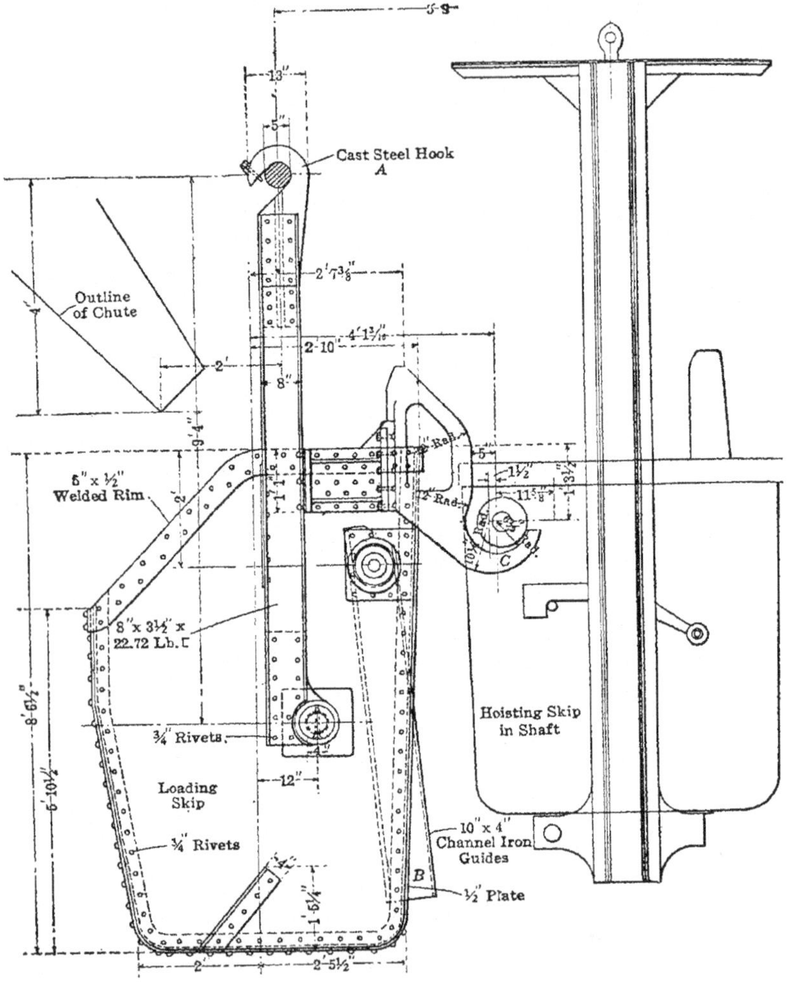

FIG. 102.—THE WHITFORD MILLS SKIP-LOADING DEVICE.

the weight of the ore opens the pocket. The chain prevents the wheels from turning too far past the center.

Skip Loading Chute.—Details of an ore chute for loading skips used at the mines of the Granby Consolidated company, Phœnix, B. C., are shown in Fig. 101. It is an improved form of finger chute, combining fingers with a sheet-iron gate for holding the fines. In the operation the sheet-iron gate is raised first by the air lift, then, as the arm is raised still higher by the piston of the air cylinder, the fingers are raised and the coarse ore allowed to escape. When the air is released the fingers fall first, catching the coarse rock, and there is a sufficient interval of time for the chute to clear itself before the sheet-iron gate is closed. The operation is rapid and the skips are filled nearly as fast as the skip tender can operate the valves.

Whitford-Mills Skip Loading Device (By E. M. Weston).—An apparatus for loading hoisting skips devised by Messrs. Whitford and Mills, respectively general manager and engineer of the City Deep, Ltd., Johannesburg, S. A., is shown in Fig. 102. It is designed to load 5-ton skips faster than could be done by means of Kimberley chutes. The idea consists essentially of a second skip in each compartment in front of the main doors of the bin at the bottom of the shaft. These skips hold five tons, as do the skips in the shaft, and are to be filled from the bin while the hoisting skips are running in the shafts. They are hung at *A* and balanced so that their movement while tipping is controlled by guides *B*, in such a manner that the hoisting skips on their descent tips them automatically by engaging the hooks *C* on either side. The loading skips themselves never project into the shaft even while tipping. In this manner hoisting could be carried on without any pause except for reversing the engines, say 15 seconds. One possible drawback to the use of the device might be the possibility of damage to the apparatus by a skip reaching the loading station with too much velocity; but as electric winding is rapidly being adopted, this system can easily be adjusted for automatic action, and steam winding engines could also be provided with one of the well-known types of automatic reversing and breaking devices.

Red Jacket Ore Pockets.—The Red Jacket shaft of Calumet & Hecla is the one shaft in the United States equipped with a Whiting hoist. This is used in raising the ore, while an ordinary drum hoist is used for raising men. Skips holding 7 1/2 tons are used and the shaft is arranged so that the pockets for the different skips are on alternate levels. These skip pockets are large enough to hold 9 tons, but only three cars, a skip load, are dumped in at a time. The pocket is lined with steel, on top of which, both on the sides and bottom, wearing plates are bolted. The steel bottom plate rests on a cast-iron bottom plate 4 in. thick which in turn rests on a bottom of 12×12-in. timbers, as trouble was experienced with the steel bottom plate when it rested directly on the timber bottom on account of the size of the boulders—some weighing a ton and a half—that are dumped from the cars into the skip pocket.

This skip pocket has a hand-operated swinging door as shown in Fig. 103. The door piece *A* is hinged at the top, the strain on the hinge seats being carried back to rock wall of the pocket pit by means of two bolts equipped with turnbuckles. A lever, *B*, leads back from this door to the main locking lever, *C*, to which it is connected by a pin joint. The door as it swings back after it is released by easing on the band brake (with which the shaft of the locking lever is equipped so as to prevent the return of the door before the pocket has emptied itself, as might be the case if there was much fine ore in the pocket) forces this locking lever past dead center so that the weight of the ore pressing against the door holds the lever against the inside stop that limits its downward motion. The outward swing of the door is limited by the locking lever coming in contact with a similar stop at the other end of its range of travel. In case that this lock-

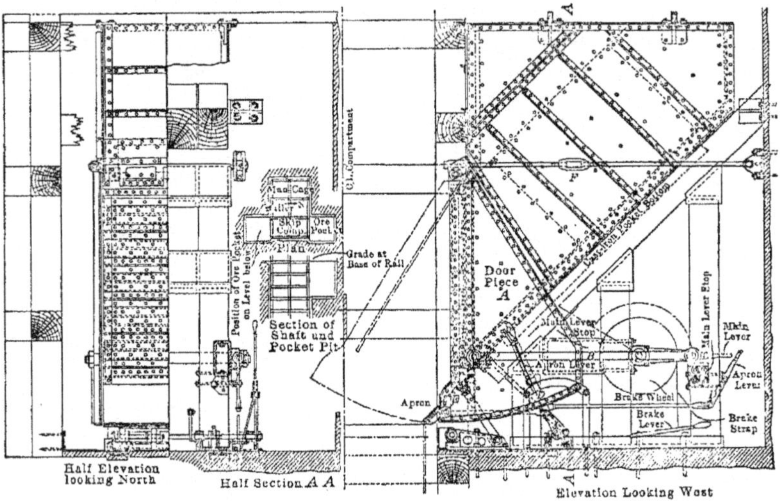

FIG. 103.—DETAILS OF A STEEL ORE POCKET IN RED JACKET SHAFT.

ing lever sticks so that a man cannot conveniently start it by a moderate lift on its outer end, the skip tender resorts to the use of a wooden auxiliary lever that is fastened to the back post of the pocket so as to aid in starting the locking lever; but this is not usually necessary. To bridge the gap between the skip and the mouth of the pocket, an apron *E*, is provided, which is thrown in by means of either of two levers *D*, shown in the drawing. This apron also is equipped with a wearing plate.

Measuring Pocket for an Inclined Shaft.—The North Kearsarge shaft No. 4 is sunk in the foot wall of the lode, and the ore is trammed to loading bins at the shaft that serve two levels. These bins are covered with a grating of crossed rails so that boulders must be broken to 18 in. to go into the bins. On the skip chutes proper a counterbalanced arc gate that closes from below through a system of togglelevers is used, as shown in Fig. 104. On the measuring pocket, doors of unique design are used. The lower one opens and closes the upper gate so they might be described as being of the clam-shell type. Owing to the shape of the levers by which the two gates are suspended the bottom gate moves up less rapidly than the top gate and therefore always closes after and over the other. This is illustrated in Fig. 104 which shows the gates open in full lines and closed in dotted lines. On each side on the lower gate is a semicircular arm in the top of which is a notch. Two hooks, connected by a bar and seated on the posts that brace the bottom of the measuring pocket,

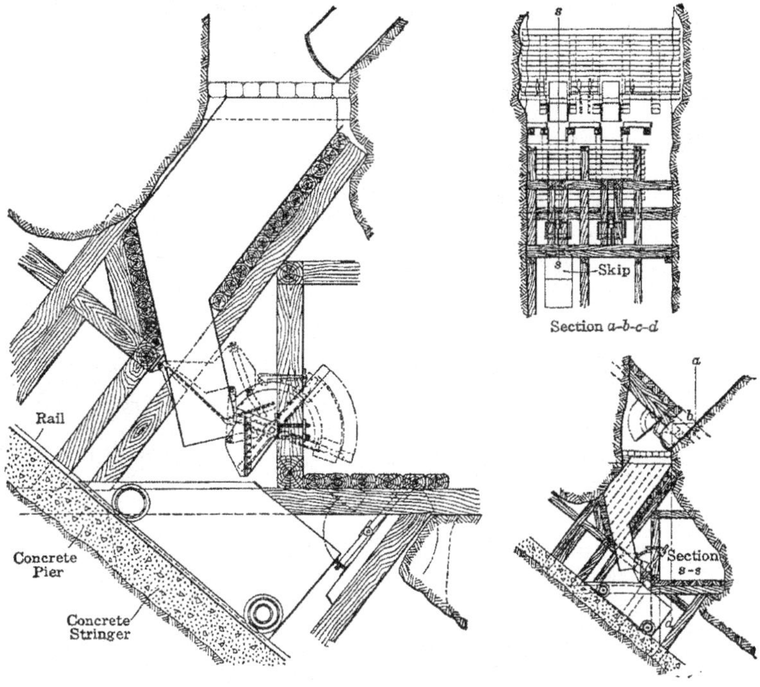

FIG. 104.—SKIP-LOADING DEVICE AT OSCEOLA MINE.

drop into these notches and latch the gates in position when they are closed. By means of a lever these hooks are raised, then the weight of the ore forces the back door backward and the front door forward, giving the ore a free passage down to the skip. Extending from the front gate are the arms carrying the counterweights, one on each side. These are adjusted so that a slight lift is necessary on the lever arm of the gates to close them, but the adjustment is so close that the jar of the skip gate as the skip leaves the bin will close the doors.

An Underground Ore Pocket.—At the iron-ore mines of the Tennessee

Coal, Iron & R. R. Co., on Red Mountain, Ala., hoisting is done in skips of 10-ton capacity. In sinking slopes without an ore pocket, the skip is lowered until it rests against a pentice of rock, and there receives ore hoisted, by an auxiliary engine from the face of the slope, in 2-ton end-discharge cars dumped by an ordinary curved-rail tipple. These 2-ton cars run over the rock pentice and, when no pocket is used, dump directly into the skip. This necessitates the use of an auxiliary steel car which can be attached to the rear of the skip and pulled out with it, or else the holding of the skip at this point until it can be filled from the small cars. To eliminate this waste of time, an underground pocket is used in the slopes of the Tennessee company, on the Ishkooda and

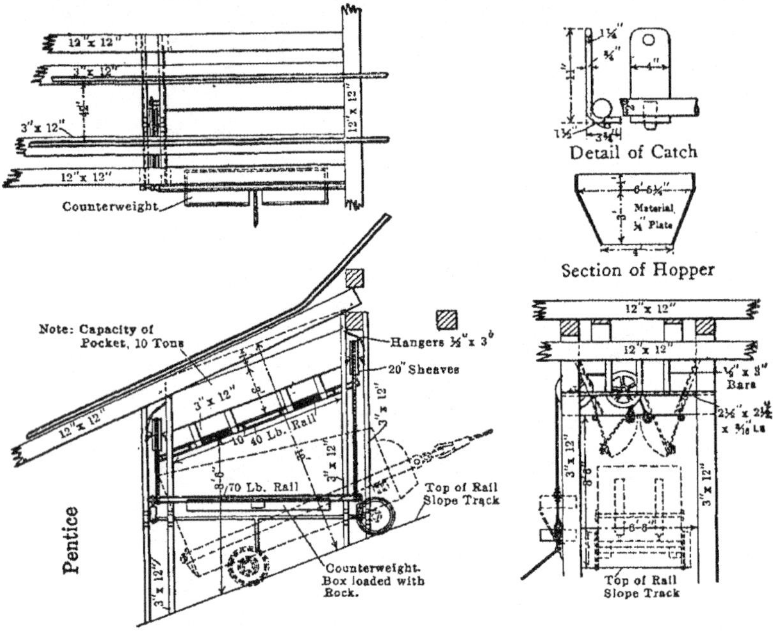

FIG. 105.—UNDERGROUND ORE POCKET.

Fossil divisions. The design and method of setting up this pocket are shown in Fig. 105. It is of 10-ton capacity, and is set so that the skip can be run under the pocket and filled directly without any loss of time. The pocket is hopper shaped, and made of 1/4-in. plate; it is fitted with hinged bottoms opening outward, as shown. These bottom doors are braced with 40-lb. rails at the ends, abutting at the center line of the bottom of the pocket. To the ends of these rails are fastened ropes which pass up and over two 20-in. sheave wheels and down on one side of the pocket framing where they are connected to a 70-lb.

rail. A latch operated by a lever, details of which are shown, holds this rail, and consequently the gates in a closed position. By releasing the latch, the weight of the ore is allowed to open the pocket bottom, but sufficient counterbalance is attached to the 70-lb. rail to swing the gates closed after the pocket has discharged. The pocket is supported from two lateral 12×12 timbers which are carried on 12×12 cross-timbers hitched into the walls of the slope.

HEADFRAMES, TIPPLES AND DERRICKS

How to Erect Three-leg Shears (By A. Livingstone Oke).—The correct way to erect three-leg shears, using a tackle and rope from a hand or power

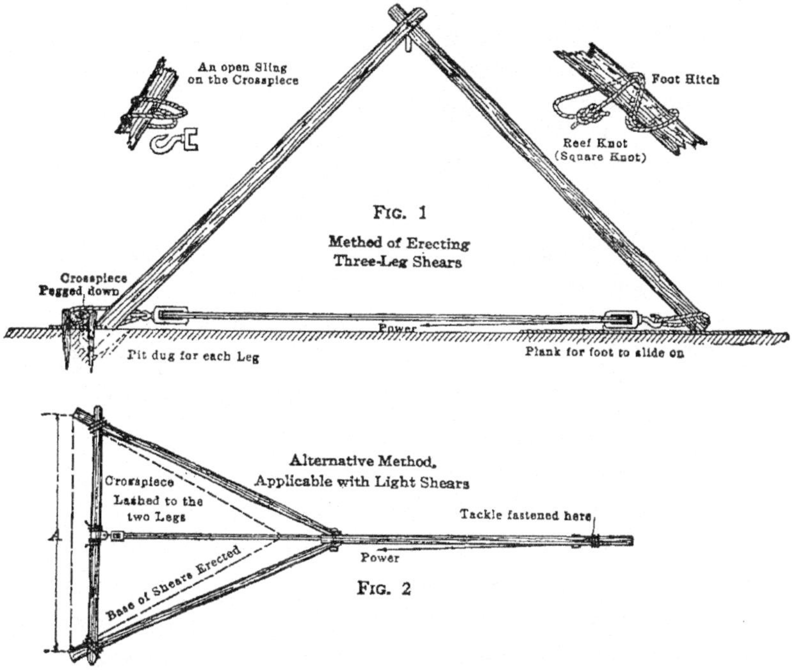

FIG. 106.—PLAN AND ELEVATION OF THREE-LEG SHEAVES.

winch, is shown in Fig. 106. The three legs are laid out first on the ground, as shown in the plan, two of them being placed with the butt ends at the distance *A* which is to be the spread of the shears when erected. On these two legs a cross piece is secured, either by lashing or by pegging down, as shown in Fig. 1. One end of the tackle is attached to the cross piece and the other end to the single leg. It is necessary to lift the center off the ground 2 or 3 ft., before applying the power. The hauling line from the tackle should come from the single leg as this is the one that slides. Boring the holes for passing the pin should be done bylaying out the three legs, as shown in the plan, with the spread *A* equal to the proposed base when erected. In this way there is no risk of the pin being bent, as the angle between these two legs remains constant and cannot be altered without bending the pin. The height of the shears may be altered by moving the middle leg nearer or further from the other two.

Headframe for a Prospect Shaft.—The sinking of a prospect shaft is often done under unnecessarily dangerous conditions. It is taken for granted that such work must be hazardous because, until ore has been found, the safety of the miners is not regarded as warranting extra expense. For this resaon much prospecting work is done with meager equipment and poorly constructed head gear. Until a depth of 40 or 50 ft. is reached a windlass may be used, but for

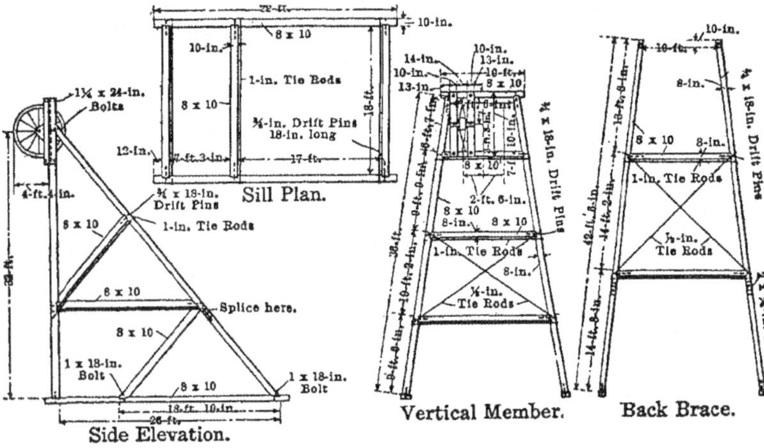

FIG 107.—PROSPECT HEADFRAME AT SAND GRASS SHAFT.

deeper work, to a limit of 200 ft., a horse whim will be needed, which requires nothing more than a tripod to support a sheave about 15 ft. above the collar of the shaft. When, however, the shaft is to be sunk to greater depth than 200 ft. it will be found more economical to use some form of power hoist; generally a geared hoist is used. A power hoist will require a substantial headframe. That the headframe need not be inordinately expensive is demonstrated by the cost of the structure shown in Fig. 107. This headframe is well suited to the purposes of an important prospecting shaft; it was designed by J. M. Fox, assistant superintendent for the Tonopah Mining Co., and was built at the Sand Grass shaft. It is a substantial structure, and insofar as the headframe has to do with the safety of shaft sinking, provides abundant security. With such headframes it has been possible to raise 100 tons of ore per day through a one-compartment shaft from a depth of 300 ft. without crowding.

The headframe is quite strong enough for prospecting work and is designed for use with a cage in counterbalance when mining of ore is started. In case more room is desired about the collar of the shaft, the lower diagonal brace can be made vertical. In the construction of this headframe, 3200 board feet of 8×10-in. timber were required. All daps were cut 1 in. deep and painted with creosote to protect them from decay. The headframe after it was erected was painted in order to preserve it from the weather. The wages of carpenters at Tonopah vary between $5 and $5.50 per 8-hour shift, yet the cost of this headframe complete and in place was but $330. The most important items were 3200 board feet of lumber at $37.50 per thousand; framing timbers and erecting labor, $120; iron work, $65. A 3/4-in. hoisting rope is used and until actual mining begins, or while the shaft is being sunk, a bucket of about 18 cu. ft. or a little over 1 ton capacity is used for raising rock.

Headframe for a Winze Hoist.—In mining operations it is frequently the case that a shoot of ore has been followed down in some part of the mine remote

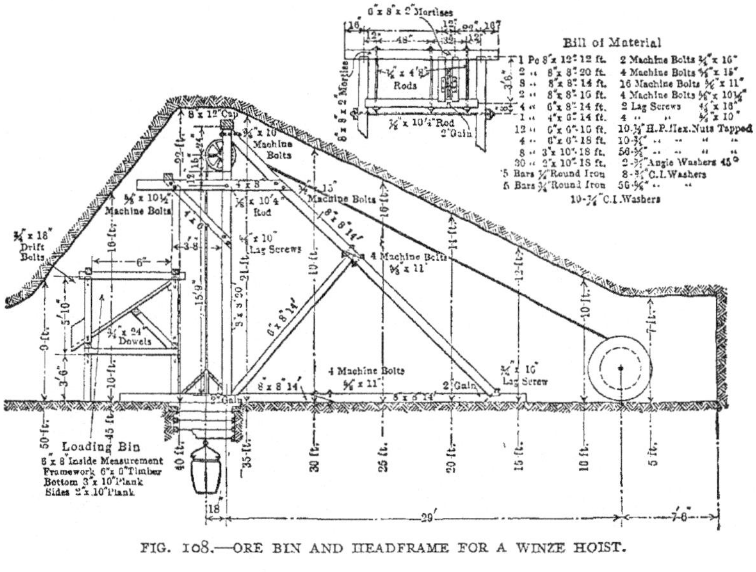

FIG. 108.—ORE BIN AND HEADFRAME FOR A WINZE HOIST.

from or not connected with the main hoisting shaft. The desire to develop rapidly the shoot and other reasons may make it necessary to raise more rock or ore through a winze than can be handled by the usual hand-operated windlass and bucket, and it becomes necessary to equip the winze with a power hoist and headframe to carry the sheave. The accompanying illustration, Fig. 108, furnished by Percy E. Barbour, is of a headframe for a winze such as was designed for use in the Copper Mountain mine in Nevada. The design follows closely the usual two-post surface headframe, but is not so high nor is it built of as heavy timber as is usually deemed necessary for a surface structure. The details are fully shown in the illustration. The station is cut out as closely as possible to just admit of the erection of the headframe and is, of course, preferably situated where the walls are strong enough to require minimum timbering. While no guides are shown in the illustration, if it is desirable to use them while sinking with a bucket, they may be supported in the same manner as if the headframe were at the surface. To receive the ore and rock raised and facilitate the loading of cars, a small box-like ore bin is built in front of the frame.

An Underground Hoist.—The accompanying illustration, Fig. 109, shows the method of arranging the hoisting equipment for a large winze that was sunk

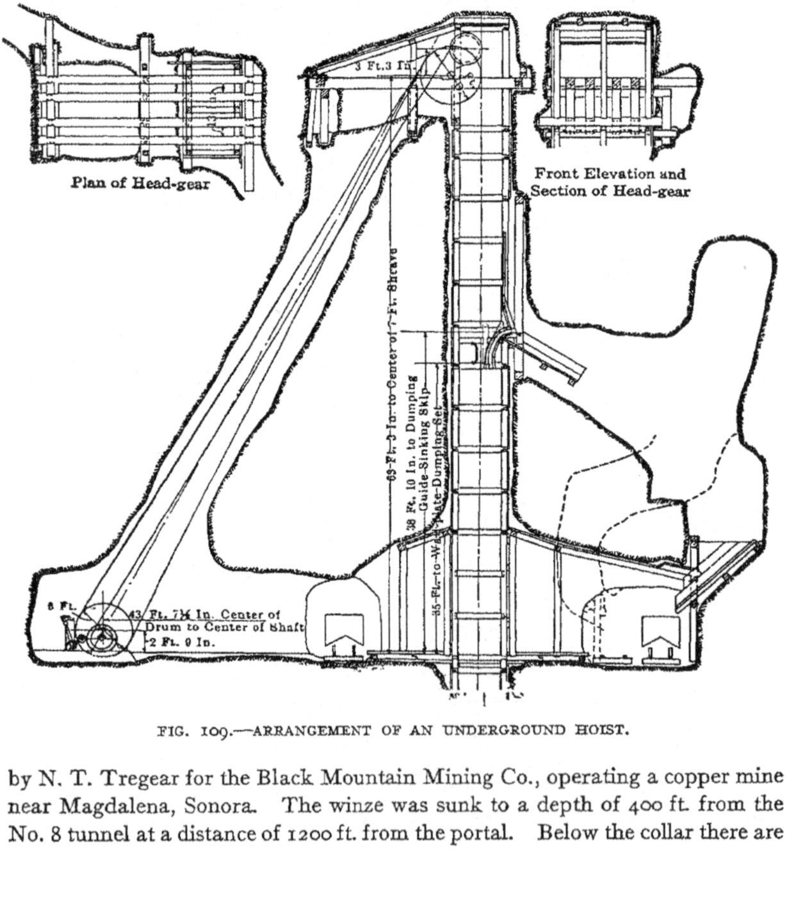

FIG. 109.—ARRANGEMENT OF AN UNDERGROUND HOIST.

by N. T. Tregear for the Black Mountain Mining Co., operating a copper mine near Magdalena, Sonora. The winze was sunk to a depth of 400 ft. from the No. 8 tunnel at a distance of 1200 ft. from the portal. Below the collar there are two compartments, each 4 1/2×5 ft. in the clear, while above the collar there are two compartments, each 4 1/2×5 ft., that were carried up to support the sheaves and to obtain head room above the tunnel for an automatic dumping skip and two ore pockets. The sheave-wheel bearers are dressed timber 11 1/2 ×15 in. and 26 ft. long; the bearing posts are 12×12 in. section, as are also the supporting bearers and dividers. The housing is made of 8×8-in. timber; the shaft sets are 8×8-in., and the auxiliary dumping set is of 7 1/2×11 1/2-in. timber; two of the collar bearers are 12×12-in., and two 8×8 in., and the guides are 5 1/2×5 1/2-in. section. The shaft sets are spaced 5-ft. centers, the station sets 15 ft. and the dumping set 20 ft. The winze is lagged with 2-in. lumber.

Details of a Wooden Headframe.—In Fig. 110 the details of the headframe of the Clermont shaft at Goldfield, Nev., are shown. This is an excellent example of the simple A-frame type of head gear. The total quantity of timber used in the construction of the headframe was 23,000 board feet. There were also used 3300 lb. of bolts and rods, and 500 lb. of cast-iron washers.

Overwinding Allowance in Head Gears.—In cases of overwinding, accidents frequently happen from a blow by the liberated end of the rope, and the rope itself may also be damaged. To prevent this, a drag rope forms a useful auxiliary. This may be of light wire rope carried on a small light drum placed near the detaching gear, the free end of the rope to be formed into a loop of such size as to allow the hoisting rope to run through freely, yet too small to admit the rope capping; this loop to be lightly fixed just over the detaching ring. When an overwind takes place, the freed end of the hoisting rope is at once held in check by the drag rope. The drag-rope drum might be provided with a brake or a coil spring inside, so as to prevent the rope running out too freely. The headgear should be of sufficient height to allow for a fair overwind in addition to the working height. As a general rule, it is suggested that the distance in feet from the underside of the sheave to the pin which connects the rope to the skip or cage standing at the point in which the journey is properly completed should not be less than the average hoisting speed in feet per minute divided by 200. For example, if the average hoisting speed be 3000 ft. per minute, the overwinding allowance would be 15 ft.

Tipple Construction in the Birmingham District.—The tipples used by the Tennessee Coal, Iron & R. R. Co. and the Republic Iron Co. at the slopes of their iron-ore mines on Red Mountain, Alabama, are different from those seen at any other slopes in the Birmingham district. These companies hoist in 10-ton skips, whereas most of the other companies use trains of five 2-ton cars. The constructional details of the tipples at a slope on the Muscoda division of the Tennessee company's ore mines are indicated in Fig. 111. The slope entry is not perpendicular to the main railroad loading-track below the tipple, this accounting for the angle at which the Nos. 1 and 2 bents are placed. The tipple carries two sets of tracks one above the other. The upper one, set at 6-ft. gage, engages the rear wheels of the skip, thus elevating it into the position

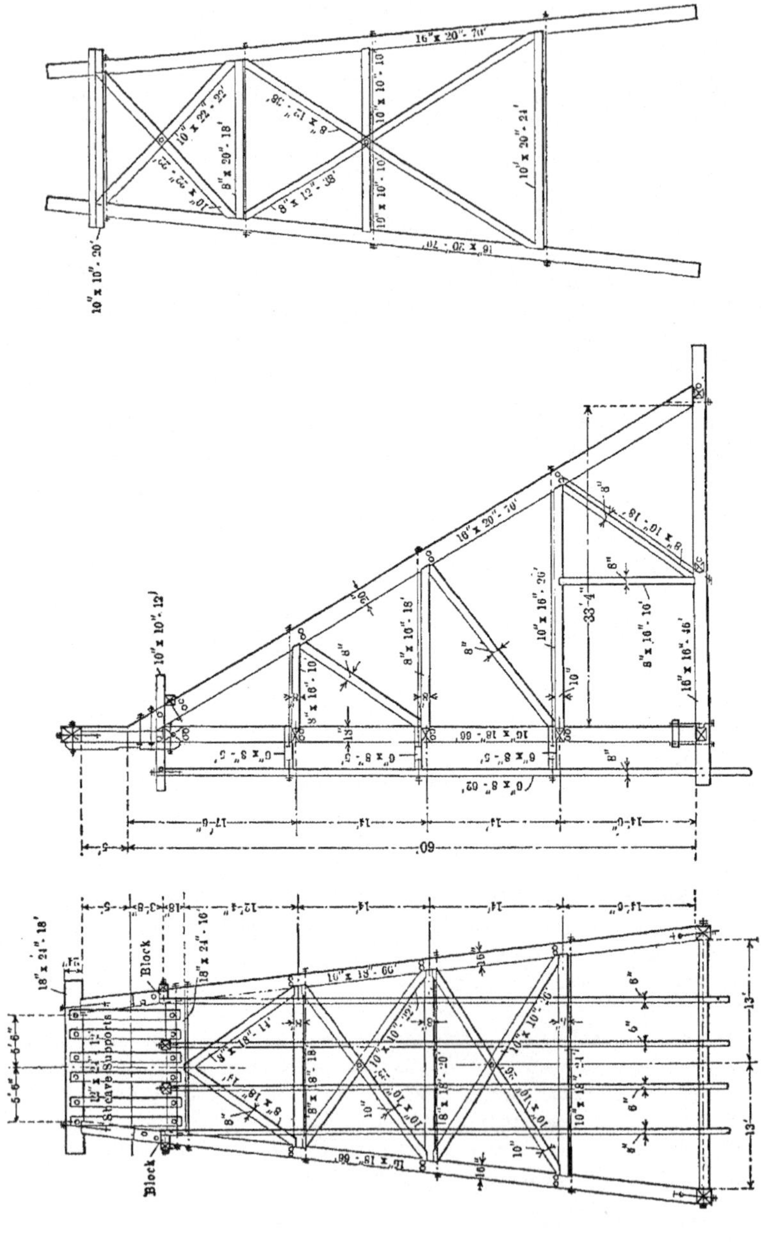

FIG. 110.—THE CLERMONT HEADFRAME, GOLDFIELD CONSOLIDATED MINES CO.

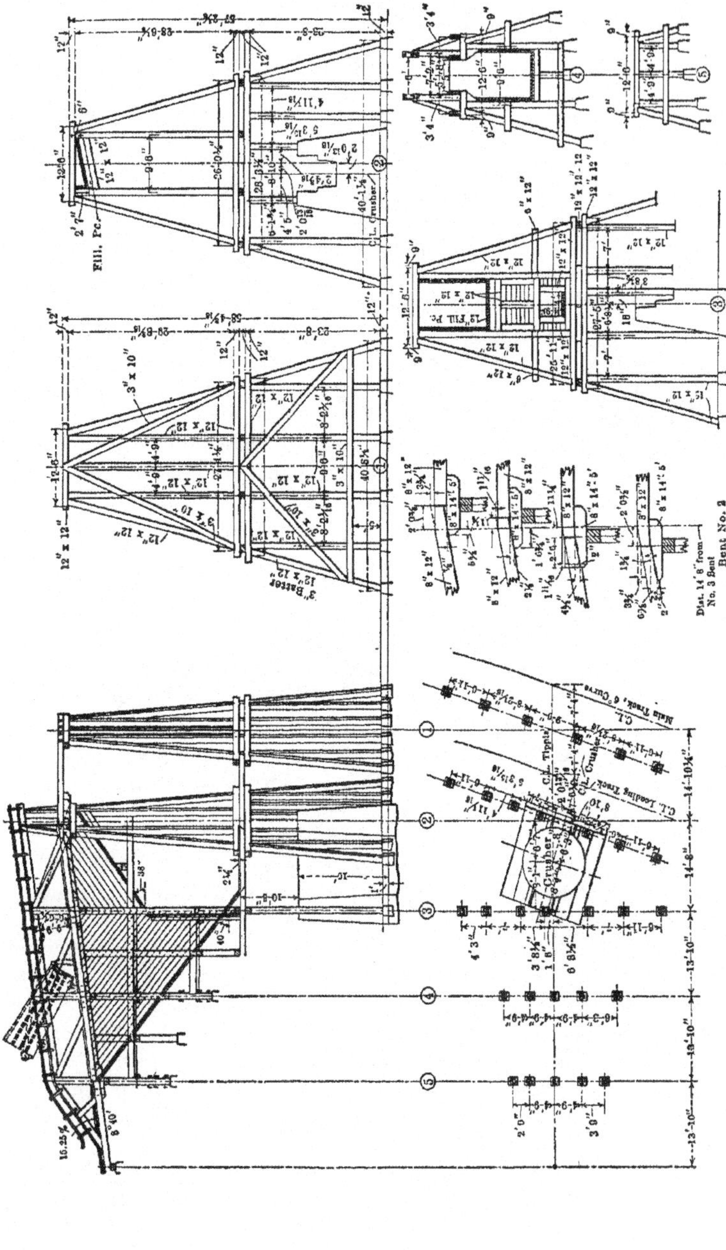

FIG. 111.—CONSTRUCTIONAL DETAILS OF TIPPLE USED IN BIRMINGHAM DISTRICT.

of dump shown by dotted line. The front wheels follow the lower tracks which are set at the regular 5-ft. spacing. The door of the skip is hinged at the top and held tightly closed during hoisting, by the bale of the skip. When the rear of the skip is raised the bale swings up and allows the door to open and the load to discharge. The ore is dumped into a bin holding about 150 tons and made long enough at the top so that the skip will not have to be dumped within close confines in order to discharge entirely within the bin. The ore bin is built with double planking on bottom and sides and is 9 ft. wide, about 26 ft. deep at No. 3 bent and has slopes to the bottom of 40 and 38°, as shown in drawing. This insures that the ore will feed freely to the gyratory crusher, which is set on a concrete base, between No. 2 and 3 bents. The crusher, a No. 8 Austin, delivers its product directly into railroad cars which are let down the track by gravity. Details of the framing of the bents, five of which are used in this particular tipple, are shown fully in the drawing. They are framed from 12×12 timbers battered 3 in. to 1 ft. and cross braced with 3×10 plank. The bents are set on concrete bases. The details of the corbels of the No. 2 bent are also given in the cut. This construction gives a strong and satisfactory tipple at a not too excessive first cost. The Tennessee company uses these wooden tipples at all of its ore mines on Red Mountain, but the Republic company has substituted steel construction for the wooden type. The general form of tipple, however, is retained.

ORE BINS

Cananea Ore Bins (By Claude T. Rice).—The drawings given in Fig. 112 show the standard bin construction that has been adopted at the mines at Cananea, Mexico. The bin has a bottom sloping at 45°, and the inside is lined with sheet iron 3/16 in. thick, in which the holes for the nails are countersunk. This slope at the bottom has been found sufficient for the Cananea ores, but it is well when building bins with a sloping bottom to bear in mind that heavy sulphide ores of copper when coming damp from the mines are apt to pack in a bin having a bottom slope of 45°. Such was the experience at the new ore bins built at the Highland Boy mine at Bingham for the use of consolidated tramways, and I understand that such also was the experience with some of the heaviest ores at the Cerro de Pasco mines in Peru. In a bin with a sloping bottom the weight is practically all thrown on the front posts, so at Cananea it is the practice to use double front posts. When a bin is designed it is the custom at Cananea to lay out the plan of the sheet-iron lining, so that when the construction is ordered the plans can be taken to the machine shop, and the lining be prepared and marked, ready for putting in place in the bin. By planning the details in advance and having all parts ready, costly delays in the erection of the bins are avoided.

Tonopah Orehouses.—In the orehouses at Tonopah special provision is

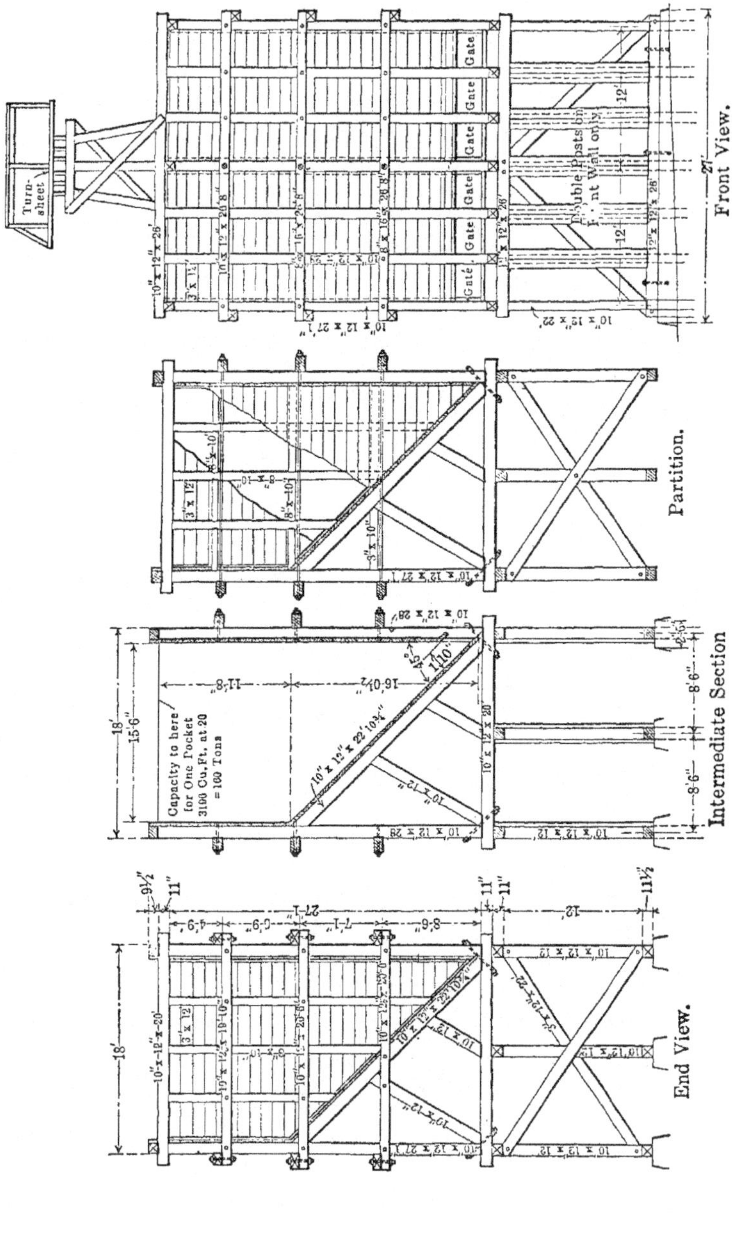

FIG. 112.—STRUCTURAL DETAILS OF THE CANANEA ORE BINS.

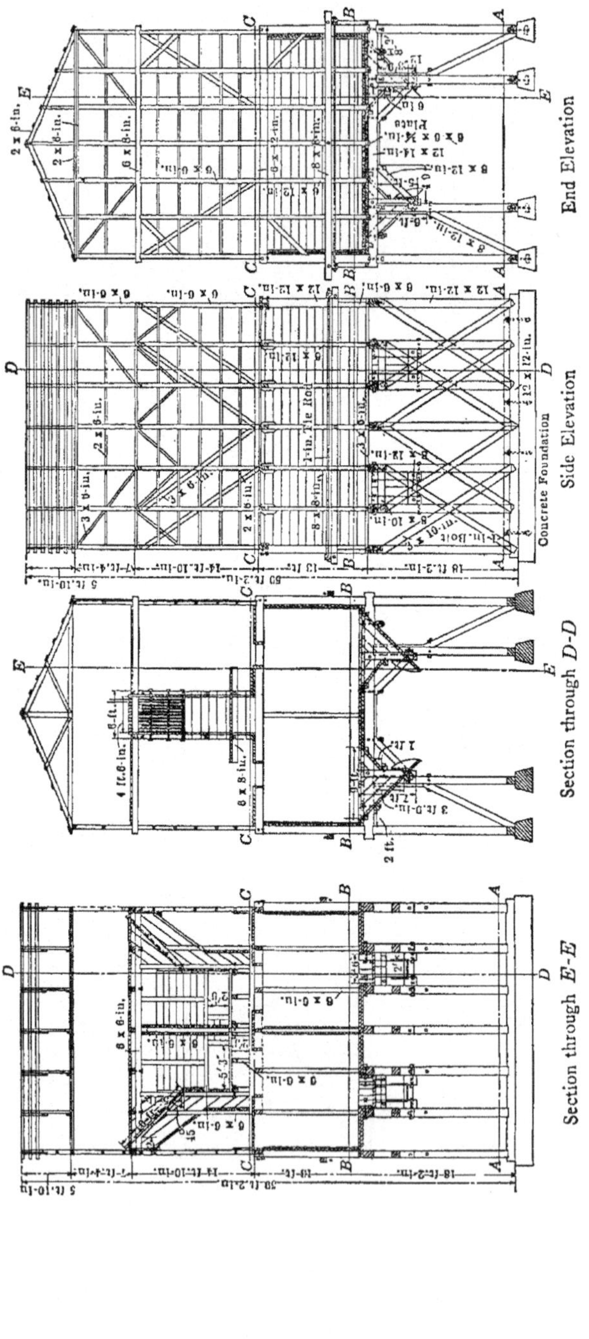

FIG. 113.—OREHOUSES OF TONOPAH MINING CO.

made for sorting the ore. The drawings shown in Fig. 113 give the details of construction of the new bins at the Red Plume and the Silver Top shafts, which were designed by J. M. Fox, while assistant superintendent for the Tonopah Mining Co. The older bin of the Mizpah orehouse was built with a sloping bottom, but that form is not so cheap in the first cost, and also requires great expenditure to keep it in repair.

The Silver Top and the Red Plume orehouses required from 43,000 to 45,000 board feet in their construction, and as they had flat bottoms, the only protection required was a small amount of sheet iron around the mouths of the gates. The capacity of each bin is about 450 tons. The cost was between $5000 and $6000 each completed, with Oregon spruce at $37.50 per thousand, and erection done by contract in a camp where carpenters earn from $5 to $5.50 per 8-hour shift. The upkeep for the last two years has been practically nothing.

The Mizpah orehouse has a bottom sloping at 40°, so that it has to be protected with a sheet-iron covering. It took 110,000 board feet to build that orehouse and the capacity is between 400 and 450 tons. As the orehouse was erected early in the life of the camp when the cost of materials was excessive, it would not be fair to use the actual cost in a comparison. However, it costs two and one-half times as much to build a sloping-bottom bin as one with a flat bottom. One of the commendable features of the Tonopah orehouses is that the grizzlies are placed so that the ore is dumped against instead of with the slope. This insures a better screening action on the grizzlies. In order to sample the fines from the grizzlies a 3-in. channel iron is carried underneath clear across the grizzlies and at right angles to the bars, but at an angle of between 30 and 35° from the horizontal, so that the fines, which are caught in the channel sampler after falling through the grizzly openings, 1 1/2 in. wide, slide down into a sample box on the floor of the orehouse.

Besides the economy with which the timber has been used in the erection of the bin, a leading feature of the construction is the commodious arrangement of the upper part so as to facilitate the sorting of the ore. The oversize from the grizzlies runs into a small upper ore-pocket, having feed holes at the bottom which permit the ore to run out by gravity upon the sorting tables as sorting progresses. Consequently, the ore with only a little scraping is in a thin layer in front of the ore sorters so that they can quickly throw the pieces of waste into a mine car on a track nearby, while the ore can be easily scraped into holes in the floor that lead to the ore bin proper. These openings are so placed that a man cannot easily walk into them.

The Tonopah-Belmont orehouse is noteworthy for the novelty, at least in metal mining, of the manner of constructing the ore bins, which is quite similar to some of the bins that have been built at eastern cement works. The floor of the bin is carried on a series of separate inner posts, while the outside posts, against which the bottom beams abut, extend to the top of the bin and serve as binding posts, as it were, to give stability. The accompanying elevations and sections (Figs. 114 and 115) show the details of this type of construction. The sorting shed construction is quite similar to the arrangement in the orehouses of the Tonopah Mining Co., except that instead of the ordinary bar construction of the grizzlies, they are made of cast-steel plates 1 in. thick with bored holes 2 in. in diameter at the top and 2 in. at the bottom face, arranged in staggered holes and with plates set at an angle of 40°, at which they clear themselves well. The ore hoppers above the sorting tables are lined with steel.

In the construction of the orehouse 73,000 board feet of lumber was used. The bin has a capacity of about 750 tons, being divided into three compartments,

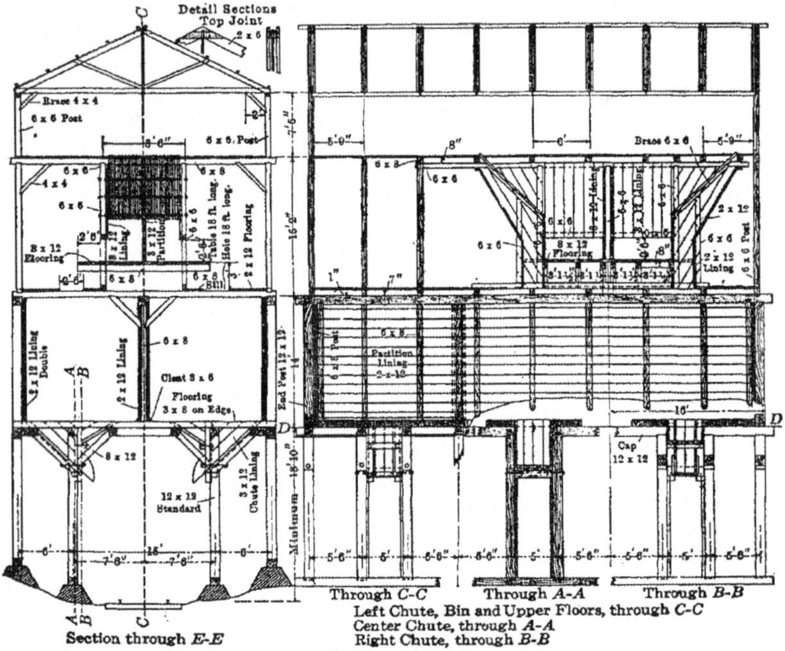

FIG. 114.—SECTIONS OF TONOPAH-BELMONT OREHOUSE.

two of 10,000, and the other of about 5000 cu. ft. capacity. The structure is 51 ft. long by 28 ft. wide by 64 ft. high to the top of the ridge pole. The main sill timbers are bolted to massive concrete mud-sills, while the bin is supported on seven bents of four posts 12×12 in. each. The outer posts reach from sill to top of bin, while the inner line of posts support the longitudinal caps on which the bin floor rests. This floor is made of 3×8-in. planks placed on edge, and is continuous across the whole width, with the outer ends resting on girts that are framed into the outer post. To increase the stability of the bin the girts and long posts are heavily reinforced by 8×12-in. cleats and sway braces.

The sides of the bin are made of a double lining of 2×12-in. planks, with joints broken, while the partitions also are made of 2-in. planks. There are six steel-lined loading chutes, three on each side of the track. These are fitted with No. 3 Bolthoff lever gates, with movable steel spouts. A clearance space of 18 ft. is provided above the rails, so that a locomotive can go under the bin. Above the bin the orehouse part is covered with galvanized, corrugated iron.

Concrete Storage Bin (By Fremont N. Turgeon).—A reinforced-concrete storage bin, interesting for its size, ease of construction and cheapness, is at one end of Witherbee, Sherman & Co.'s new magnetic-concentrating mill, at Mineville, N. Y. It is circular in section, 25 ft. in outside diameter, and 45 ft. high.

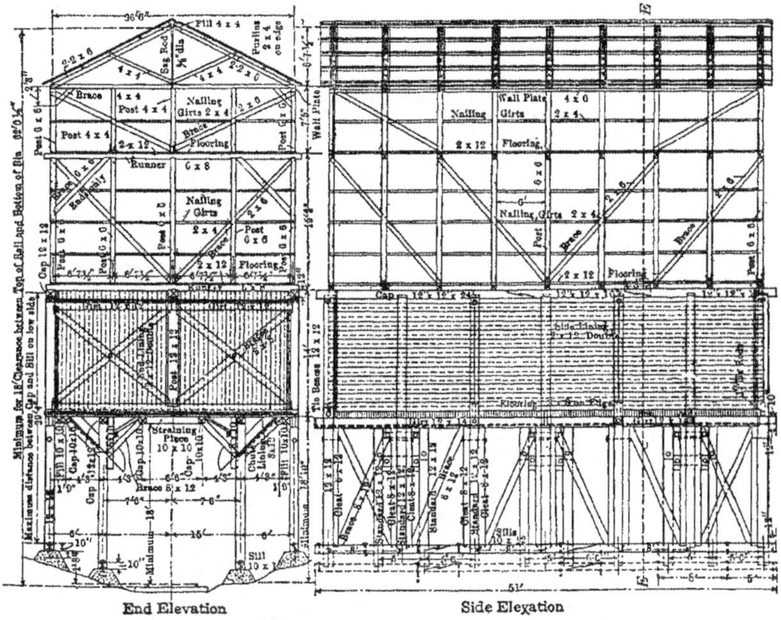

FIG. 115.—ELEVATIONS OF TONOPAH-BELMONT OREHOUSE.

It has a capacity of 1500 tons of crude ore, of which about 1000 tons will run out without handling. The foundation is of rough uncrushed mine stone bonded together with a lean mixture of tailings and cement in the ratio of 10 : 1. The walls are 18 in. thick at the bottom, and decrease 2 in. in thickness every 9 in. up to the top, which is 10 in. in thickness. The mixture of the walls is 4 : 1, tailings and cement. The reinforcing is wornout steel hoisting cable, 1 1/8 in. in diameter, spaced as shown in Fig. 116. The ends of each hoop of cable are fastened together with old cable grips. Vertical cables are placed every 4 ft. about the circumference and anchored in the foundation, and the hoop cables are fastened to each vertical cable with ties of old bell wire, and are placed 4 in. from the outside circumference.

The forms were 1-in. matched boards nailed to circular forms cut from 2-in. plank, spaced every 4 ft. vertically, and supported by short studs from the set below. The entire outside form was made first, then the reinforcing put in and fastened and supported on the outside form, the work being done from a staging inside the bin; then the first 9 ft. of inside form was put in and tied to the outside form with 1/2-in. tie rods, and the walls cast. On top of this the next 9 ft. of

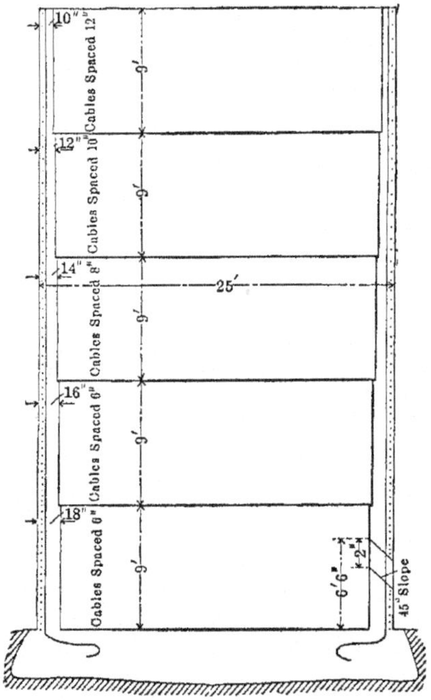

FIG. 116.—CONCRETE STORAGE BIN, MINEVILLE, N. Y

form was erected and the walls cast, and so on to the top. The concrete was raised by an elevator to the level of the top of the form being cast, and carried to the form by wheelbarrows through a small door in the outside form which was subsequently closed. The walls of the bin are thicker than necessary, to allow for the wear of the ore sliding down, as the bin is emptied daily. As may be seen, all ore does not run out naturally. To overcome this, before being used the bin was filled with crushed barren material until it ran out of the door, and ore was put in on top of this.

VIII

HOISTING AND TRANSPORTATION

Theoretical Considerations—Notes on Practice—Hoists—Miscellaneous Devices—Aërial Tramways.

THEORETICAL CONSIDERATIONS

Graphic Solution of Skip Loads (By F. W. Collins).—The pull on the bail of a skip used in an incline shaft and the load on the wheels can be determined graphically by the method illustrated in Fig. 117. In the sketch θ is the angle

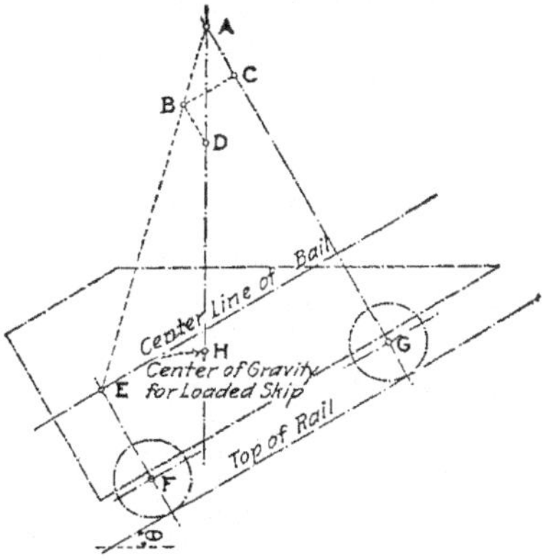

FIG. 117.—GRAPHIC DETERMINATION OF PULL ON A SKIP BAIL.

of inclination of the shaft; *AH*, a vertical line drawn through the center of gravity of the skip; *AG* is the center line of the front wheel drawn normal to the rail; *EF*, the center line of the rear wheel drawn normal to the rail; *E* is the point of intersection of *EF* with the center line of the bail; *A* is the point of intersection of *AH* with *AG*, the center line of the front wheel. After completing this construction lay off on *AH* the distance *AD*, the weight of the skip and load, to any convenient scale, and draw *BD*, parallel to *AG*, and *BC*, parallel to the center line of the bail; then *BD* will be the load on the front wheels, *BC* the pull on the bail, and *AC* the load on the back wheels. The bail need not be parallel to the rails and may be hinged at any point.

Vertical Unbalanced Loads Lifted by First Motion Hoists.—The accompanying charts were prepared by L. F. Mitten of the Vulcan Iron Works,

Wilkesbarre, Penn. The first, reproduced in Fig. 118, is plotted from the following formula:

$$Load = \frac{P \times A \times 2L \times 1\ cylinder \times 0.85}{3.1416D}$$

P = Initial pressure at the throttle.
A = Area of cylinder.
L = Stroke in feet.
D = Diameter of winding drum in feet.

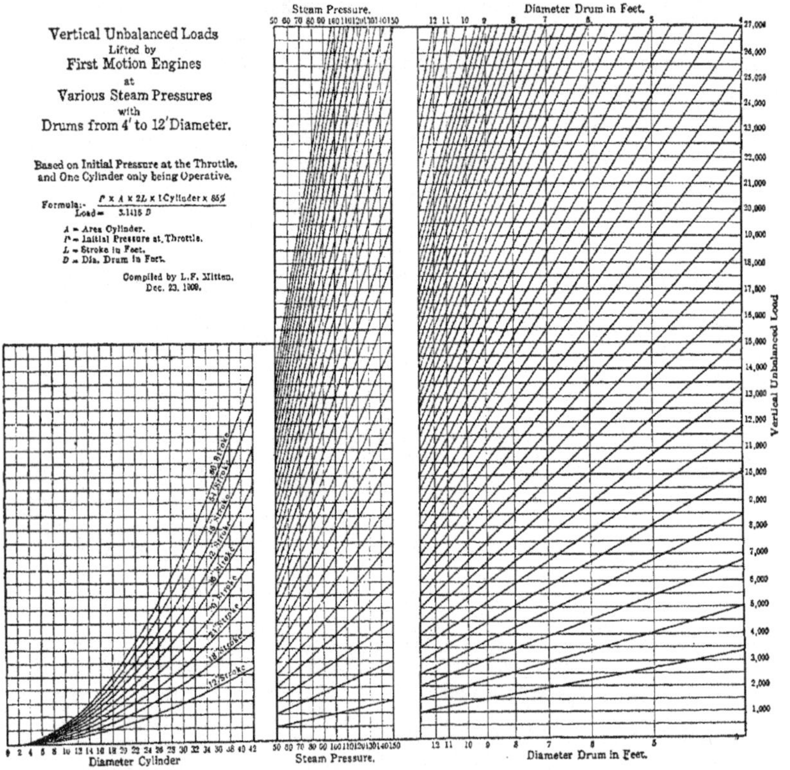

FIG. 118.—CHART FOR FINDING THE PROPER-SIZED ENGINE WHEN THE UNBALANCED LOAD AND THE STEAM PRESSURE ARE KNOWN.

Either cylinder of the pair of engines is capable of starting the unbalanced load given, inasmuch as one cylinder may be on a dead center and would therefore be inoperative. The steam pressure is taken at the full initial pressure at throttle. An efficiency of 85% has been taken, this having been found after various tests to be a fair average.

Knowing the unbalanced load and the steam pressure at the throttle, the proper-sized engine for the work may be found without any figuring whatever. It will also be readily seen that, having the size of the engine cylinders, the

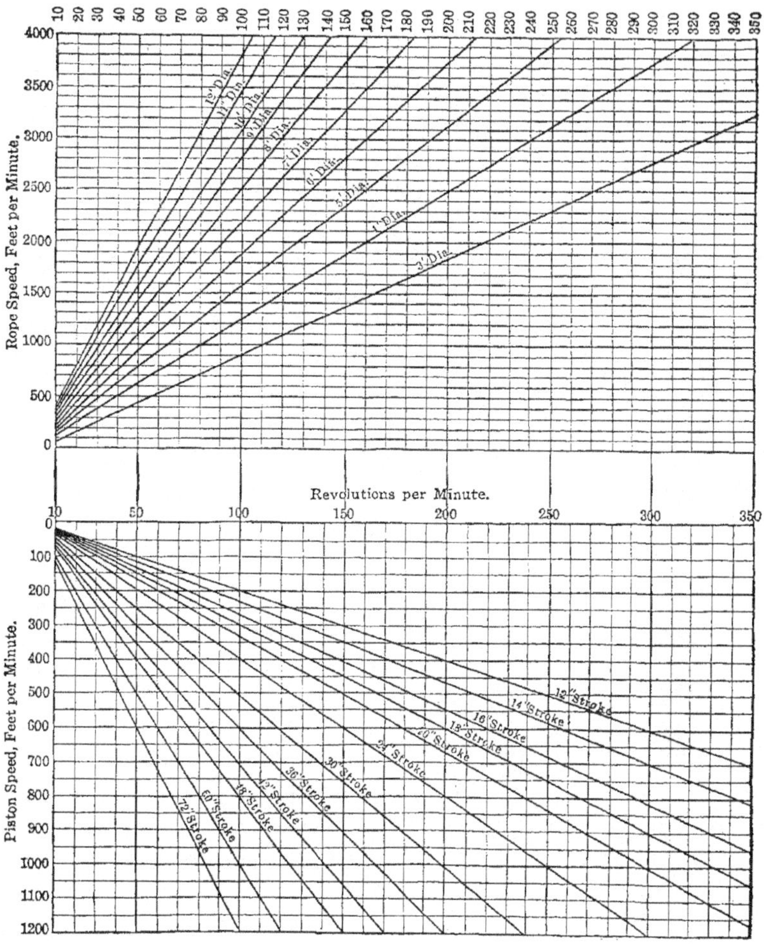

FIG. 119.—CHART FOR DETERMINING THE ROPE SPEED IN HOISTING.

vertical load that can be lifted by them may also be found for any steam pressure and various diameters of drum.

As an example, find the size of engine capable of handling a vertical unbalanced load of 12,000 lb. with 100-lb. steam pressure at the throttle. The rope manufacturers give the working load for 1-in. rope as 6.8 tons. However, in actual practice, for this load a 1 1/4-in. rope would probably be used and wound

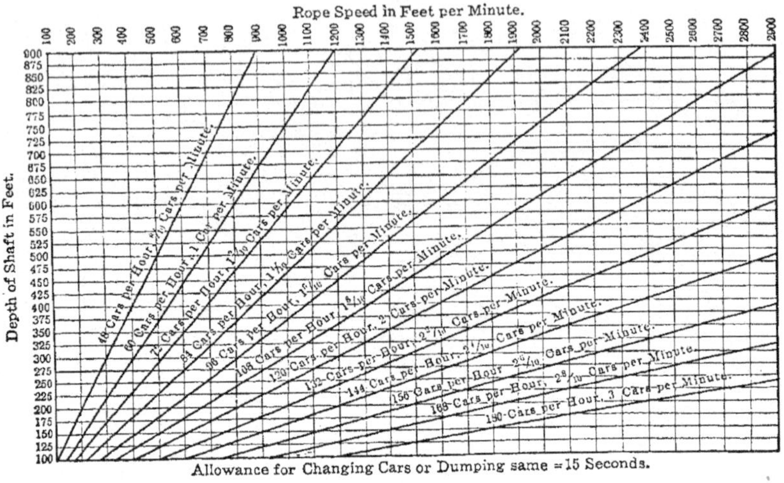

FIG. 120.—CHART TO DETERMINE NUMBER OF CARS HOISTED PER HOUR.

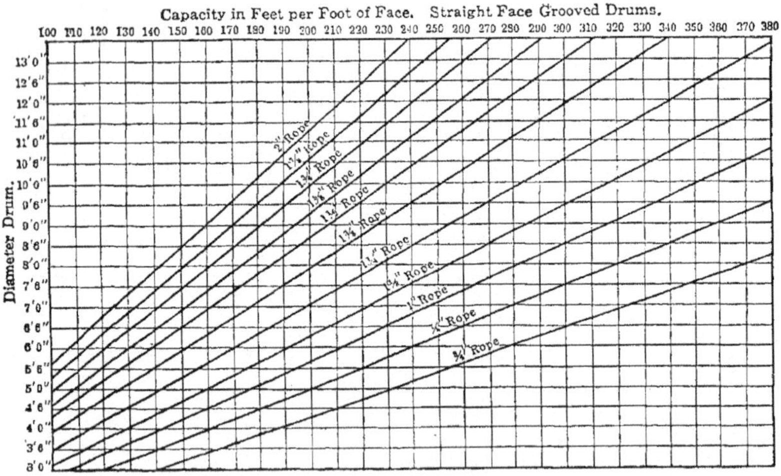

FIG. 121.—CHART FOR DETERMINING THE FACE OF A GROOVED DRUM.

on, say, an 8-ft. diameter drum. Follow the horizontal line marked 12,000 until it is intersected by the diagonal line at or about the vertical line representing an 8-ft. diameter drum; then follow the diagonal line down to left side of chart; follow horizontal line across until it is intersected by diagonal line at or about the line representing 100-lb. steam pressure; follow this diagonal line as previously described to left side of chart; follow horizontal line until proper

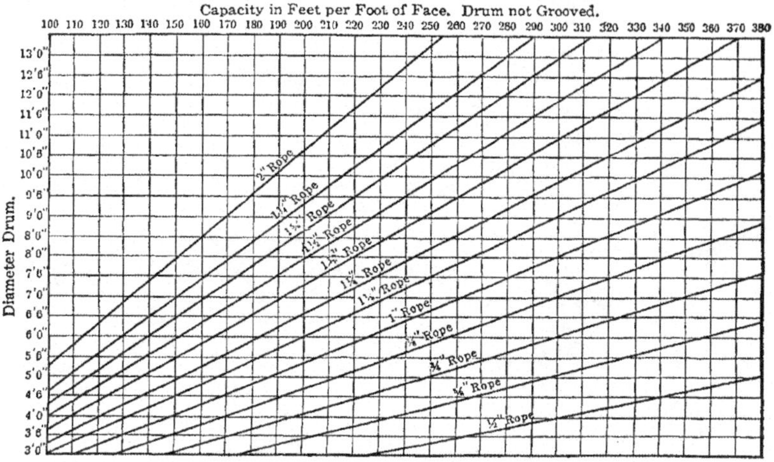

FIG. 122.—CHART TO DETERMINE THE FACE, WHEN THE DRUM IS NOT GROOVED.

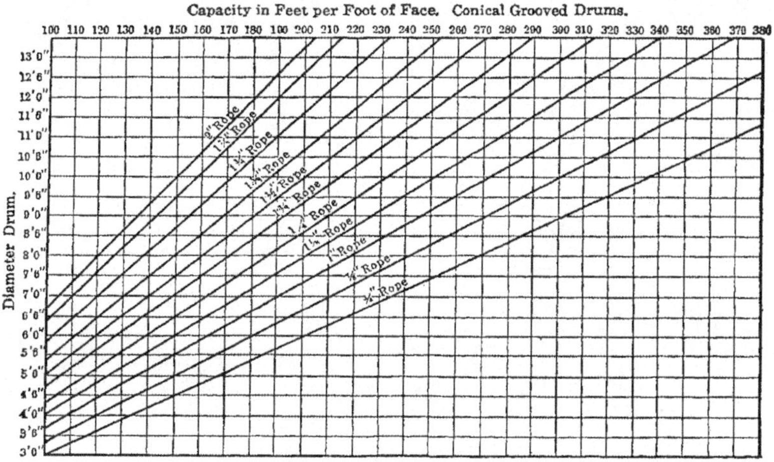

FIG. 123.—CHART TO DETERMINE THE FACE OF A CONICAL DRUM; THE DIAMETER SHOULD BE THE MEAN DIAMETER.

combination of stroke and diameter are found. For the work in hand the following engines would be satisfactory: 24×48 in. or 28×36 in.; the 24×48-in. size would probably be selected.

Determining the Rope Speed in Hoisting.—In determining the rope speed in a hoisting operation when the average piston speed, the length of stroke and diameter of drum are known, the second chart (Fig. 119) provides a handy method of calculation. As an example, let us assume an average piston speed of 600 ft. per minute, and a 48-in. stroke; then, following the horizontal line on the chart at 600 until it is intersected by the diagonal line representing a 48-in. stroke, follow this line to the upper section until it is intersected by the diagonal line representing an 8-ft. drum; by following the horizontal line to the side of the chart the rope speed is found to be 1890 ft. per minute.

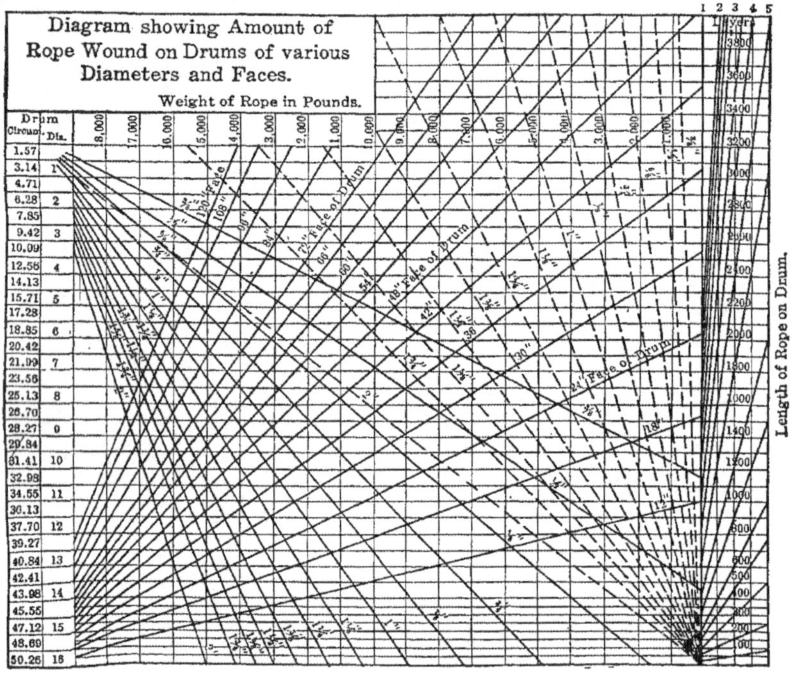

FIG. 124.

Determining the Number of Cars Hoisted per Hour.—The third chart of the series enables the quick calculation of "cars per hour," when depth of shaft and certain other factors are known. As an example, assume the shaft is 475 ft. deep, and a rope speed of 1900 ft. By referring to Fig. 120, it will be readily seen that from a shaft 475 ft. deep we can get out two cars per minute, or 120 cars per hour. This chart is based on an allowance of one quarter of a minute for changing cars and a double-compartment shaft.

Determining the Face of Winding Drums.—In determining the face of winding drums when the diameter of the drum and the rope diameter are known, the charts reproduced in Figs. 121, 122, and 123 will be found useful.

Assume that the drum has a diameter of 8 ft., and that the rope diameter is 1 1/4 in.; also that the shaft is 475 ft. deep; then referring to Fig. 121 it is found that an 8-ft. diameter drum will wind 227 ft. per foot of face. The drum required would therefore have to be, say, 2 ft. 6 in. face, which would allow a sufficient number of grooves at one end of drum for fastening the rope.

The other charts shown in Figs. 122 and 123 would, of course, be used in

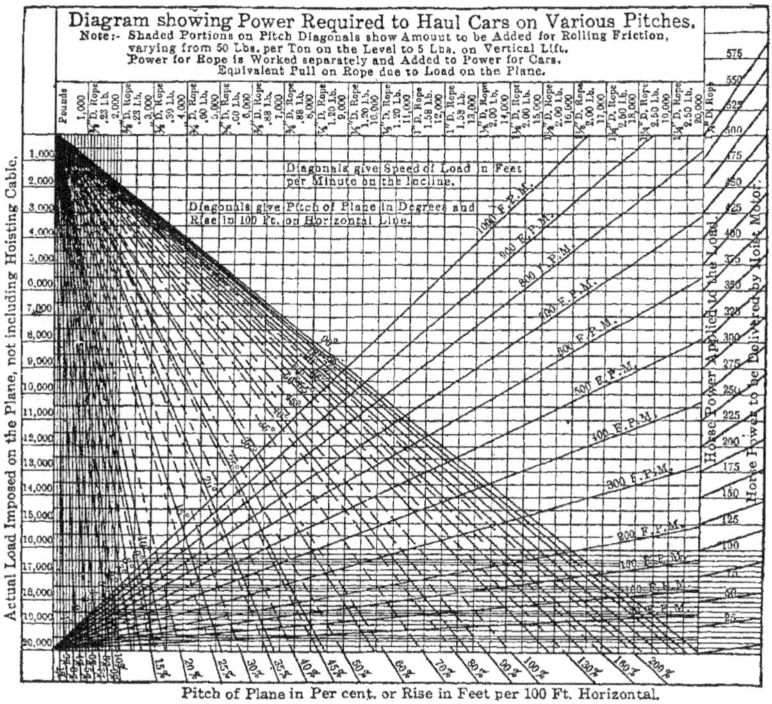

FIG. 125.

the same manner as illustrated above. The diameter used for conical drums, however, should be the mean diameter.

Determining the Amount of Rope Wound on a Drum.—In determining the amount of rope wound on drums of various diameters and faces, the chart given in Fig. 124 provides a quick method of calculation. It is, of course, essential that the circumference and diameter of the drum are known. The chart also assumes that the face of the drum is a known quantity. With these factors at hand, the method of calculation is self-evident.

Power Required to Haul Cars on Various Pitches.—The sixth chart, reproduced in Fig. 125, is intended to simplify the calculations for determining the power required to haul loads on planes of various pitches. For example, it is desired to haul two loaded cars, each weighing 6000 lb., up a plane 1000 ft. long, having a pitch of 40° from the horizontal, at a maximum rope speed of 500 ft. per minute. What is the equivalent rope pull? What is the brake horsepower required to handle the load? Two loaded cars weighing 6000 lb. each equals a load of 12,000 lb. exclusive of winding rope. Referring to the diagram, follow the horizontal line representing 12,000 on the chart until it is intersected by the diagonal line representing 40°. Directly above this point of intersection will be found the rope pull, which in this case is 7900 lb. It is also found that a 3/4-in. rope will be satisfactory and that this size rope weighs 0.88 lb. per foot. Following down this imaginary line representing 7900 lb. until it is intersected by the diagonal line representing 500 ft. per minute rope speed, one finds 120 h.p. applied to the load, or the brake horsepower required. The horsepower to be delivered by this hoist motor would be 141; this is based on an efficiency of 85% for the entire equipment. It has been found that a 3/4-in. rope would be required and also that this size rope weighed 0.88 lb. per foot; 1000 ft. of rope at 0.88 lb. equals 880 lb. By working this out as was done for the loaded cars, it is equivalent to 9 h.p., which should be added to the 141 h.p., making a total of 150 h.p.

Rope Capacity of Drums.—The rule used by the A. Leschen & Sons Rope Co. for computing the rope capacity of any size of drum, is as follows: Add the depth of flange in inches to the diameter of the drum, and multiply this result by the out to out width of the drum. This product is then multiplied by the figure below corresponding to the size of the rope used:

1/4 in.	4.16	1 3/8 in.	0.138
3/8 in.	1.86	1 1/2 in.	0.116
7/16 in.	1.37	1 5/8 in.	0.099
1/2 in.	1.05	1 3/4 in.	0.085
9/16 in.	0.828	1 7/8 in.	0.074
5/8 in.	0.672	2 in.	0.066
3/4 in.	0.465	2 1/8 in.	0.058
7/8 in.	0.342	2 1/4 in.	0.052
1 in.	0.262	2 3/8 in.	0.046
1 1/8 in.	0.207	2 1/2 in.	0.042
1 1/4 in.	0.167		

This rule applies, of course, to a drum on which the rope is to be wound in successive layers up to the full height of the flange.

NOTES ON PRACTICE

Flat Rope vs. Round Rope.—A correspondent asks about the comparative results in practice of flat and round wire ropes. He says that he has had the bestsul rets from the flat rope, and considers it the safer because of being open to closer inspection. His superintendent contends, however, that round cable is the better, and cites Lake Superior and South African practice. Our correspondent asks, What are the manufacturers' claims? With respect to this matter a leading manufacturer of wire rope informs us that flat rope is now used only in exceptional cases, there being but little demand for it, owing to its greater cost.

The effect of wear in round wire rope shows first in the outer layers, and anything radically wrong can therefore be readily detected. Round wire rope exposes less surface to atmospheric oxidation than flat rope, and the core wires, which are saturated with grease, are less likely to suffer from oxidation or corrosion from the action of water, and especially water containing acid, so that as far as safety is concerned round rope is considered to be superior to flat rope. The superiority of round rope in point of safety is generally recognized by engineers. This explains why round rope is so generally used at Lake Superior and in South Africa. Some engineers go so far as to say that the use of flat rope ought to be forbidden. A round rope may have a good many broken wires and still be safe, owing to the tight binding of this kind of construction; whereas in flat ropes the binding is much looser and broken wires quickly become a serious danger.

Remarks on Hoisting Ropes.—The latest Prussian statistics on shaft hoisting ropes were exhaustively and critically discussed by Professor Herbst, of Aix-la-Chapelle, in a series of articles in *Glückauf*. His general conclusions are as follows: The protective effect of lubrication has not been plainly proved in dry shafts. This observation suggests that the present lubrication processes for wet ropes leave room for improvement, although it is certain that all now-known lubricative agents rapidly disintegrate in shafts where the water is salty or sour. Future experiments in this direction may provide a remedy. Galvanizing or coating with zinc does not appear to have a really protective effect in wet shafts, the reason probably being that zinc coating has but little power of resistance to salt water. It is also suggested that the wires have suffered in the galvanizing process, for, although it has been proved by Winter and others that the process, when properly and carefully executed, does not unfavorably affect the ropes, it is also well known that it often reduces the tensile strength of the rope by 50% or more. The hauling efficiency of ropes in dry shafts stands in the proportion of 100 to 60 or 70 to that in wet shafts, a fact which, in view of the high price of ropes, means a substantial economic advantage for dry shafts. Tensile strength between 160 and 180 kg. per square millimeter does not unfavorably affect the flexibility or hauling strength of the ropes, while ropes of more than 180 kg. per square millimeter have given substantially lower efficiency figures. The greater or less strain, as expressed by a higher or lower safety factor, put upon ropes has had no influence upon their consistency. It may, therefore, be assumed that the advantages of a higher factor of safety are neutralized by its disadvantages that is, greater rope thickness combined with reduced flexibility and greater dead weight.

Uses for Old Hoisting Cable.—As old hoisting cable has had most of the stretch taken out of it, it makes good reinforcement for concrete work. At the Red Jacket shaft of the Calumet & Hecla company old hoisting cable is unstranded and the strands are also used on the underground-haulage systems for the wearing ropes. In removing the grease from the cable, burning was tried, but it took the temper out of the wires and the strands would untwist, so now only a part of the grease is taken off the cables before they are unstranded.

The unstranding is done in lengths of 600 ft. A block and tackle is fastened to each end of the cable and it is stretched so that it will be clear of the ground. In attaching the blocks to the cable a clamp is used consisting of a bar with a hinged top piece which is tightened on the rope by a bolt at the other end. The block and tackle at each end is fastened to a swivel so that the cable can twist in either direction as it is being unstranded. Two strands are unstranded at a time and each of these strands is fastened to a block so that they can be kept tight as they are being unstranded. These blocks are fastened at some distance from each other as well as from the main cable, so that neither one of them will interfere with either the other or the main rope while it is being unstranded.

About 2000 ft. of cable can be unstranded in a day, and about 12,000 ft. of single-strand rope obtained. Both 1-in. and 7/8-in. hoisting rope has been unstranded for the haulage systems. This old rope has been found to work quite satisfactorily for the hauling rope, but good rope must be used for the tail rope if a return rope from the same engine that does the hauling is used, as the unstranded rope will not readily pass through the pulleys. Strands of 1 3/8-in. rope have also been tried, but they were found to be too stiff. In case a rope breaks or a broken wire begins to ball up on the rope, the individual wire or the ends of the broken rope are heated so as to take out the temper, and then the ends are tied together and hauling is continued until the rope can be spliced properly.

Gravity Planes at Cheever Mine (By Guy C. Stoltz).—The Cheever Iron Ore Co., operating at Port Henry, N. Y., trams the concentrates, resulting from magnetic separation, by gravity planes to the loading chutes of the Delaware & Hudson switch on the shore of Lake Champlain. Topography favored the installation of two planes, the first plane being 700 ft. long with a drop of 55 ft., and the second about 2000 ft. long and a drop of 193 ft. The grade is not at all regular. The tracks conform, wherever possible, to the surface of the ground. Three 30-lb. rails are laid at 3-ft. gage on each plane and four rails with the spread for turnouts are laid at the half-way points. Side-dump steel cars of 4 1/2-ton capacity are used. A trip of two loaded cars is released on the slight down grade at the storage bin and on their downward journey to the first turntable they pull the two empty cars, attached to the other end of the cable, to the loading bin. At the turntable the loaded cars are deflected about 60° and attached to the free end of the cable for the second plane and on their downward course pull up two more empties. Sheaves with brakes are installed at the top of each plane. At the terminal of the second plane the cars are delivered to a turntable and trammed by hand to the several loading chutes. It is intended to replace the first turntable by a steeply banked curve, which will increase the capacity of the system and lower the surface-tramming cost by almost one-half.

Car Stopping Devices on Gravity Inclines.—It is of great importance to have, at the upper end of every gravity plane, a device to regulate the admission of cars, one at a time, to the plane, and at the same time protect the men working at the bottom. Fig. 126 illustrates three different types of appliances used in Germany to accomplish this.

The device shown in Fig. 1 consists of a pair of stops, one at the extreme top and the other a distance of 2 m. down the incline; both are raised into effective position by cams keyed to axles which lie underneath and across the track. The movement of the axles is controlled by levers connected in such a way that a single motion of the hand lever will raise one stop into position and simultaneously drop the other out of position. The first motion of the hand lever drops the upper stop, permitting the car to start down the incline. The car is blocked by the second stop, until a motion in the opposite direction lowers this stop, allowing the car to pass down the incline, and raises the upper stop into position to retain the next following car.

The apparatus shown in Fig. 2 consists of an axle about 1 m. long, lying below and parallel to the rails, and supported in this position by two journal boxes. To each end of the axle is fastened an arm, at 90° to one another, of such length that the extreme end of each arm will reach out and rest upon the top of the adjacent rail, thus forming an obstruction to the wheels of the cars. When one rail is blocked, the other is free, so that to permit the cars to pass one at a time it is only necessary to rotate the axle through a few degrees alternately to one side and the other. The top tender does this with his foot.

The type of which two views are shown in Figs. 3 and 4 consists of a heavy, square beam pivoted at its ends and extending across the top of the incline at a sufficient height to permit the loaded cars to pass beneath it. At one end of the beam is a single-notched ratchet engaging a pawl, which prevents the former from rotating beyond a certain point. Two strong arms are fastened to the square beam in such a way as to block the passing of a car on either track so long as the pawl holds. The latter can be released by pulling the handle on the end of the cord, which is within reach of the top tender, allowing the car to pass. As soon as it has gone far enough, the arms fall back into their first position, and their impetus carries the notch in the ratchet to within reach of the pawl, when the device is ready for the next car. It is apparent that the apparatus interposes no obstruction to the passing of a car coming up hill.

The danger to be apprehended in the device last described is that, if two

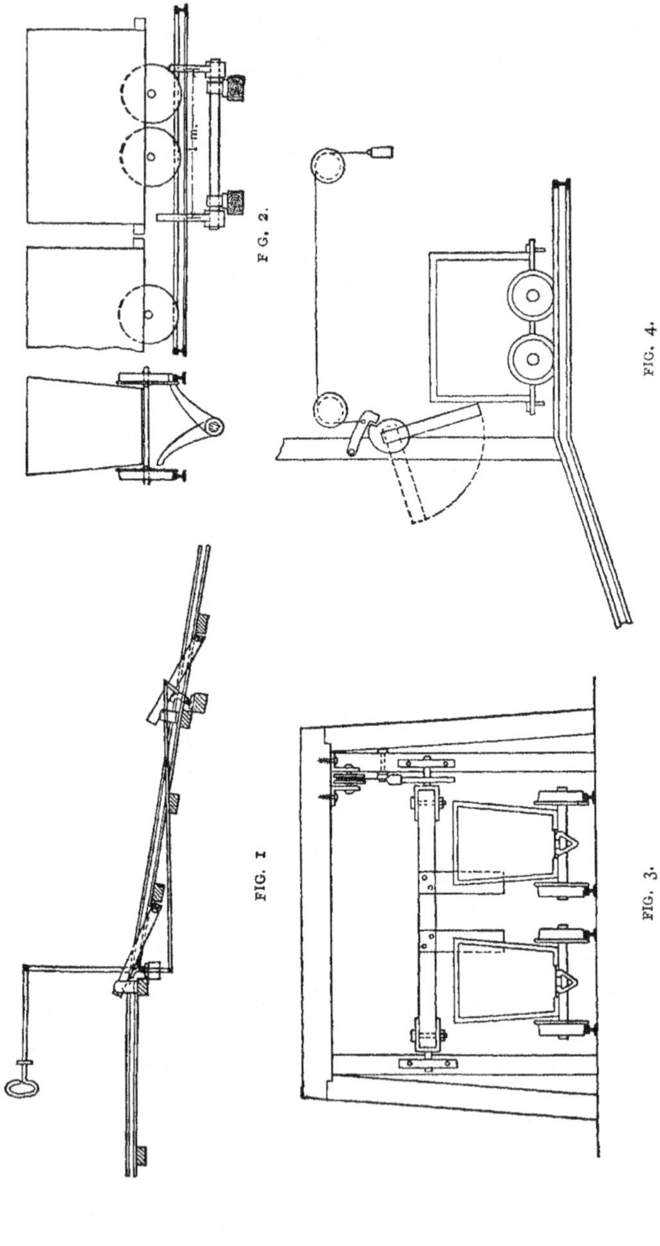

FIG. 126.—SOME GERMAN CAR-STOPPING DEVICES.

cars should follow one another closely, in passing over the knuckle, by the time the first car had gone far enough to release the restraining arm, the second car would be so far advanced as to prevent the ratchet from establishing connection with the pawl, and the second car would race the first one down the hill.

Tail Rope Haulage Operated by Skips.—At the Republic mine, Michigan, where an unbalanced ore skip is used, the descending skip furnishes power for hauling empty ore cars. The tram has a length of 900 ft., and has a sufficient down grade for the loaded cars to run out by gravity. The car is of the double-truck type, side dump, weighs 6500 lb. and carries 3 1/2 tons of ore. A tail rope is attached to the car and is connected with a winding drum on the axle of the sheave. This drum is 8 ft. in diameter, the same as the sheave. A friction clutch is used to operate the drum. There is also a brake on the drum to control the speed of the outgoing cars. Both brake and friction clutch are operated by a wire rope from the station level. An indicator is also used to show the exact position of the car on the track. The car dumps automatically. Of course in this particular case the skip remains at the surface until the car has reached the bin and is ready for its return trip. As the skip descends, the friction clutch is thrown in, and the empty car is drawn back to the shaft ready to receive the ore when it comes up. This scheme works excellently with a shaft that is not working at its full capacity or where the stopping of the skip for a few moments does not interfere with the output. A second drum is now being installed at the same mine. This one will only be 4 ft. in diameter, as the distance for tramming the ore is less. In a system of counterbalanced skip a similar scheme is used, except that the ascending skip also brings in the empty cars. In this case the loaded car goes out as the skip descends and then returns as the skip comes up. The drum is at the station level and is operated by a rope drive from the sheave. The drum, as before, is provided with friction clutch and brake. This adds a little extra strain on the hoisting cable, but not enough to seriously affect its working.

An Underground Haulage System (By Albert H. Fay).—The problem of handling a large tonnage of ore underground is usually a serious one, especially in the matter of cost. The system that is now being installed between the 700 and 800 levels by Witherbee, Sherman & Co., promises to be one of great importance in handling the magnetite ore in their mines at Mineville, N. Y. The installation is expensive and could be used only by mines handling a large tonnage for a number of years. Up to the present the mining has been carried on by working on large faces of ore 50 to 75 ft. high and several hundred feet wide. The ore was shot down to the foot of this face and then shoveled by hand into the mine cars which were pushed to the shaft by hand. The ore is heavy and in most cases it requires at least three to five men to handle a single car.

The scheme now under way consists of a haulage way installed 60 ft. below the present working level. This underground passage is in the form of an ellipse, with a circumference of about 1000 ft., and opens an orebody 500 ft. wide, 60 ft. deep, with an indefinite length. At least 1,000,000 tons of ore are now blocked out to be handled by this installation. Along this drift are at least a half-dozen raises, inclined at about 45°, which will be used as mill holes. The mining will be carried on by stoping down, setting the machine drills near the raises and shooting the ore down to the loading platforms. The large number of raises will give ample space for a number of machines and as the work progresses it will give still more room. As the ore passes down through a raise, it falls upon a loading platform built of concrete. This platform is 4 1/2 ft. high, 10 or 12 ft. long, 8 or 10 ft. wide according to the condition of the ground, with a loading chute 3 ft. above the top of rail. It is built with a slope from the back to the front as well as a slope from each end to the center. The chute will be covered with sheet iron.

At the shaft an 800-ton storage bin has been built, the bottom of which is 46.5 ft. below the top of the car tracks. The tipple is 26 ft. long and will dump three cars at once. It is operated by an electric motor and revolves upon trunnions. The bin gate and the loading chute are operated by air hoists. An auxiliary tipple at the left will dump only one car at a time and is to be used only when tramming by hand in case the electric motor haulage system is out of commission. This is also to be used as a waste pocket when it is necessary to dispose of waste from the same loading station. Between the shaft and the bin is a rock pentice which serves as a support for the auxiliary tipple and at the same time forms the front wall of the ore pocket.

The cars are of three tons' capacity and will be handled in trains of nine cars each. The motor truck has two 25-h.p. motors operated on 220-volt direct current. The haulage track is 45-lb. rails. While lighter rails could be used, practice has demonstrated that the heavy rails are better as they are not easily broken by heavy pieces of ore falling upon them. They are also more solid, require less ties and give a better track. The motor in passing around the track gathers up the loaded cars and pushes them in front. When the tipple is reached, three cars are dumped. These are then pushed through the tipple and three more dumped until all are empty. The motor then goes around the tipple on a side track and couples the nine empty cars on behind. When the first loading station is reached, the loaded cars are picked up, and one, two or three empty cars left in their place as may be desired. In this way the motor will be in operation all the time and with an 800-ton storage bin it will be possible to keep the hoist working up to its full capacity without the loss of time which was usual when the cars were operated by hand power. When this equipment is completed, it is expected to be able to handle 1000 tons per 10-hour shift. The ore is hoisted in self-dumping skips. I am indebted to S. LeFevre, chief engineer, Witherbee, Sherman & Co., for the above information.

An Underground Hoisting Station (By S. A. Worcester).—Fig. 127 shows the layout of a winze hoisting station in a mine in southwestern Colorado.

The winze is nearly square in section and is divided into three compartments. The largest compartment is rectangular in section and is used for hoisting ore in cars of 2200-lb. capacity, two cars being placed tandem on the single deck of the cage. The 150-h.p. electric, two-reel cage hoist occupies a large room excavated in hard rock at the west side of the station; it is not shown in the illustration. The flat rope from one of the reels runs over the large upper sheave at the left, thence down to the cage. The flat rope on the other reel runs over the lower sheave supported on an A-frame thence down the shaft to an overbalance weight. This weight is made of several sections and is similar to the ordinary elevator weight. It is so weighted that the work of the motor when raising the weight is the same as when raising the cage with its maximum load.

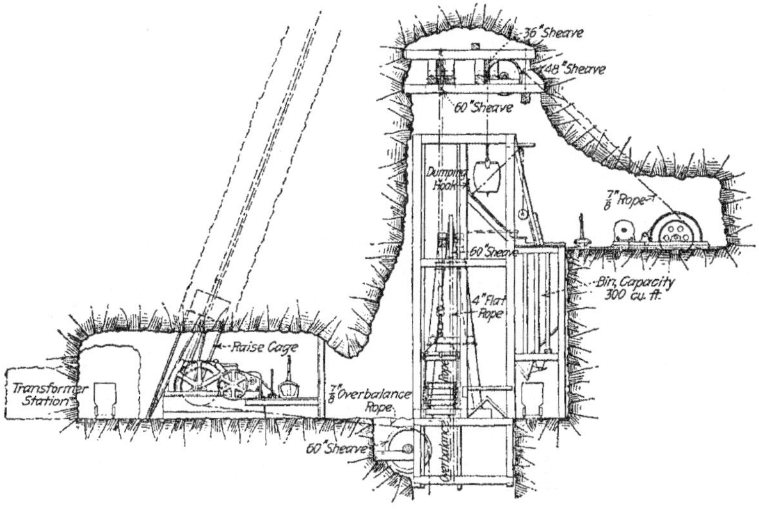

FIG. 127.—WINZE HOIST STATION IN A COLORADO MINE.

The two other compartments are nearly square in section. One is used as a pipe and ladder compartment, the other is lined with planks throughout and is used only as a bucket hoistway for sinking operations. This bucket is raised by the 75-h.p. electric motor shown in the upper room on the right-hand side of the illustration. When the bucket has been raised above the level of the floor of this room the counterbalanced door is lowered and the dumping rope shown at the left of the bottom of the bucket is hooked into the ring at the bottom of the bucket. Upon lowering, the bucket turns over and dumps its load into the bin below from which the ore or rock is drawn into cars and hauled 2400 ft. along the vein then 2200 ft. through an adit to daylight; thence it is conveyed 11,500 ft. by a Bleichert tramway to the mill. The operating levers of the hoist are placed near the winze so that the hoist engineer can attend to the dumping.

At the left of the illustration a third room is shown, excavated in rock, in which is shown a 50-h.p. two-drum hoist. Above this room the position of an inclined raise is shown. The hoist rope passes up the raise for 500 ft., over a sheave, thence down to the cage. The 7/8-in. rope on the drum passes over a sheave at the collar of the winze, thence down the cage compartment to an overbalance weight that runs by the side of and is similar to the weight used on the cage hoist. The raise hoist is used for raising men, timbers and supplies to the upper levels. Direct current for all these hoists, for other hoists in the mine and for the mine locomotives is supplied from a storage-battery plant at the surface. The battery is charged by rotary converters and a booster, the current being generated at a hydro-electric power plant several miles from the mine.

Catenary Hoisting Cable.—There is an unusual installation of a hoisting cable at the Republic mine, Republic, Mich. In order to utilize a central power plant, it became necessary to have a cable operate across a small lake a distance of 1800 ft. before the headframe of the shaft could be reached. The shaft is about 800 ft. deep and is inclined at an angle of 70°. In constructing the cable across the lake, the towers for supporting the cable would be very high if an attempt had been made to keep the cable in a straight line from the drum to the shaft. A catenary curve between the two places was figured out on the basis of the breaking load of the cable. Towers were erected at intervals of 100 ft. entirely across the lake, the one near the center being only 10 or 12 ft. above the level of the lake. The catenary is so flat that the cable has no tendency to lift off the pulleys and at the same time the friction on the pulleys is less than it would be if the cable were worked in a straight line. A large saving also resulted from the construction of smaller towers.

Double Hoisting Cables.—At the Beust shaft of the Deutschland mine at Hasslinghausen, Germany, the slipping of the cable is prevented by using a double rope running on double-grooved sheaves. Thus the bearing area on the packing in the sheaves is increased. The cables are attached to the cage by a drawbar to which the cables are fastened by turnbuckles. These turnbuckles permit the strain on the cables to be equalized. Bolts prevent the moving of the turnbuckles by the twisting of the ropes.

Hoisting Cable Run through a Drill Hole.—A 6-in. drill hole from the surface penetrated a body of ore which later was stoped out. Upon sinking the drill hole ore was struck at a lower level and a winze sunk with the drill hole as a center. In order to work the winze to any depth it would have been necessary to install a hoist underground, which was not practical. A hoist was therefore erected at the surface and the cable operated through the drill hole. Ore was hoisted to the main working level and then trammed to the hoisting shaft.

Rapid Hoisting with Wire Guide (By Hugh C. Watson).—A remarkable feat of hoisting is the one now in operation at La Ojuela mine in the State of Durango, Mexico. By means of wire guides and an unbalanced, first-motion hoist, a bucket holding about 1800 lb. of ore is filled, hoisted, dumped and returned to the bottom of a 1700-ft. shaft in 2 minutes 10 seconds. This is not record time, but ordinary hoisting speed. It has been done in 2 minutes flat.

Wire-rope guides are not the best that can be used, but they have certain advantages, especially when working a mine on a prospecting basis, where first cost is one of the essential features. The principal advantages are small cost, ease and speed with which they can be installed and shifted, that they will work in any sort of a vertical shaft, and, considering the speed attained at Ojuela, they do not seriously limit the capacity of the haulage system.

Six men are employed in connection with the hoisting plant: One is the hoist man, two are topmen and three are fillers. The topmen do nothing but close the doors, dump the bucket and open the doors for the down trip. Of the three men at the bottom of the shaft, one sits above the ore-bin door with a bar to see that the door does not get stopped up. One stands directly in the shaft to hook and unhook the buckets as they are pushed to him by the third man. The empty bucket is caught on a low truck set to receive it between the guides. The man in the shaft unhooks it from the crosshead, the third man gives his full bucket a push that shoves the empty bucket and its truck to the far side of the shaft, the shaft man then hooks the full bucket to the crosshead and gives the signal to hoist. The bucket now at the bottom is filled while the other bucket is being hoisted and dumped.

The bucket itself is of the ordinary type attached to the bail slightly below center, and kept upright by means of a link and ear. The doors on the shaft are of the type known locally as "doghouse." They are simply two doors which, when closed, form a gable over the mouth of the shaft on which the ore slides into a bin on either side. The guides are 5/8-in. four-strand, galvanized-steel, wire rope. They are attached at the bottom to a cable that is stretched across the shaft below the track level. This cross cable is anchored on each side of the shaft to a 1 1/2×18-in. eye-bolt, which has a split point and wedge to keep it fixed solidly.

On top, these guides pass over two small sheaves and down to a drum that serves to hold an excess of rope and is also fitted with a counterweight, a crank handle and dog, so that the guides can be easily and rapidly tightened. This is only one of the many ways that these guides can be fastened; in fact, on this same mine there are at the present time seven shafts that have wire guides, and each one has its own system, each taking advantage of some peculiar condition existing in that shaft.

The crosshead is made in the shape of a triangle, of 6×6-in. pine, and at the lower side two lead guide blocks are used of such a form as to be easily changed and solidly attached to the triangle. The guides last indefinitely, but the guide blocks have to be constantly replaced. On shaft No. 4, which hoists about 200 tons daily, the guide blocks last four days. When lowering men, four accommodate themselves on the triangle, four stand on the edge of the bucket and two, sometimes three, climb into the bucket. The principal disadvantage of this method of hoisting is that it offers absolutely no way of attaching safety devices, and that it takes a larger shaft for the same size of bucket. The principal advantages, other than those already mentioned, are that no timber is required in the shaft, the system can be used either with the balanced or unbalanced method of hoisting, and that by this method a bucket can be passed through a cave or a big stope, where it would be costly to put timbers. This last is the principal reason for its adoption at Ojuela.

Concrete Chute Bridging a Level.—Concentrating hoisting at a few levels is a practice that is growing in favor at many large mines. At the Osceola, No. 13 shaft of the Calumet & Hecla company, the ore from five levels, each 100 ft. apart, is delivered at the level through a chute in the shaft pillar. This chute is inclined at an angle of 40° from the horizontal. The ore is drawn from it into a car of 7 1/2 tons capacity, the flow of ore being controlled by a chute gate of the hinged type. As this chute was cut after the levels were driven, it was necessary to continue it from the opening in the roof of the drift diagonally across to the similar opening in the floor, near the opposite wall. This was done by bridging the gap with a long, reinforced-concrete box without ends. The bottom of the box was made 24 in. thick, and was reinforced with old 30- and 40-lb. rails. Large blocks of rock from the vein walls, some of which were 16 to 18 in. diameter, were imbedded in the concrete in making the bottom, as it was believed that this material would withstand abrasion by the ore better than concrete alone; a 1:2:5 concrete mixture was used. The sides of the box were made 12 in. thick at the floor and tapered to a thickness of 6 in. at the roof. The cover or top of the box was made of concrete 12 in. thick, and it, as well as the sides, was reinforced with old rails. An opening in one side of the box at the floor on each level afforded means of dumping ore into the chute. A grizzly was placed over the opening, and was held in place by a 12×12-in. timber, to which the grizzly bars were screwed.

A Cheap Mine Road (By S. H. Brockunier).—Recently I had occasion to build 2600 ft. of a side-hill road to the Erie mine, Gaston, Calif. I had only a week to complete the road and the problem was further complicated by fallen trees 3 or 4 ft. in diameter, and several hundred feet of swampy ground, so I decided to place main reliance on dynamite. The first day only two pairs of men were put to work, one pair at each end of the road, in order to see how much they would accomplish and how many men would be needed. In this way I estimated that eight men would complete the road in seven days; as a matter of fact the ditching of the swamp detained them a day longer and eight days were actually consumed in the construction. The men were given

40% dynamite, fuse, caps, augers, and bars. They were told to put 5-ft. vertical holes along the upper side of the proposed road and loosen the earth with dynamite, regulating their charges so as to move the earth as much as possible toward the downhill or lower side. In this way not a pick was used on the entire road and shoveling was reduced to a minimum. When a fallen tree was encountered it was bored for a charge of dynamite, blasted in two, and pushed out of the way. The swamp was corduroyed with 8-ft. slabs blasted from fallen trees. These slabs or poles should have been 12 ft. long, because if shorter they rock too much in soft mud. Cutting and splitting trees with dynamite is an easy method when compared with axe, saw, hammer and wedges and the necessary men to handle such tools. The total cost of this road was, seven boxes of dynamite, $49; fuse and caps, $13; labor, $202; total, $264. This is at the rate of 10 cents per lineal foot for an 8-ft. road. The day the road was completed a 6000-lb. load was drawn over it with ease and it has been teamed over steadily ever since.

HOISTS

Snatch Blocks Applied to Hoisting (By Stephen L. Goodale).—At the Bristol mines, at Pioche, Lincoln county, Nev., a large amount of rich ore was

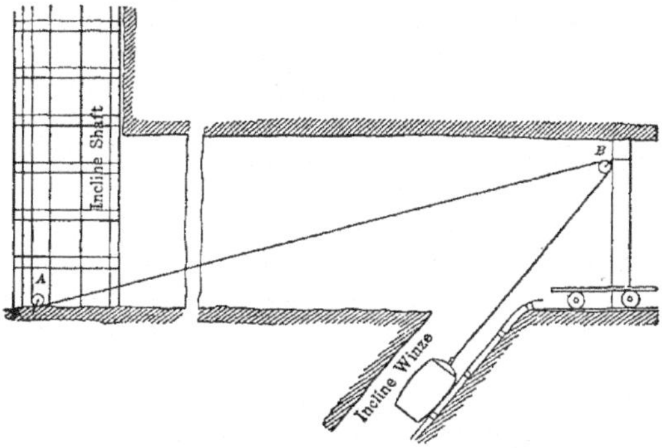

FIG. 128.—SNATCH BLOCK APPLIED TO MINE HOISTING.

taken out from an oreshoot or chimney close to the Gipsy shaft, and as the work got away from the shaft a small electric hoist was installed. This was found to be an expensive arrangement, especially as the electric drill, on account of which the dynamo was primarily installed, proved unsatisfactory, and the dynamo had to be driven for this hoist alone. This meant an extra man at $4 per day to drive the dynamo, also the hoisting engineer below and the hoisting engineer at the top of the Gypsy shaft, each of whom got $4 per day.

To replace this, two 12-in. snatch blocks were placed at the 450-ft. station—one in the floor of station at *A*, as shown in Fig. 128, close to the shaft, and one at *B* hung from a well braced stull in line with the winze. The bucket was hoisted from the winze and lowered to a truck on the station level. The hoist rope was run out to get slack and taken off the snatch blocks; the slack was again taken up and the truck run to the shaft. The snatch blocks were placed, as shown in the diagram, to allow the main hoisting rope being carried around the snatch blocks and down the winze. This arrangement could be worked rapidly and lessened the number of men, cutting out the high-priced engineer at the May Day shaft and replacing the $4 man at the 450 level with a $3 man, whose duty it was to manage the hoisting from the winze and the placing of the hoisting rope around the snatch blocks. Frequently also during a shift this man was able to get out several hundred pounds of high-grade ore near the shaft. It might seem that there would be danger of overwinding and pulling out a snatch block, considering that the bucket had to be stopped within 8 in. of a given point on the 450 level, and that the hoisting engineer had to rely largely on a bell signal. Again, it was difficult to maintain a mark on the rope for the engineer's guidance. However, no trouble was experienced. While this cannot be considered an ideal arrangement for mining a large deposit, still it worked satisfactorily for prospecting more than 100 ft. below the level of the snatch-block station.

A Simple Form of Lift.—A coal lift used at some of the steam plants on the Mesabi range is shown in Fig. 129. The coal is dumped on the ground outside

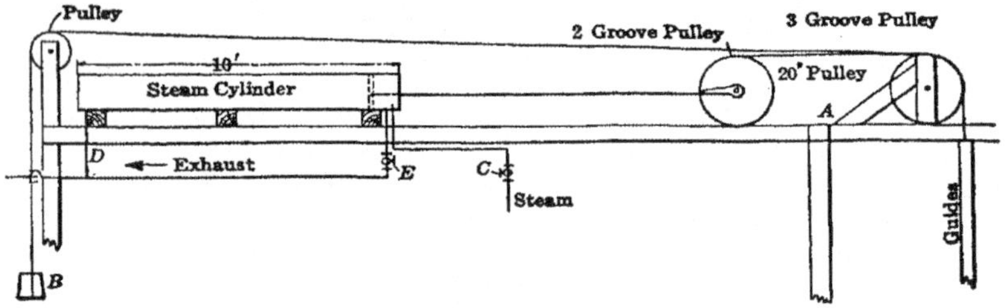

FIG. 129.—SKETCH SHOWING PISTON ARRANGEMENT FOR COAL LIFT.

the boiler house. It is then loaded by hand into 1-ton cars and trammed to this lift and elevated to the bunkers. In most of the newer plants, where it is possible, the coal is discharged directly from the railroad cars to the bunkers, thus saving the extra handling with the lift. The device is operated by a steam cylinder about 10 ft. long by 12 or 14 in. in diameter. At the end of the piston rod is a double-grooved sheave over which two 3/4-in. cables operate. One end of these cables is fastened at *A*, so that in this way when the piston moves 10 ft. it will lift the car 20 ft. The car platform works between guides and is balanced by a counter-weight *B*. Steam is turned on at *C*, the exhaust *D* being open, forcing the piston along and lifting the car of coal. To lower the car, shut off the steam and open the exhaust valve *E* and the weight of the car will operate the device by gravity. The area of the piston must be such that the product of the area, steam pressure and distance shall be in excess of the load, multiplied by its distance. If these are equal it gives a balanced system and no movement takes place. The amount of steam consumed is small, simply enough to fill the cylinder. The steam and exhaust valves may be at any convenient place, not necessarily as shown in the diagram.

Cable Drum for Lowering Timber.—One of the best cable drums for letting mine timber and lagging down a shaft is shown in Fig. 130. This

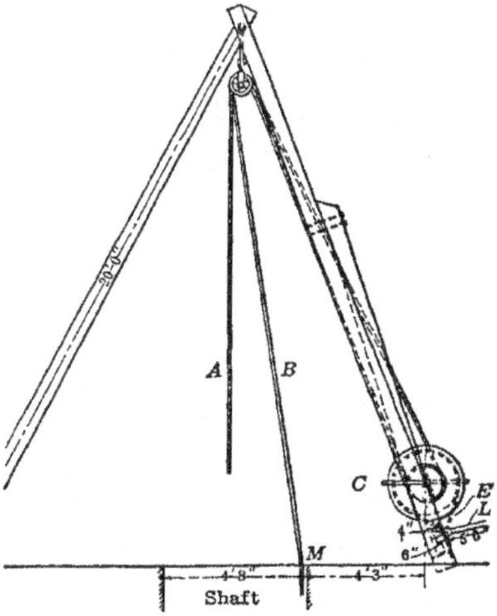

FIG. 130.—DEVICE FOR LOWERING TIMBERS IN IRON MINES.

form is in use by most of the underground mines of northern Minnesota. The timber or lagging is loaded on a small car and pushed to the edge of the shaft. A double slip noose is placed around the timber and the rear end of the car raised up so the load drops into the shaft. It is allowed to drop slowly down the shaft by the friction band *E*, controlled by the lever *L*. As the rope *A* is unwinding, the rope *B* is being wound up and is kept to the side of the shaft by a guide on the collar of the shaft. The end of rope *B* reaches the collar as the load strikes bottom; and another load is then attached to rope *B*, which again pulls up the unloaded rope with chain. The drum is of sufficient length to allow for any reasonable length of rope. The larger the load handled more turns of rope are necessary to hold it. A 3/8-in. wire rope is used in most cases and will wear for years. The friction band *E* is made of strap iron 4 in. wide and 1/4 in. thick. The shafts are lined with plank placed vertically, so that the bundle of moving timbers does not catch in the shaft timbering.

A Portable Winch.—A portable winch is an extremely useful piece of machinery at any mining operation. At the Republic mine, Republic, Mich., an ordinary hand winch is mounted upon a heavy frame which in turn is mounted upon trucks for a standard-gage track. A 7 1/2-h.p. electric motor is also mounted on the same frame and connected by belt to the pinion shaft which operates the drum. A friction clutch is used to throw the drum in gear. This winch can be moved to any point where there is a car track and is easily anchored by fastening to the rails, or by means of chains to stakes in the ground. Where electric power is available, this arrangement is quite satisfactory, as power can be obtained from any point along the line. The entire apparatus is not so heavy but that it can be moved over smooth ground without the aid of rails. This one is used where a temporary hoist is required, and also in the erection of trestles for car tracks on stock piles.

Combination Timber Hoist and Winch.—The design of a combination drum for lowering mine timbers and a winch for hoisting is shown in Fig.

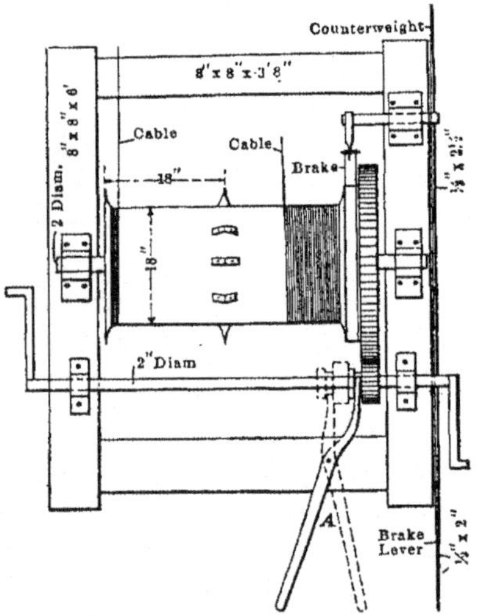

FIG. 131.—TIMBER HOIST AT HEMATITE MINE, ISHPEMING, MICH.

131. The apparatus here described is used at the Hematite mine, Ishpeming, Mich. The drum is 18 in. in diameter, 3 ft. long and is mounted upon a heavy frame of 8×8-in. timbers as shown. On one end of the drum is a brake wheel and band, also a cog wheel into which a small pinion meshes. This pinion may be thrown out by means of a lever *A*, and the timbers lowered by the use of the band brake only. The drum is divided into two sections, upon

which are placed two cables. As one cable is run out with the lowering of the timber, the other cable is being wound up ready to receive a second load of timbers. In the event any of the timbers are too heavy for the brake to control their descent, the pinion may be thrown in and the crank employed. The winch may be used in hoisting pieces of machinery.

Interchangeable Arrangement for Steam and Electric Hoist.—At the Gold Cliff mine of the Utica company at Angels Camp, Calif., the hoisting engine is simply arranged for the use of either electric or steam power. The hoist was originally built for steam power, but it is more economical now to use electricity as a motive power, so it has been rigged for direct connection

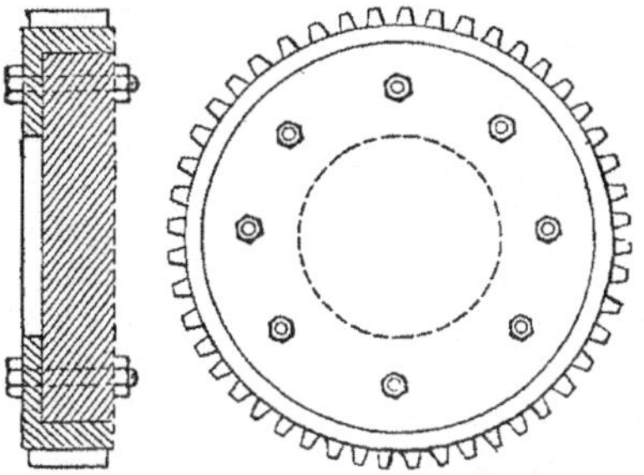

FIG. 132.—INTERCHANGEABLE ARRANGEMENT FOR STEAM OR ELECTRIC HOIST.

to a motor. When electric power is to be used for driving the engine, the connecting rods to the steam cylinders are taken off and a specially constructed rim with ratchet gearing fastened to the crank, the rim engaging the pinioned drive pulley on the motor. The crank on the engine is a solid wheel. A wheel of larger diameter, the size desired to secure the proper hoisting speed, is turned down so as to fit flush against and partially over the crank, the projecting edge forming a rim or tire about the latter. Both the crank and the auxiliary wheel are drilled for tapered bolts by which they are fastened securely to each other. Fig. 132 illustrates the details of construction. The rim can be slipped over the crank and bolted to it in a few minutes; then, by disengaging the connecting rods on the engine, the hoist is ready for electric driving. A rawhide pinion is used to reduce noise and friction. This arrangement permits a satisfactory interchangeable driving of the hoist without making any serious alteration of the plant.

A Cone Friction for Mine Hoists.—Friction hoists are used at mines because they are usually the simplest and cheapest hoists that can be built. When two drums are mounted on one shaft for hoisting in balance, it is often desirable to work one drum while the other is at rest and one of the methods of transmitting motion from the engine to these drums is by means of a friction gear. The friction hoist finds its widest application in mining where heavy loads are raised while the engine is using steam and light loads are lowered at a speed controlled solely by the brake and friction gear, no steam being used in the engine. Most friction hoists are not built with reversing engines unless occasional heavy loads are to be lowered. Hoisting through winzes where rapid return of the empty bucket is a requirement is a typical use of the friction hoist.

There are two types of friction gear more widely used than others for driving hoisting engines, the band and cone frictions. The band friction consists of two semi-circular bands of wrought iron that carry wooden shoes and which can be tightened to grasp the drum in a manner exactly similar to the operation of a band brake. The cone friction consists of a ring, in shape like the frustrum of a hollow cone, which is bolted to the part of the hoist actuated by the engine through gears. This cone engages a similar larger cone bolted to the drum. Where one such ring is used on the drum the gear is termed single cone friction; if there are two rings, one of which engages the outside and one the inside of the engine cone, the gear is called a double cone friction. The engine cone is therefore male and the drum cone female. Contact is made and broken by shifting the drum laterally upon its shaft. Such lateral movement of the drum varies from 1/32 to 1/8 in. and is effected through a hand lever operating shifting devices that vary in construction in hoists of different manufacture.

In the earlier forms of friction gear two metal surfaces were employed but when any slip occurred the amount of heat developed was excessive. This form was succeeded by that in which one member was made of wood and the other of iron or steel. In the double cone hoists the male cone is made of wood and the female cones of iron as it is easier to replace the male cone when worn out; the wooden cone of course wears more rapidly than the metal. Some manufacturers make the wooden cones so that the friction surface is against the grain of the wood as such a surface wears longer; the surface parallel to the grain gives a better hold and as replacement of the cone is easy some manufacturers prefer to use this surface and change cones more frequently.

Single cones can be adjusted as they wear but this is compensated in part by the greater amount of wear they are subjected to over double cones. The object of making the gears in the form of a truncated hollow cone is to get sufficient surface of contact and reduce the amount of end thrust. The best angle, for the face of these cones, as determined by experiments, is 30°; that angle is best for quick release and brings the least pressure against the mechanism for shifting the drum on its shaft.

An improvement in wooden cones is shown in Fig. 133. The two surfaces of the wooden male cone that engage the iron surfaces of the female cones are bored, staggered as shown. Into the holes cork cylinders, a little larger in diameter than the holes, are forced under pressure so that they bulge above the level of the wooden surface. The convex surfaces of the cork insets are then planed flat but allowed to project about 1/32 in. above the wooden surface of the cone. A peculiar fact noted in the use of these insets is that while both wood and cork surfaces eventually wear the corks always protrude about the same amount until the cone is worn out—due, no doubt, to the squeeze to which the cork is subjected by being forced under pressure into a comparatively small hole.

In the cone with cork insets, the holding power on the drum is increased about 100%. This means that the hoisting engineer has only to exert about one-half the pressure on the lever to hoist the load. The cone withstands the effects of heat better than the plain wooden cone, in fact the cork will not burn so readily as the wood and as the drum does not slip as much as with wood, the cone is more durable. Oil and water do not cause slipping to anything like the extent that they do with a plain wooden surface. The cone also takes hold and lets go more quickly.

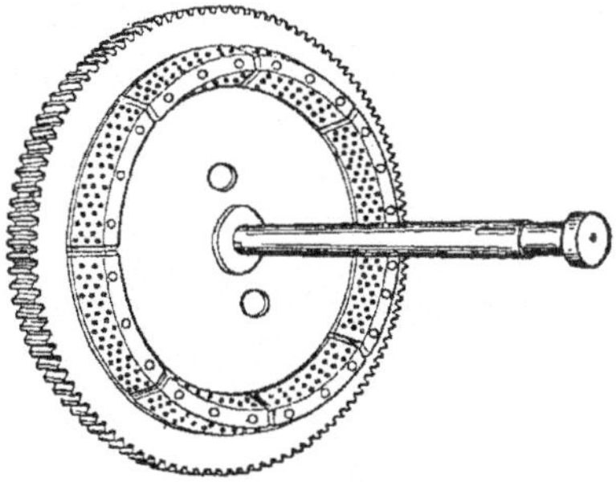

FIG. 133.—A HOIST FRICTION WITH CORK INSETS.

The efficiency of a friction cone has nothing to do with the elasticity of the materials, but with the character of the surfaces in contact. The good results obtained by the use of cork insets seem to indicate that a yielding surface is an advantage, but in the light of more recently developed friction devices this is not proven to be the case in all instances.

Deep Sinking with Gasoline Hoists.—At the Boston & Ely property at Kimberley, Nev., prospecting has been done by sinking shafts far beyond what is generally thought to be the range of gasoline hoists. Of course, it probably would have been more economical to use larger hoists driven by electricity, but the work shows what can be accomplished with gasoline engines by those who understand them. By means of a 15-h.p. Fairbanks-Morse gasoline hoist having a rope speed of 200 ft. per minute, the shaft, which was timbered with 4×4-in. sets, was sunk to a depth of 840 ft., using a bucket and crosshead that gave, with the weight of the rope, a dead load of one ton. This 840 ft. was the limit of the rope and so, when a new rope was ordered, a 40-

h.p. gasoline hoist of the same make was installed, but with the old hoist a speed of 50 ft. per month was attained in sinking. With the 40-h.p. hoist the shaft was sunk to a depth of 1126 ft., it having been enlarged from one and a half to two compartments below the 1100-ft. level. In sinking to that depth a bucket and crosshead were used, but cleaning out broken rock became too slow; it took 4 1/2 minutes to hoist and return a bucket when the shaft was that deep. A cage and car were therefore substituted, and an air-driven hoist installed on the 1100-ft. level. Then by means of a bucket, which dumped into a bin on that level, the shaft was sunk below that point without trouble, as sinking could go on independently of the surface hoist, at least up to the capacity of the bin. From the bin the rock was loaded into a car and hoisted to surface on the cage. This made a gross load of 3500 lb. on the hoisting rope, which was a 5/8-in. Hercules steel rope, weighing 62 1/2 lb. per foot.

Steam Hoists for Shallow Mines (By Sven T. Nelson).—Several years ago, in the iron-ore fields of northern Minnesota, a large number of mining companies were hoisting from comparatively shallow depths with primitive slide-valve engines of the poorest fuel economy. By shallow mines are meant those from 200 to 1200 ft. deep. The average load handled, exclusive of the rope and skips, is 5 tons, and ordinary service requires a speed of 700 to 1000 ft. per minute.

Some 30 years ago corliss engines or engines with an automatic cut-off were just being introduced for hoisting purposes. These new plants were chiefly in the Lake Superior copper district. They consisted largely of regular mill engines, purchased from the corliss engine builders; some were furnished with gear and pinions and some were not. The drums were built up of wood at the mine.

With these exceptions, the hoists then in use were equipped with slide-valve engines of the commonest type. Crude as were the first corliss hoists, and numerous as were the objections and jibes cast at them on account of their "trappy" and "complicated" mechanism, the reduction in fuel consumption which they secured by means of the automatic cut-off was so great, that the slide-valve engine was soon crowded from the field.

The iron companies of the northern Michigan field were also impressed with the fact that corliss hoists did their work on from one-third to half the fuel required for the slide-valve pattern. The first engines installed at any of the iron mines with a detachable valve gear for automatically closing the steam valves were not of the corliss type, but of the same type as an engine that is still built at Fitchburg, Mass., by the Brown Engine Co. A modified type of the Brown engine was adopted by one of the hoisting-engine manufacturers of that time, and several of them installed. These engines were found to be satisfactory and were economical on low steam pressures. However, they did not lend themselves so successfully as the corliss type to the constant increase in steam pressure, which took place with improvements in boiler manufacture.

From this time the corliss engine became the standard for deep hoisting practice throughout the Lake Superior region and the western mining fields as well. By deep mines are meant those ranging from 1000 to 5000 ft. in depth.

With the knowledge and experience gained in designing hoists for deep mines at their command, the engineers attacked the problem of securing hoisting economy for shafts of more moderate depths.

It is out of the question to use first-motion corliss hoists for this work, on account of the limitations in speed of corliss valve gear requiring engines unduly large for the service required. In shafts only a few hundred feet in depth after the load is accelerated, but a few revolutions of the engine will be made with the automatic cut-off in action, so that a direct-acting corliss hoist would be not only needlessly high in first cost, but more extravagant in fuel than a slide-valve engine of the simplest type.

Corliss-geared hoists were tried for some of these shallow mines, but as in the case of first-motion plants, engines disproportionately large had to be furnished, to keep the number of revolutions as low as possible. Difficulty was also incurred in certain fields, in securing engineers who would and could take proper care of a corliss engine; so that mine managers continued to use the old-fashioned, plain slide-valve hoists, with their excessive cost for fuel.

After much study, the type of hoist known as the automatic slide-valve hoist was worked out, and has now been in satisfactory use for several years.

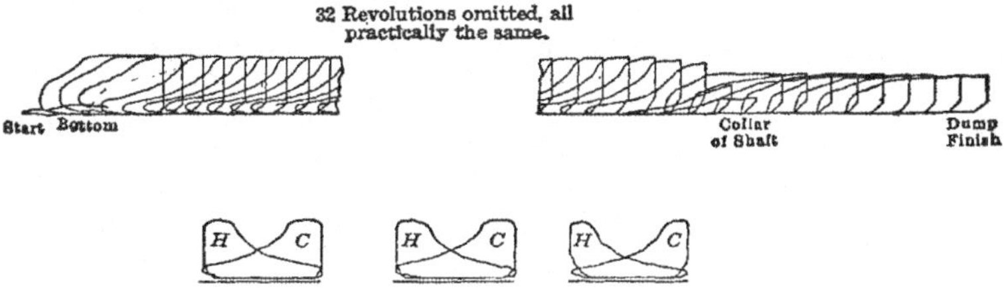

FIG. 134.—CONTINUOUS DIAGRAM FROM AN AUTOMATIC CUTOFF HOISTING ENGINE.

That this design fulfils the conditions will be seen readily from a study of the continuous indicator diagram, shown in Fig. 134, taken from the hoist of this pattern at the Webb mine of the Shenango Furnace Co., at Hibbing, Minn.

The hoist consists of two slide-valve engines each 16×18 in. geared to a single drum 6 ft. in diameter by 6 ft. long. It takes steam at 145 lb. boiler pressure and runs at 140 r.p.m., hoisting a load of 5 tons of ore in addition to the weight of the rope. Two skips are used in balance, so that their weight is offset. The diagram represents the entire trip from the start at the bottom to the dump. Each diagram indicates one revolution of the engine. The gear and pinion ratio is four to one, so that the engines make four revolutions to one of the drum. As there are 57 diagrams the depth of the shaft is 269 ft. and the load is hoisted in 14 revolutions of the drum.

In starting from the bottom, it should be noticed that the first diagram takes steam three-fourths of the stroke, the second about five-eighths of the stroke, and the third about one-third of the stroke; at this point the acceleration is completed. From there on, up to the point referred to as "collar of the shaft," the engine is cutting off at about one-fifth. At this point there is a drop in pressure, as indicated by the diagram, and the length of admission of steam to the cylinder. When this point is reached, the engineer closes the throttle partially to slow up for the dump; this puts the automatic cut-off out of action, in precisely the same manner as the dashpots cease to drop when a corliss engine is being retarded. From the point called "collar of the shaft" to the point referred to as the "dump and finish" the regular slide-valve action is secured, just as would be the case for the entire distance from top to bottom were it not for the automatic cut-off.

The three individual diagrams, below the continuous diagrams, were set aside from the continuous diagram so that users of engines not familiar with continuous diagrams can tell from these the action of the valve gear and the steam distribution. The slightly jagged appearance of the lines is due to the high speed at which the engine was running and a small amount of water in the indicator pipes, which caused the indicator pencil to chatter.

It should be noted that in hoisting engines, hand adjustment of the point of cut-off is out of the question. It would require an engineer's constant attention to give his engine steam for the entire stroke when starting the load, to set the valves at the proper point of cut-off when the load is under full motion, and to lengthen the cut-off again, at the end of the trip. This is obviously impossible, nor would it be possible for the engineer to set the cut-off at its most economical point each time, owing to variations in steam pressure and load.

The engines of the Webb mine hoist and of others of this type, are of the plain, double, slide-valve pattern, and the valve gear places no limit on the speed at which they can be run. The mechanism controlling the automatic cut-off is so arranged that no extra thought or action is required of the engineer. When the throttle lever is pulled, the first 2 or 3 in. of its movement opens the main throttles, admitting steam during the entire stroke to start the load from the bottom, as shown by the first cards. As the lever is pulled back, it admits steam to an auxiliary valve mechanism and cylinder. The piston of this cylinder actuates a crank and shaft which in turn moves a vertical rack at the rear end of each cylinder. These racks engage pinions, one at the outer end of each cut-off valve stem. The admission of steam to the auxiliary cylinder therefore automatically places the main valves in the position of shortest cut-off, and this action is shown in the cards as above described. At the end of the trip, the reversal of the lever, to close the main throttles, admits steam to the opposite side of the piston in the auxiliary cylinder, and the cut-off is restored to its first position.

The hoists are built with drums ranging from 6 to 8 ft. in diameter, and with engines from 14×14 in. to 16×24 in., adapted for steam at 150 lb. pressure.

The hoisting speeds range from 700 to 1000 ft. per minute and the loads from 5 to 7 tons.

MISCELLANEOUS DEVICES

Sheave Supports for Underground Hoists.—At the Red Jacket shaft, it is necessary to work the lower part of the Calumet & Hecla company's ground by means of a blind shaft or winze that starts from the 5700-ft. level of the Calumet No. 2 shaft. There an electric hoist is installed which is controlled

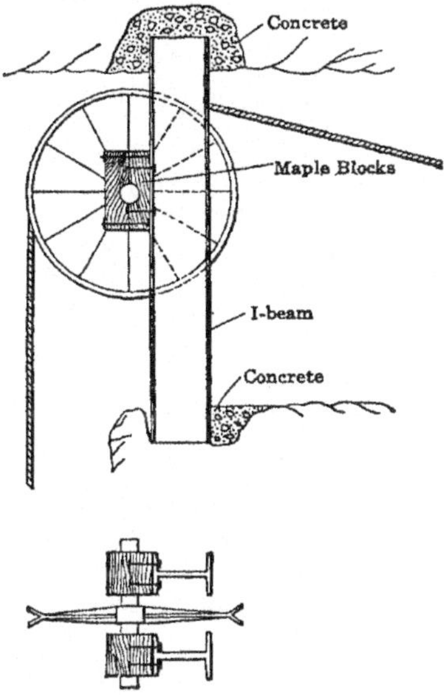

FIG. 135.—SHEAVE SUPPORT IN SHIFTING GROUND.

by the Ward Leonard system of wiring. The ground around the station has a tendency to move and so it is necessary to mount the sheaves so that the settling of the ground will not cause trouble. This is done as shown in Fig. 135. Two I-beams are put together as posts and anchored in hitches so that they will stand the strain coming upon them. To these posts are bolted maple blocks, long enough to extend beyond the steel posts and take the side thrust of the sheave. The sheave shaft with the sheave wheel loosely mounted on it is then bolted tightly in its seat in the maple blocks; maple blocks are also used as cap pieces. In this way of supporting a sheave there is only one babbitted bearing to maintain, while the main feature is that, in case either post should move relatively to the other, the wooden blocks will adjust themselves to the change and any serious trouble be promptly remedied. In case the movement is great, new blocks can be put in cheaply and the shaft lined up again. If two babbitted bearings are used in this ground that has a tendency to settle and move, endless trouble arises.

Arrangement of Sheaves at the Tobin Mine.—In Fig. 136 is shown the arrangement of sheave wheels that had to be adopted at the Tobin mine, Crystal Falls, Mich., in order to use the hoisting engine without turning it around. The old shaft was in ground that will eventually be mined and the new shaft was sunk 1000 ft. deep in the foot wall and back of the engine room. The drums are 8 ft. in diameter, 8 ft. long and operated by a Nordberg engine.

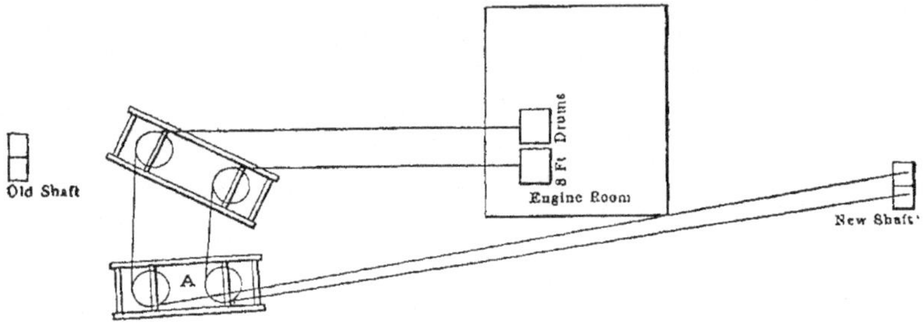

FIG. 136.—ARRANGEMENT OF HOISTING PLANT AT THE TOBIN MINE.

The sheave wheels which are anchored in front of the engine room are 8 ft. in diameter and held in place by 12-in. framed timbers weighted down with rock. While these additional wheels add to the friction losses, no trouble has been experienced with this plan. Three-ton skips are hoisted from the bottom of the shaft in 30 sec. The wheels at *A* are inclined, to conform with the slope of the rope to the top of the shaft house.

Rope Guard for Idler.—Trouble is often experienced in keeping hoisting ropes on their idler wheels, especially where the angle to the sheave wheel is very great. This can be overcome by means of the simple device shown in Fig. 137 which is in use in many mining districts. The wheel *B* slides horizontally on its shaft to allow for the wind on the hoisting drum. Unless there is ample pressure exerted on the idler wheel by the hoisting rope, it will jump off; and especially is this so when there is any slack in the rope. Ropes are supposed to keep in place by the weight of the rope on the wheel, the idler wheels being placed in a straight line with the sheave wheel and the drum or, if anything, slightly above that line. In using this device it should be placed in a straight line with the drum and the sheave wheel. The slotted wheels, *C* and *D*, are held in place by two straps of iron, one placed on each side of the idler wheel *B*, as in sketch. The bearing edge of the upper wheels should be about 3/8 in. above the hoisting rope.

Rope Idlers for Inclined Shafts.—In the conglomerate shafts of the Calumet & Hecla company where, owing to the necessity of not cutting into the foot nor the hanging wall, the shaft must follow the lode in all its changes of dip, the rope idlers or rollers are subjected to severe wear, and some of them have to be replaced once per shift, while the average life of an idler in the whole shaft is not over a week. It is important, therefore, to make the idlers in a cheap and simple way. They are maple logs turned down to a diameter of 16 in. and a face of 24 in., in which three grooves are turned along which the cable may run as the idler is shifted in the shaft. Through the center

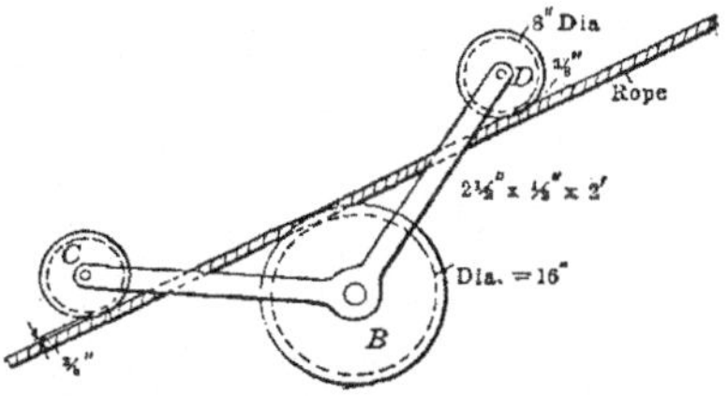

FIG. 137.—IDLER WHEELS FOR HOISTING ROPES.

of the idler a 1 1/2-in. hole is bored for the axle. Formerly an iron pipe was inserted in the idlers for a thimble and the idler revolved on a fixed axle. This arrangement is the cheaper and is good enough for shallow shafts where the hoisting speed is not great, but there is too much friction to use such idlers with high rope-speeds as it is impossible to grease them properly. The idlers now generally used have a fixed axle that rests in two bearings on the idler frame which is wedged in place in the shaft between the cedar ties of the

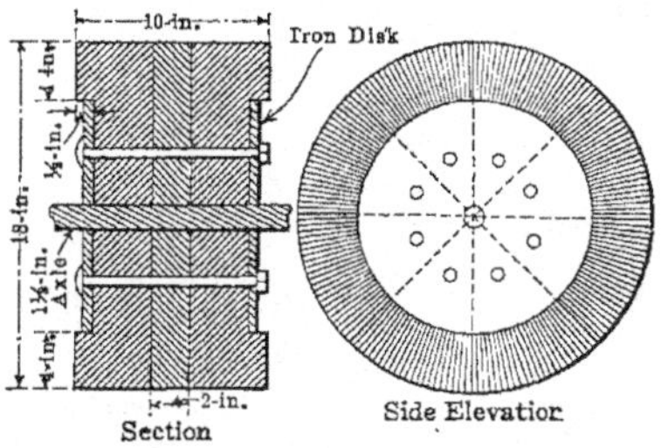

FIG. 138.—IDLER FOR SKIP ROPE IN SHAFTS.

skip tracks. These frames are set in the shaft so that the rope passes through the right-hand groove of one idler, the center groove of the next and the left-hand groove of the next. By changing the idlers from one frame to another they may be completely worn out.

This type of idler is used in the amygdaloid mines of the district, except at the shafts of the Osceola Consolidated, where a built-up idler is used. This idler, as the amygdaloid shafts are sunk at a regular dip, does not have to be made so wide to insure the rope staying on the idler, while as the wear is not so great as in the conglomerate shafts, the idler part is built up of segments so that the wear will come end-on instead of sidewise with the fiber of the wood. These built-up idlers are 18 in. diameter and 10 in. wide. There are three layers of octant segments held together by flanges and bolts through the apex of each, as shown in Fig. 138. The outer segments are sawed from 4×6-in. maple or beech planks, being cut so that the grain runs approximately length-

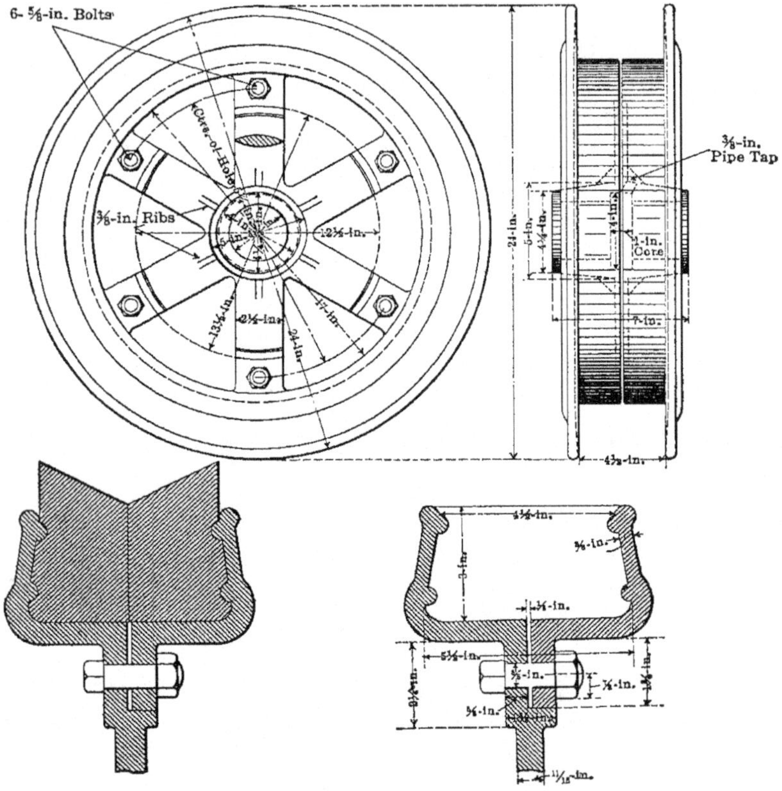

FIG. 139.—IDLER FOR HOISTING ROPES USED AT CHAMPION MINE.

wise, while the inside segment, the one that takes the wear, is sawed from a 2×6-in. plank of the same material. It is, therefore, possible to replace the segments without throwing away the whole idler when only a few of the segments in a layer are worn. The outer segments are made with a shoulder 1/2 in. deep turned in them to receive the flanges so that the bolts that run through the idler will not extend beyond the sides to wear and cut the rope in case it should slip off the idler. These idlers have been in use several years. They last much longer than do the other types, and as they are lighter, they get up to speed when the rope comes on them quicker than do the others. In the conglomerate mines they cannot be used, because the wear is so great that it does not pay to try to increase the life of the idlers by complicating the design.

Idler for Hoisting Rope in Inclines.—In the shafts of the Copper Range company in Michigan which have a dip of about 70°, the hoisting ropes are carried on wood-lined idler sheaves at intervals of about 33 ft. The idler, as shown in Fig. 139, is made of two malleable-iron castings bolted together to grip the wood lining pieces in the jaws of the wheel. On the jaws are cast two beadings for cutting into the wood and holding it tightly. The wood

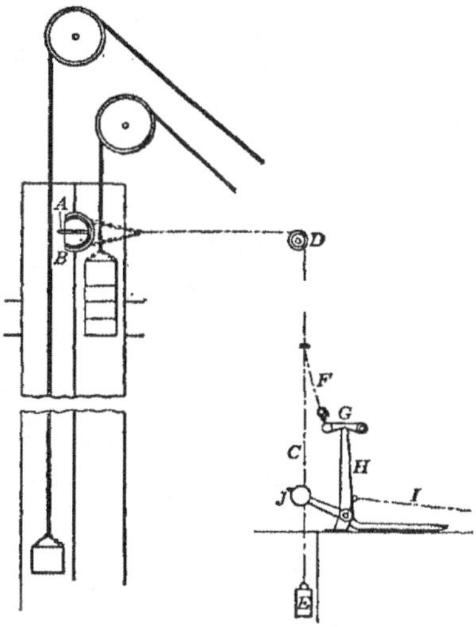

FIG. 140.—DEVICE TO PREVENT OVERWINDING.

filling pieces, which are sawed so that they take the wear on the ends of the fibers on the wood, are in two pieces arranged so as to stagger the joints between the segments of each half of the lining. These segments are cut so as to have a chord of about 6 in. across their outer face. The wood pieces should be sufficiently thick so that when the nuts are brought tightly home on the bolts the two pieces of the frame will not quite touch. Then the bolts can be tightened so that the pieces of wood will be held securely even if they shrink. Owing to the fact that the hoisting cable does not have much side play in these shafts, a face of 4 1/2 in. on the idlers is sufficiently wide. The rim of the casting is much wider than the spokes so the bolts that hold the frame pieces together are well within the protection of the rim and, in case the rope should slip off the idler, it would not come in contact with the nuts on the clamping bolts. The bearings for the idlers rest on I-beams carried on concrete pedestals from the bottom of the shaft, and the whole frame is in turn protected from injury by the skip by means of wooden buffer pieces running along over the main I-beams and in turn carried on I-beam crosspieces.

Device for Prevention of Overwinding.—The device shown in Fig. 140 is installed at a German shaft to prevent overwinding. At a point about 30 cm. above the highest normal position of the cage is pivoted an axle, carrying at its center a lever *A*, which projects out into both hoisting compartments, and at its end the pulley *B*, to which a length of chain is attached at two points, as shown. The middle point of this chain is connected to a wire rope which, after passing over the pulley *D*, hangs down a suitable distance and is kept taut by the weight *E*. A short piece of rope *F* connects the first rope with an end of the latch *G*, which engages the top of the lever *H*. The rope *I* attached to *H*, leads to the throttle valve of the engine, and also to the valve of a steam-actuated brake. A slight upward pressure against either end of the lever *A* thus lifts the latch *G*, allowing the weight J to act through the rope *I* on the engine. The same effect could be produced electrically, though possibly not with equal certainty, by establishing contact through the guide shoes of the cage, and providing a magnetic release for the latch *G*.

Device for Cleaning Flat Wire Cables (By M. J. McGill).—An arrangement that I have used successfully for cleaning hardened grease, dirt and rust from flat wire cables is illustrated in Fig. 141. The contrivance consists of a box made in four parts to be hooked and stapled together, in which is fastened a U-shaped steam fitting through which the cable is passed. The bottom of the box is first placed on a platform above the collar of the shaft, the cage or skip being lowered just far enough so as to clear all the cable fastenings. A fitting *B*, made as shown in the drawing, with two 1-in. pipe legs with 1/16-in. slots and capped ends, is placed with one leg on each side of the cable, the slots being turned a trifle downward. The top parts of the box are then placed and hooked to the bottom parts. An opening just large enough to admit the fitting, so as to make it as near grease tight as possible, is made in the side of the box. A steam line is coupled to the fitting *B*, and the steam turned on, while the cage is slowly dropped. As the cage drops, the cable passes through the steam fitting, and being subjected to the scrubbing action of live steam is readily cleaned. The time required to clean thoroughly depends on the condition of the cable. A few minutes after applying the steam, grease and dirt will begin rolling over the blocks *C C*, which are placed across the inside of the box to strengthen it and to divert grease toward the nipples at the end of the box, from which it falls into buckets placed to receive it.

In three hours' time I have thoroughly cleaned 1500 ft. of the worst-looking cable that one can imagine. Some cables of equal length that were not in such bad condition only required 1/2 hour. This apparatus has proved superior to any I have seen or read of for thoroughly cleaning flat wire cables. I always lubricate the cable immediately after cleaning.

Mine Signal Switch.—A mine signal switch designed by A. H. MacGregor, Palatka, Mich., is shown in Fig. 142. The principal feature in this switch is that it is strong and not likely to get out of order as does a more delicate one under the rough usage of the miner. The parts are mounted upon a hardwood board, 1×8×16 in., and inclosed within a box as shown by the dotted line. The switch lever is of 3/16×1 1/4-in. steel upon which a copper contact is soldered. The lever is pivoted at *B* and is held in position by a bar *A*, which prevents any side movement. The copper bar *C* is in contact with the lever at

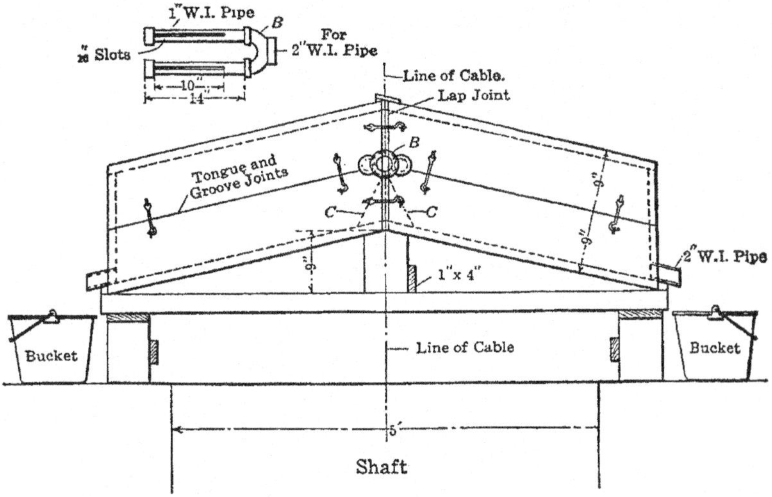

FIG. 141.—CLEANING DEVICE FOR FLAT-WIRE CABLES.

all times. The circuit is completed with the contact *C'*. A No. 10 tension spring breaks the circuit as soon as the operator releases the handle *D*. The switch is placed about elbow high so that it requires some effort to operate it. In this way the contacts are positive and distinct, and there is no fluttering as is the case when a switch is in such a position that it can be operated rapidly. The board and handle *D* are covered with an insulating paint. The device has been in use over a year and is now installed at several of the Pickands-Mather mines.

Another type of spring switch used in connection with an electric signal system is shown in Fig. 143. It is built in a metal box 6×10×5 in. deep with a hinged door on one side. When giving signals the operator catches hold of the wooden handle that hangs below the box, pulls down, and the spring in the box breaks the circuit as soon as the operator releases his hold. The signal system, which is used in connection with this switch, flashes a light in addition to ringing the gong.

Electric Signals for Underground Tramways (By W. S. Grether).—Fluorspar associated with galena occurs in a vein near Rosiclare, on the Ohio

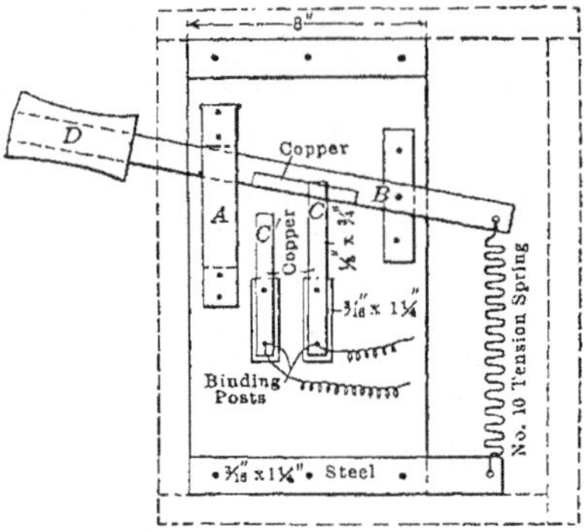

FIG. 142.—SIGNAL SWITCH AT BALTIC MINE, MICH.

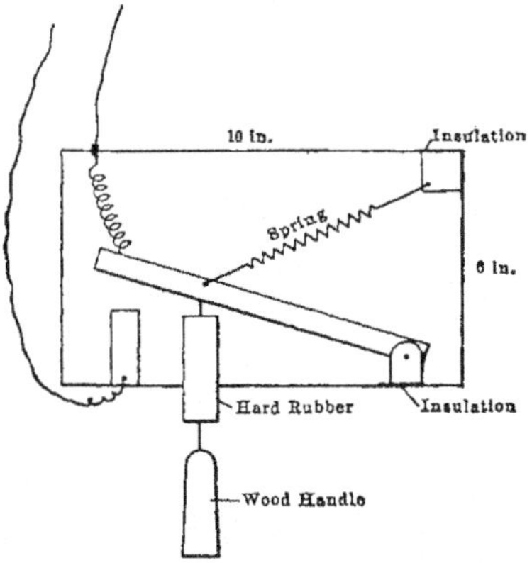

FIG. 143.—SPRING SWITCH FOR ELECTRIC MINE SIGNAL.

River, in southern Illinois. The vein has been prospected for a length of 2 or 3 miles and to a depth of 500 ft.; it is from 3 to 20 ft. wide. In mining the ore is cleanly broken from the foot and hanging walls and not more than 10% of the ground broken is waste. The drifts are tortuous and not of uniform width which makes electric haulage impracticable on the 235-ft. or main working level.

The ore is mined by overhand stoping and the 1-ton tram cars are loaded from wooden bins situated 20 ft. apart along the single track in the drift. The cars are moved by men to the three-compartment shaft sunk midway between the ends of the vein; mules are used to pull the empties, 10 cars per trip, back to the loading bins The excessive grade of the drift, 1 1/2%, prohibits hauling the loaded cars with mules. Ore is trammed from either end of the mine, one end being called the south, the other the north workings. The caging at the shaft is often delayed on account of wrecks, lowering timbers, etc. The trammers in the ends of the workings perhaps 2000 ft. away are not aware of these conditions and formerly continued to bring cars to the shaft, making matters worse on account of the limited switching facilities near the old station. To overcome this difficulty a simple electric signal system was installed.

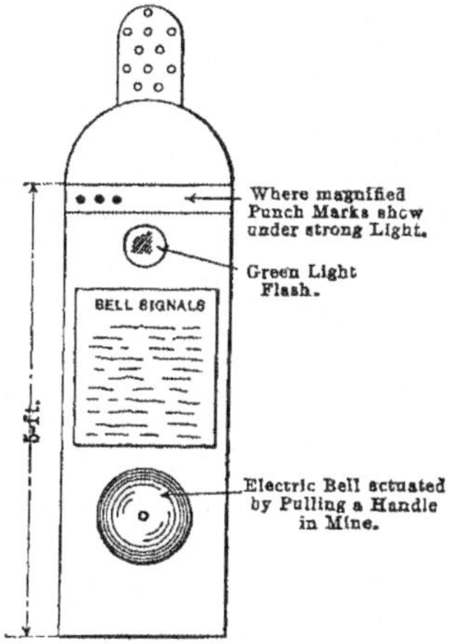

FIG. 144.—A GERMAN SIGNALING DEVICE.

A switch and two 110-volt, 16-c.p. lamps, one green and one white, are placed on the south side of the shaft, and similar equipment is installed in the south workings near the loading bins. Similarly, switches and red and white lights are placed on the north side of the shaft and in the north workings. All switches are left open except when signals are being sent. When the cager wants cars at the shaft he closes the switch three times. The trammer, when ready to move cars to the shaft, answers with three signals. Likewise when the trammer wishes to push cars to shaft, he signals three flashes to the cager; if the cager is ready he flashes three in return. In this way blockades at the shaft are avoided and the trammer may utilize his time while waiting for signals in oiling cars or cleaning the track.

An Electric Signal Device (By P. B. McDonald).—Two German signal devices for hoisting, the general scheme of operation of which is shown in Fig. 144, have been installed in the Negaunee mine hoist house. When a man in the mine pulls the signal handle, two things happen in the engine room. The electric bell rings and a strip of paper resembling a stock exchange ticker tape is punched with a hole. By means of a strong magnifying device and mirrors, a magnified reflection of the small hole punched in the strip of paper appears in a horizontal aperture. Thus if the hoist engineer is not sure of the number of bells which were rung, a glance at the black circles tells him, also the strip of paper serves as a permanent record in case a dispute arises as to the signals that were given. In addition to these electric signal devices the old style mechanical bell will probably be installed, for use when the electrical apparatus is out of order. The green light flashes only when the paper strip of the punch-mark mechanism is exhausted.

Hand Bell Signal Wiring (By Guy C. Stoltz).—The disadvantages of hand signaling are: The difficulties presented in counterbalancing the long line

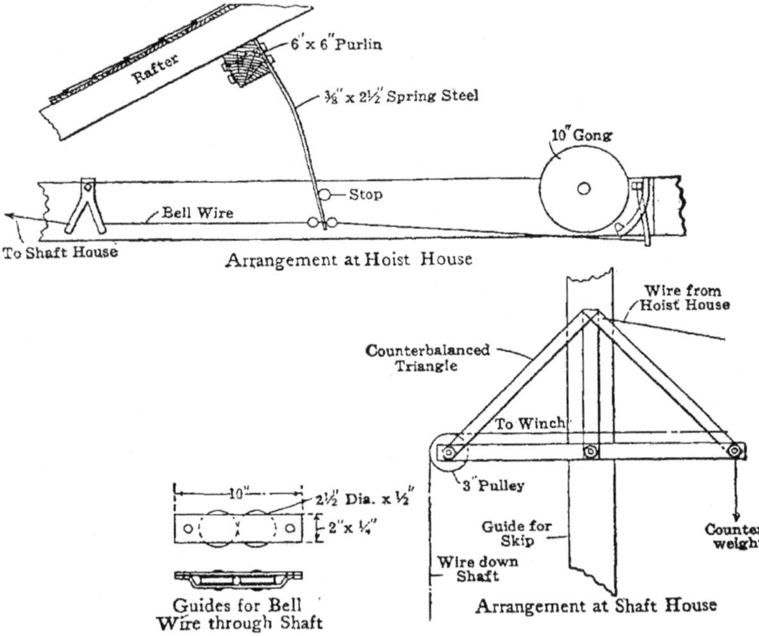

FIG. 145.—ARRANGEMENT OF SIGNAL-BELL WIRING AT PORT HENRY, N. Y.

of bell wire necessary to reach the lowest levels in the shaft; in guiding the wire through the shaft over the several angles to the stations and to the hoist house with the least friction; and in keeping the system taut to eliminate all possible lost motion. The system installed must be positive and hand ringing should accompany every electric bell or light-signaling installation. In Fig. 145 a satisfactory method of rigging is shown. The strap of spring steel introduced before the gong does away with any lost motion to the gong and keeps the wire taut to the counterbalanced triangle at the headframe. Here the wire is kept on a winch and the necessary length is guided over a small pulley to prevent kinks, and down the shaft by cranking the winch. A grip is clamped to the wire after the required length has been unreeled, and this, bearing against the triangle, makes the wire fast. In this way the wire is kept one length with no splicing and as longer wire is required on sinking, the winch has a supply. Hand-holds for signaling are clamped to the wire at each station. The wire is guided through the shaft by passing between sets of 2 1/2-in. pulleys placed in an iron frame. This frame is secured to the shaft timbers. The counterweight is attached to the triangle in the headframe and is varied in amount as required.

AERIAL TRAMWAYS

The Solution of a Cableway Hoist Problem.—At Mine 21, Mineville, N. Y., the greater part of the ore in the open pit was hoisted and carried to the gondola cars on the surface by means of a Lidgerwood traveling suspension cableway. The towers on opposite sides of the pit were 400 ft. apart, the greatest hoisting depth 300 ft. vertically, and the carriage was usually carried out about 200 ft. where the bucket was dumped by hand. Some time after installation and as hoisting depth increased the scoop on being raised began to rotate in mid air, twisting the fall block cables and throwing out most of the ore. The problem as to how to guide the scoop in its upward journey required some little attention. Many schemes were suggested and many were tried. A leading rope manufacturer recommended trying his non-twist rope. This was given a trial but did not help matters in the least. The master mechanic finally devised the following scheme. An 18-in. arm was attached to the fall block frame and from this a 5/16-in. wire rope was run over two 12-in. sheaves (spaced 10 ft. apart center to center and hung by a frame to the traveling cable) and then to an overhanging light counterweight. The rope and counterweight proved to be a satisfactory guide. In fact the rope alone was sufficient to steady the scoop during the ascent, the only part played by the counterweight being to keep the small rope taut so it would travel well over the auxiliary sheaves. Besides preventing the twisting of the block this arrangement also serves, to some extent, to prevent the load from swinging.

Turning Device for Tramway Track Cables.—The companies that erect tramways instruct that the track cable be frequently turned so as to equalize the wear, but this has proved to be a direction easier to give than execute. For instance, at the United States tramway at Bingham, the tramway men tried, without success, for over a year to turn the cable.

The directions usually given by manufacturers are for twisting the cable by means of stilson wrenches. Sections of the cable can easily be turned, but the difficulty is to make the cable stay in the new position, for if not held it gradually works back to the old position. To obviate this difficulty, Joseph Ruttle, foreman of the Highland Boy tramway, Bingham, Utah, has devised a method of turning and holding the cable that is certain in its operation. The device for accomplishing this, known as the Ruttle turning strap, has been in use some time, and it is probably as much due to its use as to any other one cause that the old Highland Boy tramway was noted for the long life of its track cables.

The turning strap consists of an iron strap 2 1/2 in. wide, made of No. 12 band steel that is clamped to the track cable by means of two T-head bolts, which have their flat heads turned toward the passing buckets. This band steel is continued to form an arm 12 in. long, and then a 3/4-in. round rod is bolted to the end of this arm between two nuts working on a right- and left-handed threads. In order to prevent the outer bolt from working off and allowing the arm of the clamping strap to swing around and catch on the bucket, a cotter pin is inserted in a hole drilled through the end of the rod. This rod is made long enough to pass through a detaining brace, or loop, which is made by bending double a 3/4-in. round rod. The iron loop is just wide enough for the arm of the turning clamp to move freely back and forth, with the stretch of the cable, and is made 3 ft. long, so as to provide for that much stretch. The detaining brace, or loop, is fastened by means of two 3/8×4-in. lag screws to the timbers of the tower, the rod being flattened to 3/8 in. where it comes in contact with the tower timbers. The turning straps are put on the track cable at each tower.

Whenever it is observed that the track cable is wearing, or about once in two weeks, the cable is turned one-eighth way around by means of stilson wrenches, the clamping bolts on the turning clamps having been previously loosened. Then the clamps are again tightened on the cable. Needless to say, the twisting must be done in the direction of the twist of the cable, or else the strands will be unlaid.

Cable Clamp for Tramway (By Claude T. Rice).—Tramway cables have to be frequently stretched, especially in the early period of the operation of a tramway. This is a troublesome task, owing to the design of the cable clamps in common use. The clamp furnished by the tramway companies is made of two iron plates, with a shallow central groove to increase the contact and consequently the friction of a cable when the clamp is tightened. At the end, the two plates are given an outward bend, so as to keep apart the ropes that are fastened to the plates of the clamp. Two different strands of the block and tackle system are attached to the twisted lengths in the ends of the two clamping plates. Thus there is little if any clamping effect obtained from the pull of the ropes, the entire pressure of the cable being obtained from the tightening of the six bolts that hold the plates together.

On the traction cable this design of clamp holds fairly well, but with the

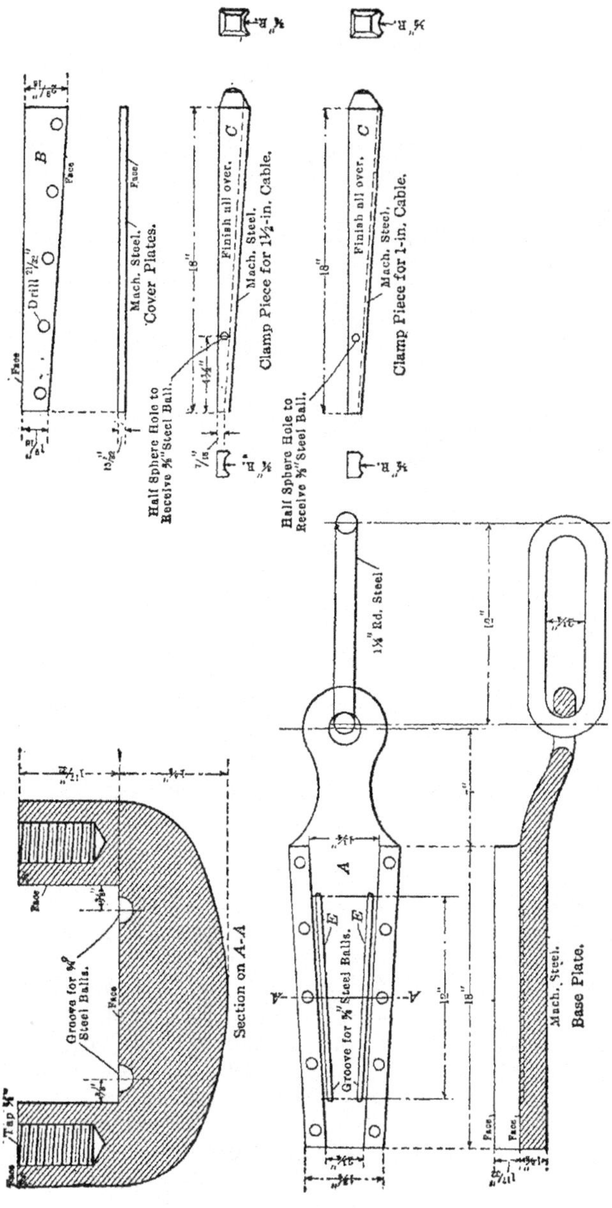

FIG. 146.—DETAILS OF THE RUTTLE CLAMP FOR TRAMWAY CABLES

heavier track cables it does not give satisfaction. Joseph Ruttle, foreman of the Highland Boy tramway at Bingham, Utah, has devised the clamp that is now in general use at Bingham. This clamp is simple in design, effective in its operation, easy to put on the cable, even by one man, and can be cheaply made in any machine shop. This clamp is caused to grip by the pull of the tightening tackles and so is correct in principle. The tramway man has only to give the grip pieces a slight blow to cause an initial grip on the cable. In case the clamp should be put on at a point where the cable is larger than elsewhere, and there should be any slipping, the clamp adapts itself to the diminishing diameter of the cable and grasps the cable more securely than ever. This device is unpatented and has been in use on the Bingham tramway for at least a year.

The Ruttle clamp consists of a base plate *A*, referring to Fig. 146, with raised sides, to which are bolted two cover plates *B* that partly cover the top of the base plate. The sides of the base plate taper toward the rear end, so, as the clamping pieces *C* move along the channels formed by the base plates, they are forced nearer together and grip the cable more securely. The base plate has a curved neck, to which the blocks are attached by a ring. The clamping pieces have the face toward the cable turned to the diameter of the cable on which the clamp is to be used, but by having other sets of grips the clamp can be quickly changed so as to be used on larger or smaller cables. The faces of the grips are segments of a cylinder, so they adapt themselves to the cable when it is considerably worn.

In order to hold the grip pieces in the channels in which they travel, a hemispherical pocket is bored in the grip piece to receive a 3/8-in. steel ball, while in the base plate a cylindrical groove is cut under each cover plate, in which the steel ball may travel. This groove is made long so the grip pieces may spread far enough apart at the head end to facilitate placing the clamp on the cable. When the grip piece is in its farthest position at the other end of the groove, its two faces are considerably nearer together than the diameter of the rope. The angle given to the back raised edge of the base plate determines the length of the groove, while the angle of the back edge and length of the base plate itself are determined by the length of gripping surface that is necessary to give a secure hold on the cable.

Oiling Tramway Track Cables.—The machine used to oil the track cables on the Balaklala aërial tramway, at Coram, Calif., consists of a carrier frame arranged to carry an oil tank having a capacity of 10 gallons. This carrier frame is fastened to a set of trolley wheels turned upside down, so as to secure a space between the wheels for the oil jet directly over the track cable. There are two gear wheels, one of which is mounted on the extended axle of one of the trolley wheels and the other secured to the trolley frame and carrying a wheel for the belt running the rotary oil pump which is mounted on the tank. By these gear wheels the speed of the oil pump is regulated.

By means of a cock in the pipe leading to the oiler the supply of oil to the cable is controlled. After a rain the supply is so regulated that enough oil is fed to run down and cover all the cable, while at other times the supply is only enough to cover the upper side. When the valve has been properly regulated to give the right amount of oil, the carrier is gripped to the traction rope and the automatic oiler is sent out over the line. This oiler works satisfactorily at the Balaklala mine where it was designed.

Oiler for Tramway Buckets.—It is necessary to oil the wheels on the carriers on tramways, especially new ones, frequently. Moreover, the amount of oiling necessary seems to bear some relation to the roughness of the topography of the country across which the tramway is built. In the early stages of the life of a tramway two miles long, it is necessary to oil the buckets every trip, and later at least twice a day, while with tramways four miles long the bucket carriers have to be oiled at least every other trip. This oiling can be done at either end of the tramway, but generally it is far more convenient to do the oiling at the loading station, while the bucket is being loaded.

The usual arrangement is to have a tank of oil placed at some convenient point above the loading-track station so that enough fall is given to the oil to insure a quick feeding of it to the carrier wheels. The oil is led through a small rubber tube to the metallic tip used for insertion into the oil holes on

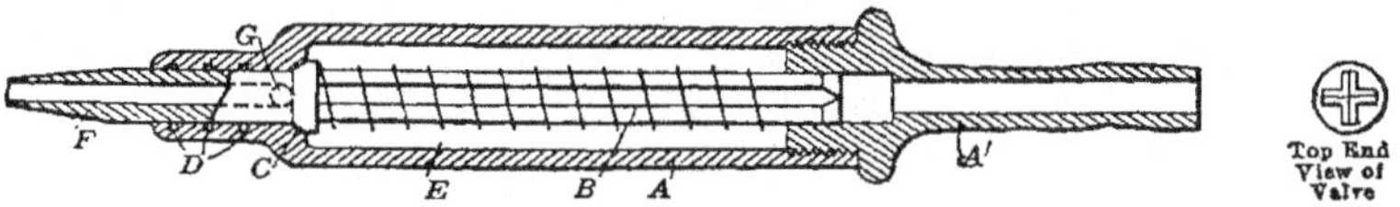

FIG. 147.—NONLEAKING OILER FOR TRAMWAY BUCKETS.

the wheels of the carrier carriage. To shut off the feed of oil, a spring clamp or pinch cock is generally used on the feed tube. Besides being inconvenient this is wasteful, for most of the oil in the tube below the point of clamping drops out while the tip is being transferred from one oil hole to another. This waste of oil is not only a source of needless expense, but it increases the danger from fire, and in time makes the loading station a greasy, sloppy, disagreeable place.

To obviate these objections, W. H. Cole, master mechanic at the Highland Boy mine, Bingham, Utah, devised the oiler shown in Fig. 147. The Cole oiler is attached to a feeding hose just as the tip is in the crude device commonly used. But in this oiler the oil cannot be turned on until the oiler has been inserted to the bottom of the oiling hole, and the feed is shut off the instant that the pressure is taken off the oiler. As a result there is no wasting of oil and the accompanying disagreeable features are eliminated. In addition, the oiling can be done quickly and easily.

The oiler is turned out of two pieces of brass A and A', to form the shell, while the interior mechanism consists of a spring valve *B*, working upon a valve seat *C*. Leak about the valve is prevented by three oil rings *D*, in which no packing is used, as whatever oil may ooze down the valve rod forms its own natural packing at these grooves. The barrel is 1 in. in diameter, 4 in. long and the surface is milled to render it easy to grip. The surface of the piece *A'* that fits into the rubber hose is corrugated to aid in attaching the oiler securely to the feed hose. The valve has a play of 3/8 in., and the stem is fluted, the upper end of the stem being sharpened so as to assist the feed of the oil into the oil chamber *E*, when the valve is opened. The discharge hole in the valve tip *F* is bored to 5/32 in., connection with the oil chamber being through the hole *G*, bored at right angles to the discharge hole.

When the tip of the oiler is pressed against any resisting surface, as the bottom of the oil hole on the carrier wheels, the valve is forced back against

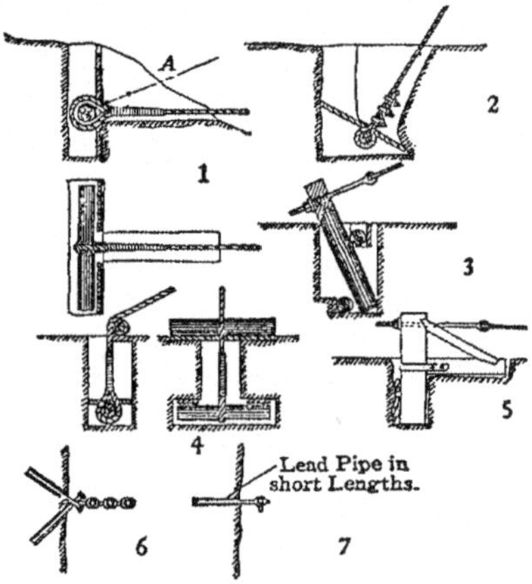

FIG. 148.—WIRE-ROPE ANCHORAGE.

the shoulder, and oil flows from the oil chamber in the oiler through the tip. When the pressure is removed, the spring forces the valve back on its seat and the transverse openings in the body of the valve are closed. This oiler can easily be made at a machine shop at small expense. It has been in use at the Highland Boy mine since the starting of the tramway which conveys the ore to the Tooele smeltery of the International company, and has been found to work admirably.

Anchoring Wire Ropes (By A. Livingstone Oke).—The diagrams shown in Fig. 148 may be found useful to illustrate the methods that may be adopted for anchoring wire ropes. In Fig. 1 is shown in section and plan how to secure the end of the rope in soft soil, such as a gravel bank. A trench is dug in the form of a T, its size depending on the load to be put on the rope. With this arrangement, which is one of the more generally employed methods, the angle of lead of the rope must not exceed that shown by the dotted line at *A*. If a steeper lead becomes necessary the modification shown in Fig. 2 may be used, in which case the trench is undercut as shown, and the planks inserted normally to line of lead. In Fig. 3 is shown a method of anchorage to a post and is to be recommended where some means of tightening the rope has to be allowed for. Fig. 4 shows a method which admits any desired angle of lead to the rope; it consists of a pit with its lower sides undercut to admit inserting the cross timber. Fig. 5 is an adaptation of the method shown in Fig. 3, where the lead is horizontal. Fig. 6 shows a method particularly adapted for more permanent work in mines. Fig. 7 is the better method, the end of the bar being split for a wedge and the whole calked with lead; for vertical upward pull cement may be used.

IX

SKIPS, CAGES, CARS AND BUCKETS

Mine Buckets—Ore Cars and Skips—Mine Cages—Special Carriers—Unloaders.

MINE BUCKETS

Drill-steel Bucket.—A bucket for handling drill steel in stopes can be easily and cheaply made as follows: The essential parts, as shown in Fig.

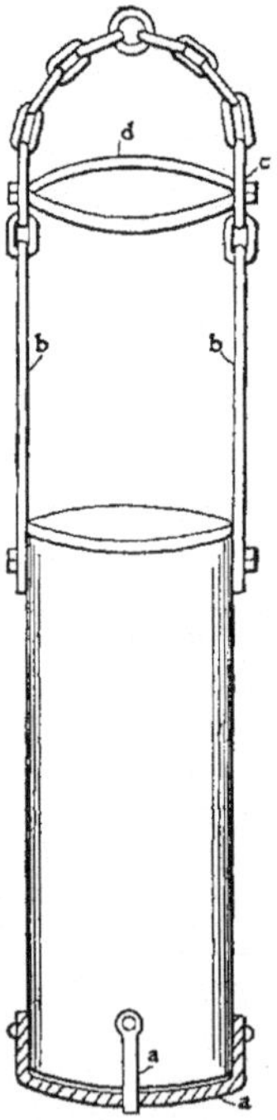

FIG. 149.—BUCKET FOR DRILL STEEL.

149, are the bucket, handle and a ring *d*. The bucket is made of 3/16-in. sheet steel, 30 in. deep and 10 or 12 in. in diameter. The bottom is made of heavier steel, and reinforced by straps *a* riveted over the bottom plate. The ring *d* is of the same diameter as the bucket and is made of 1/2×2-in.

iron or steel. It is fastened by two rivets on each side to pieces *c*, 4 in. long, 1/2×2 in., which form one link in the handle and hold the ring in a horizontal position. The straps *b* are also 1/2×2 in., and are 20 in. long. This, however, should be regulated by the length of steel to be hoisted. The ring is fastened to the long straps by a link, and three or four links above the ring will be sufficient to fasten to the hoisting rope. The bucket may be made light or heavy, according to the work to be done. A bucket of this description is being used by the Vermont Copper Company.

Tram Car for the Prospector (By Guy C. Stoltz).—In the Gowganda district, Ont., where a real tram car is a luxury to the prospector, he has

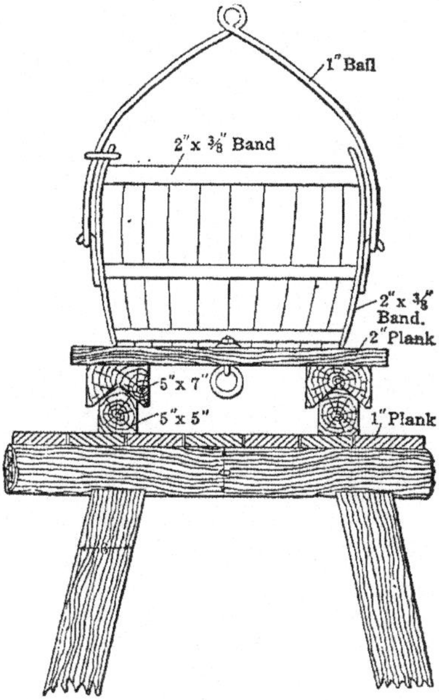

FIG. 150.—BUCKET AND TRAM PLATFORM.

evolved the substitute shown in Fig. 150. The half barrel, strengthened by iron strips, is fitted with bails and used as a bucket. It is hoisted by windlass from the prospect shaft, and at the surface is swung out on a movable platform and detached from the winding rope. The platform, 3 ft. square, is made of 2-in. plank, fitted with notched runners, and rests on the inverted V guide rails, which are spiked to a floor, covering the trestle bents. The windlass men draw the bucket of waste rock to the dump at the end of the trestle by a rope attached to the ring shown on the platform. In summer axle grease and in a freezing temperature water is applied to the runners and guide rails.

The Mineville Ore Bucket.—Buckets used in Mineville, N. Y., for hoisting from an open pit or a large winze present certain advantages over the usual cylindrical type in the manner of loading and dumping. They are of the stone-boat type. The bucket can be drawn to the foot of the stope of ore, where the low sides and large filling area facilitate the loading by hand shovels. Preparatory to hoisting, the chain which is attached by U-bolt *A* (Fig. 151) to the bail is made fast to the lip of the bucket by passing the hook on the free end of the chain through a 3-in. ring *B* welded in the eye of a strap of flat iron which is riveted to the bottom of the bucket. The hook is bent at such an angle that after passing through the eye it can be forced back

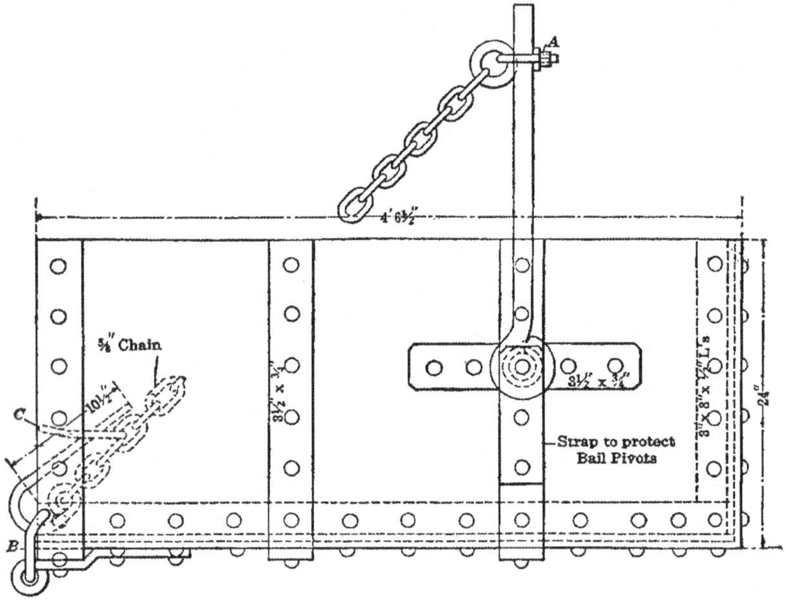

FIG. 151.—IRON-ORE BUCKET USED BY PORT HENRY IRON ORE CO.

against the chain and locked in this position by a slip-ring *C*. The bucket after being locked is hoisted vertically to the top of the pit and then carried out horizontally on the traveling cable to the loading tracks, or in the case of a winze near the top of the headframe the bail is hooked by an auxiliary rope and as the bucket is lowered it is carried out to the dumping chute. By knocking the slip-ring with a shovel, the hook is released and the bucket discharges its contents. The bucket is 24 in. deep, 4 ft. 6 1/2 in. long and weighs 1100 lb. The body is made of 1/2-in. steel plate, with the corners reinforced by 3×3×1/2-in. angle iron.

Joplin Bucket Cars.—With a few exceptions, the hoisting in the lead and zinc mines of southwestern Missouri is done in buckets—"cans" and "tubs," as they are called in the district. This permits the use of narrow-gage track, so that sharp curves can be laid around the pillars and through the "drifts," as the openings between the pillars are called. Because, as a rule, the tubs hold only about 800 lb., 1250 lb. being the maximum capacity, a light track can be used that is cheap to install and quickly and readily changed. In fact, in many of the mines the track, made of 8-lb. rails, is lifted up bodily, ties and all; the ties are usually 2-in. planks. This system is flexible, and admirably adapted to the conditions prevailing in the district. The hoisting system is excellent, and because of the simplicity of the hooking and tramming routine that has been evolved, these tubs can be hoisted at an average rate of one in 35 seconds without great exertion on the part of the men. Indeed, more than 1000 tubs have been hoisted in one shift of 7 1/2 hours by a single engine, and with the hoistman

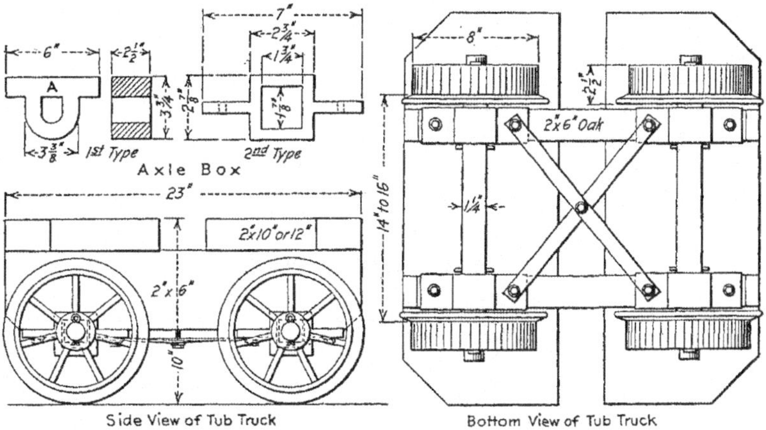

FIG. 152.—BUCKET CARS USED IN THE JOPLIN DISTRICT.

dumping the tubs himself. At some of the mines, cars have been used underground, but it is questionable whether they are more economical. The only advantage of the car over the tub is that it is more difficult for the shovelers to build "windies" in them. In the vernacular of the district, tubs loaded with boulders in such a way as to leave the maximum free space between them, are called "windies."

Many excellent features are embodied in the construction of the tub cars. For instance, the wheels are mounted loosely upon the axles, and the axles are loosely attached to the truck by sleeve bearings, so as to admit of lateral as well as up and down movement of the wheels in respect to the truck. This is an important feature, because of the lightness and temporary nature of much of the track that is used in these mines. By the use of the sleeve attachment, the four wheels of the car stay on the rails, no matter how rough the track is. It is impossible for the truck to run on three wheels at rough places in the track which, with trucks of the ordinary type of axle, is a common cause of derailment. The looseness of the attachment and the open bearing make the use of oil impossible, so cocoa butter is used to lubricate the axle and the sleeve bearings.

There are two types of bearings used on these tub cars, and likewise there are two types of axles. Some of the cheaper trucks are equipped with round axles and sleeves of the first type shown in Fig. 152, but the round axles, which are from 1 1/4 to 1 1/2 in. in diameter, turned down slightly at the ends, frequently break just at the points where the holes are drilled to receive the pins that limit the side motion of the axle in the sleeves. At one mine, instead of using two pins, a piece of old pipe is put over the round axle to form the shoulders that limit the side play, and this pipe is fastened to the axle by a rivet through the axle halfway between the sleeves where the bending strain is least. But probably 90% of the axles in the district are of square section, turned down for 3 in. at each end to receive the wheels. These axles seldom break. The truck proper consists of two 2×6-in. pieces of oak fastened together by two crossed iron straps on the under side at the bearings, while the deck of the truck consists of two 2-in. planks, 10 to 12 in. wide, nailed to the two 2×6-in. pieces. The gage ranges from 14 to 16 in., but the latter is coming to be regarded as the standard.

ORE CARS AND SKIPS

Wooden Ore Car.—A simple type of ore car used by the Sulphur Mining & R. R. Co. at its mine near Mineral, Va., is illustrated in Fig. 153. The

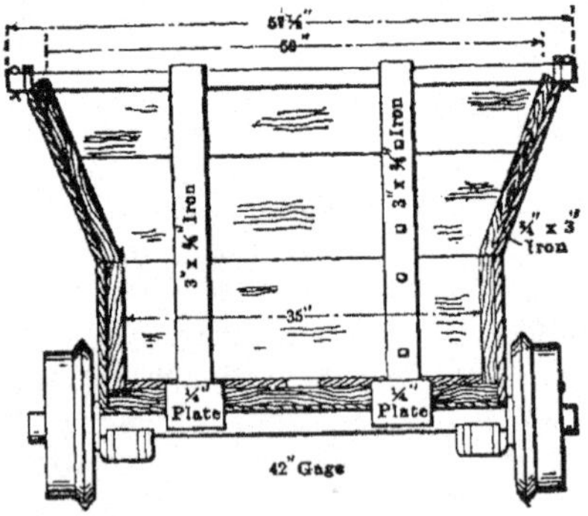

FIG. 153.—SULPHUR, MINING AND RAILROAD CO.'S ORE CAR.

side, end and bottom boards of the cars are of 1 1/2-in. white oak, braced with 3/4×3-in. wrought-iron bars. The cars are designed for a capacity of 41 cu. ft.

As the cars are dumped by a revolving tipple no arrangement is provided for unloading. The car is strong, easily constructed and is to be commended for use at the mine where lumber and labor are cheap. The wheel base is 24 in., length of car, inside, 60 in. and outside dimension 64 1/2 in. The 1/4-in. plates are on timbers that extend 4 in. beyond the end of the car and serve as a bumper.

A Joplin Car for Boulders (By Claude T. Rice).—In the mines of the zinc and lead district of southwestern Missouri the ore is hoisted in buckets, then dumped on grizzlies spaced 4 1/2 to 6 in. apart, and the barren portion of the oversize is sorted out and taken to the dump. The headframes are of the derrick type in which the hoist is placed at a height of 50 ft. or more from the ground. Therefore, it is not desirable to use a large car for tramming boulders as it would require a heavy trestle from the derrick to the boulder pile. More-

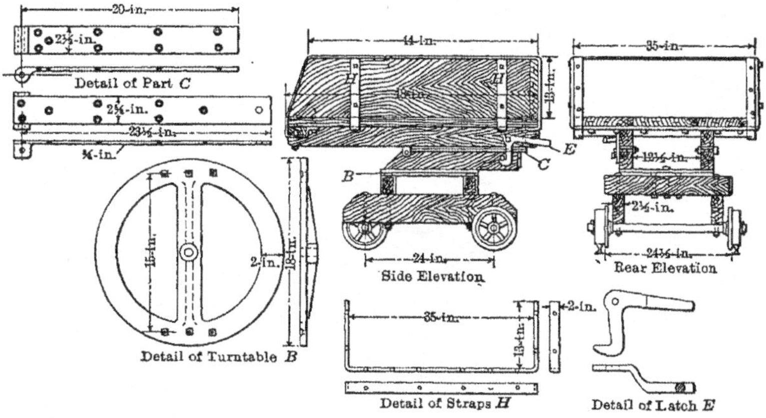

FIG. 154.—CAR FOR TRAMMING BOULDERS, AS MADE AT JOPLIN.

over, because the life of most of the mines is short, a cheap construction is desirable. As will be seen in Fig. 154, the car is an end dump one having a turntable deck carried on a wooden truck. The wooden box of the car, which is open at the front end, is hinged to the turntable. The edges of the side boards are bound with strap iron and are reinforced by the bands *H*, passing under the body of the car and up the sides. The box is prevented from dumping by a simple hook on one side of the box. The loaded car holds about 1000 lb. of boulders.

Tram Car for Stope Filling.—The accompanying drawing, Fig. 155, shows the details of construction of cars used in the stopes at the Copper Range mines at Painesdale, Mich., where a waste-filling system of mining is used. These cars hold a ton of waste and have the long, low bodies characteristic of copper-country mine cars. The long body is of advantage in spreading the filling, as it allows the filling to be dumped somewhat farther to one side than could be done with a car with a short body. The height that the shoveler has to lift the ore is also reduced. The body of the car is fastened to the truck by a hinge carried on a turntable on the truck frame. The wooden truck is now being made of fir. Some of these cars are equipped with Peleter bearings as are cars on the main levels, while others have ordinary bearings. The bolt passing through the hinge can be taken out easily when the car is to be raised into or lowered out of the stope, and the body dismounted for handling it through the chutes.

Tram Car with Automatic Door.—A tram car with an automatically opening and closing door has been constructed under the direction of A. J.

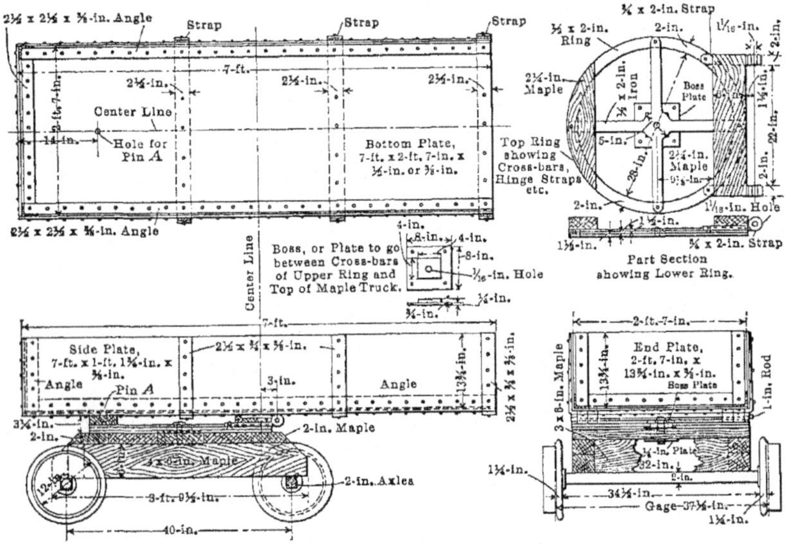

FIG. 155.—TRIMOUNTAIN CAR FOR STOPE FILLING.

Cummings, superintendent of the Cheever Iron Ore Co., operating near Mineville, N. Y. Previous to the use of a car rigged with a door in this manner, a door was used which required the tram man to open it before entering the tipple. If the door would not open, as was often the case, the loaded car had such momentum that it would enter the tipple and turn to the dumping angle, thus making it difficult to open the door. Trips were attached to the tipple to open the door, but nothing could be rigged conveniently to close the doors mechanically. Open-end cars, designed by Koppel, were used, but these required extra care in loading large lumps of ore at the open end to prevent the fine ore from rolling out on the tram tracks. It was necessary to have the car fitted with a door in order to load to full capacity. The Koppel cars were then rigged with the automatically operated doors and these have been entirely satisfactory. Iron plates are tapped and riveted near the top and center of the sides of the car and to these lugs are screwed. Two arms of flat iron are attached to the lugs and extend out to the front end where they are split and riveted to the door. On the horizontal center line of the door a strip of flat iron is riveted and the ends are swedged to 1-in. diameter to receive rollers which extend beyond the body of the car. The door is kept in position by resting on two supports formed by splitting the flange of the channels at the end of the car and bending them to the proper angle. As the car enters the tipple and dumps, the rollers carrying the door are guided in a horizontal course by riding on 4×5-in. maple pieces bolted to each side of the tipple frame.

Side Dump Mine Car (By Claude T. Rice).—Where ore is hauled in trains, the cars must be of the side-dump type to be economical. Side-dump cars are also better adapted to dumping into raises at the side of the tracks, or into the skip pockets in a shaft station or into the bins at the surface. The lifting

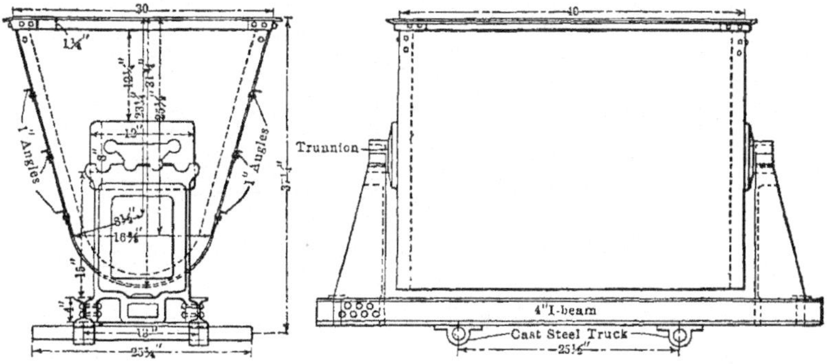

FIG. 156.—SIDE-DUMP CAR USED AT NORTH STAR MINE, GRASS VALLEY, CALIF.

of the car and its load a sufficient height to raise it out of the "keeps" and the swinging of the whole load around so as to dump the car over the side is wasted energy.

Side-dump cars have been in use at the North Star mines at Grass Valley, Calif., for several years, the first having been designed by Gerald Sherman, who also introduced them at the Copper Queen mines at Bisbee, Ariz. Fig. 156 illustrates the details in the construction of the cars in use underground at the North Star mines. Roller bearings are used on the latest cars and the wheel base is only 16 in., on account of the sharp turns in the drifts. The car has a capacity of 20 cu. ft. and holds, as loaded at the North Star mine, about 1800 lb. The cars weigh 750 lb., are easily made and cost $40 each. The frame of the car consists of two cast-steel end supports riveted to two 4-in. I-beams by two rows of rivets on each side. The trunnion is also made of cast steel and is riveted to each end of the car body. The car is reinforced by three 1-in. angle irons on each side. No catch is necessary to prevent the car from upsetting while being hauled by a mule or pushed by a man, but were these cars to be made up into trains to be hauled by electric motors, a catch would have to be put on them to keep the car from turning over because of the greater speed at which curves are taken.

Cradle for Dumping Mine Cars.—At the Copper Range mines in Michigan the bodies of the mine cars are fastened tightly to the trucks so that cradles

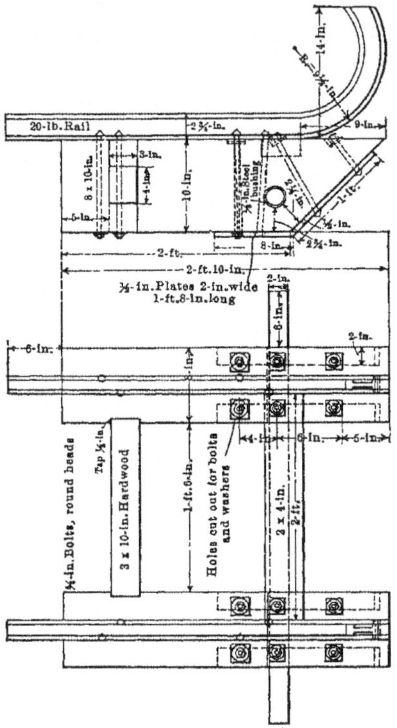

FIG. 157.—MINE-CAR DUMPING CRADLE.

must be used to dump them at the shaft. In dumping on a level, as when filling a stope, the rails at the edge of the raise coming up from the stope below are bent down and then turned up into a hook to catch the wheels. In this way the body is allowed to drop enough so that the rock will slide out of the car. The details of the cradles which are made to template are shown in Fig. 157.

Two strap-irons protect the front of the cradle timbers where they hit the floor of the station when the cars are being dumped. The axle on which the cradle turns is provided with collars to limit the side play of the cradle, while the axle is carried in strap bearings on two 6×6-in. timbers that extend back under the ties of the track so that they are securely anchored in place, and through them the cradle itself. Thimbles protect the cradle timbers from wear by the axle, and the rails on the cradle are turned up at the end so that the car cannot be pushed off into the shaft. In dumping, the front of the car either strikes the station floor or else the top of the skip if that happens to be raised a little too high, so that owing to its long body, the car cannot go over far enough to overbalance itself. Such would not be the case if the car had a short body like most mine cars used in the West. In order to prevent accidents from men stepping on the cradles, the cradles, when installed, are always tested, even though made alike and according to template, and a cradle must be able to hold the weight of a man standing on the far edge without dumping. Otherwise a cradle might sometime throw a man into the shaft when he happened to step on it. The car rests on the cradle in such a way that a little lift is necessary to dump it.

Calumet & Hecla Ore Cars.—The Calumet & Hecla is one of the few companies in the Lake Superior copper district, that uses ore cars having doors. At the other mines the cars are open at the front end and closed at the back. The men pile boulders at the front to hold in the finer dirt and this is a waste of time.

All the Lake Superior cars are built with the body resting directly on the axles so as to allow them to be made with the sides as near the ground as possible. The usual capacity is 2 1/2 tons, hence the cars are made with long bodies, especially where the tracks have the usual 3-ft. 4-in. gage; a 4-ft. gage is used at the Calumet & Hecla mines. Owing to the wide gage the cars have to have a short wheel-base compared to their length so as to allow them to make the turns without cramping. The front axle is bolted directly to the body of the car while the rear axle is attached to the forward axle by two distance straps that slip loosely over the axles. The front hole in the strap is made circular to allow the axle to turn in it when the body of the car is being raised to dump, while the back hole is made square to go over the square part of the back axle. In the center of the back axle there is a lug that fits into a hole in the bottom of the body so as to prevent side swing while the car is being pushed along the track.

The front door of the car is made to swing up and is carried from a cross rod in the ends of two horns that extend above the car so as to afford plenty of space for rolling in boulders. The door at the other end of the car is made to drop down so as to allow perfect freedom in handling the boulders. This door only comes two-thirds the way to the top. As the rear door is sometimes used as an apron to aid in loading boulders, it is secured to the car by four straps instead of three, as in the case of the front door.

These doors are locked by a rather ingenious device. To the doors are fastened two bolts with nuts on them, a link intervening so as to give flexibility, as shown in Fig. 158. By means of the nut, which is a tight fit on the locking rods, the length of the locking pin is adjusted so as to give the proper tension when the bolt is down in the keeps. The keeps into which the locking bolts drop are made of cast iron and are bolted to the sides of the car. The back

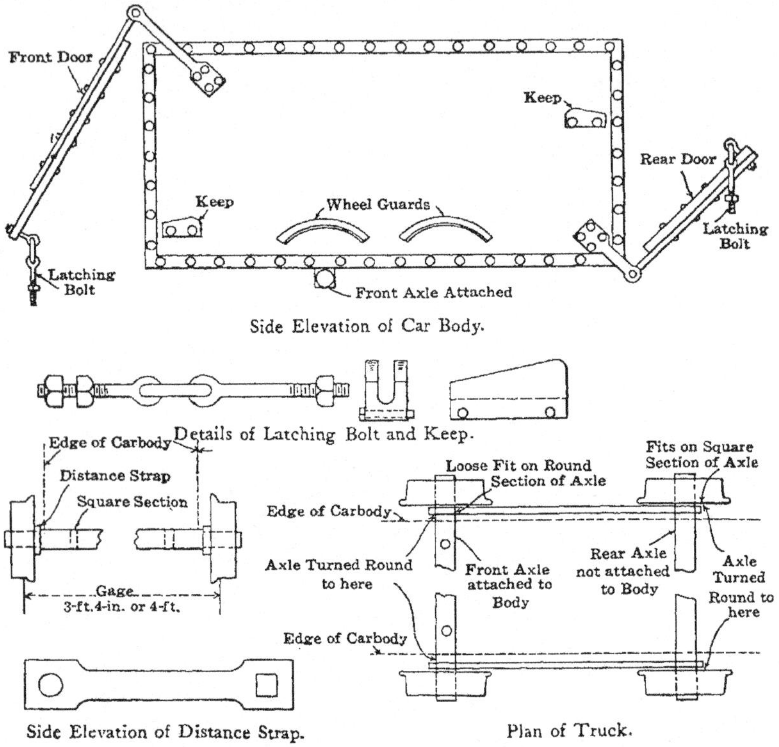

FIG. 158.—TWO-DOOR CAR FOR TRAMMING BOULDERS.

end of these keeps is slightly rounded so that the bolt is tightened as it is shoved down in them, while the bottom part is straight and designed to be perpendicular to the locking bolt when it is down in the keeps. A slight knock on the end of the locking bolt loosens it so as to open the gate, but the bind against the keep is sufficient so that no jar can loosen the door while the car is being trammed. The accompanying sketch shows the keep in detail. To prevent

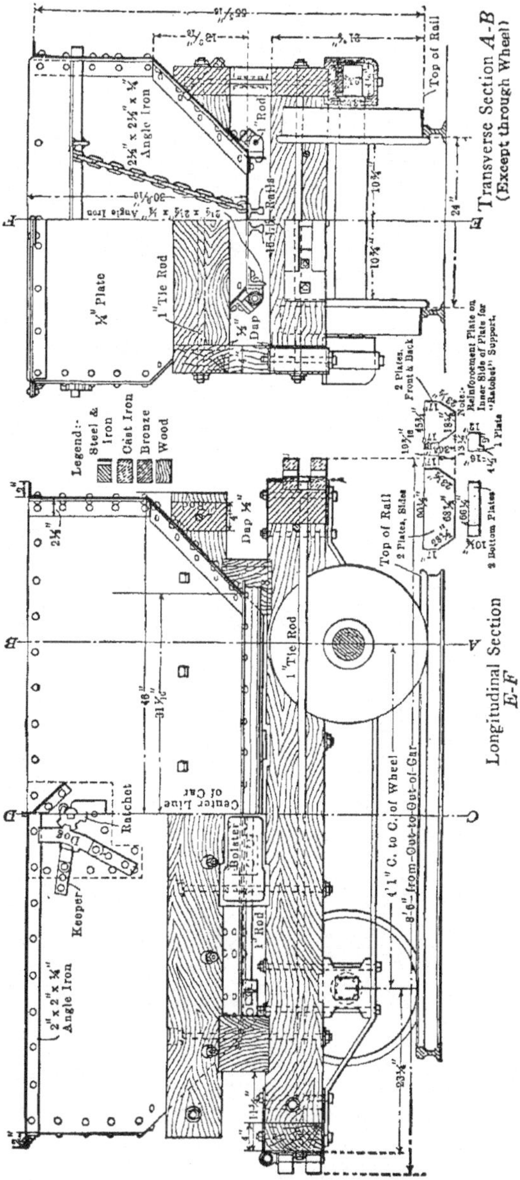

FIG. 159.—ORE CAR USED IN MAIN TUNNEL OF MORNING MINE, IDAHO.

fine rock from dropping upon the axles, wheel guards made of 2-in. angle iron are bolted to the side of the cars. The wheels are loose on the axles being held on by keys.

Cœur d'Alene Mine Car.—In Fig. 159 is shown a type of mine car used in the Cœur d'Alene district of Idaho. This particular car is used at the

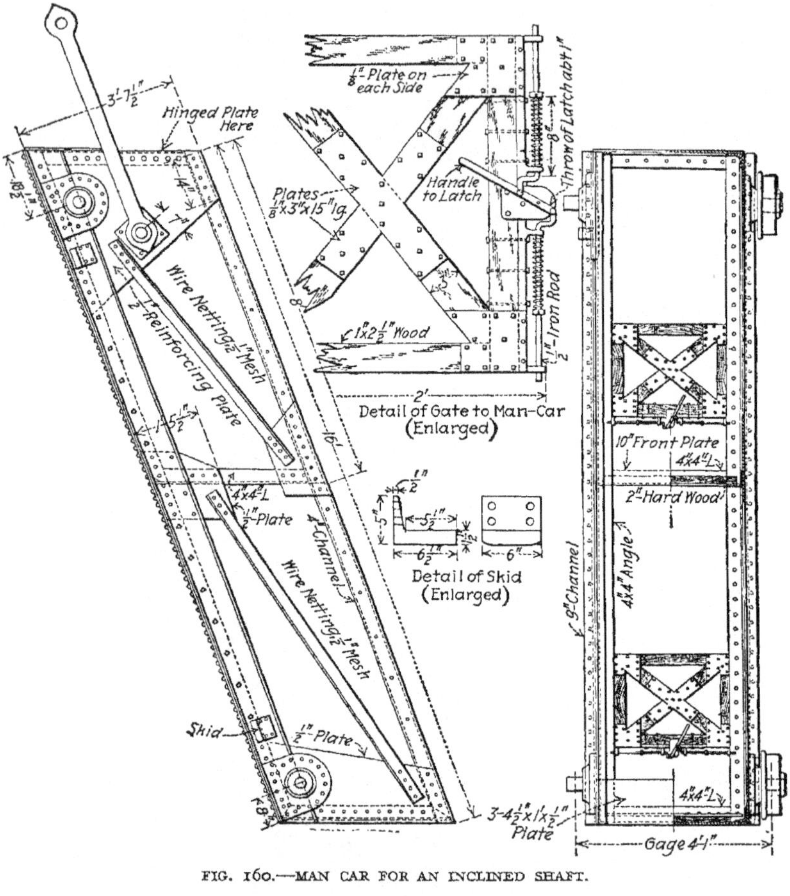

FIG. 160.—MAN CAR FOR AN INCLINED SHAFT.

Morning mine, of the Federal Mining & Smelting Co., at Mullan, and is similar to the Bunker Hill & Sullivan cars. The capacity of the car is 68.3 cu. ft., and it is used in the long adit tunnels in which ore from the mine is conveyed to the mill or shipping point. The advantages of this type of car are in its strength and durability, with which is combined simple structural details and rapidity of operation. The body of the car is made of 1/4-in. steel plate, reinforced with angle iron. The patterns of the plates forming the body of the car are shown in the drawing. This body, with sloping bottoms, as shown, rests in a cradle made of 4×8-in. wooden beams set on end, and supported from the axles. The car axles are heavy, being 3 1/2 in. in diameter, turned to 2 3/4 in. at the bearings. The wheels are 20 in. in diameter, running on a 24-in. track. The overall dimensions of the car are: Width, 51 1/4 in.; height, 55 5/16 in.; length, 90 1/2 in. The bottom of the body of the car consists of two sections of steel plate hinged along the sides of the sloping bottom of the body. Each flap is weighted along its center with 16-lb. rails and supported by chains from a winding shaft across the body of the car. The car is dumped by simply loosening the ratchet gear controlling this winding shaft. The weight of the ore and the rails on the bottom doors serve to swing them open. A few turns of the winding shaft suffices to close the car again. The handle used for this purpose is kept at the point at which the cars are to be dumped. One man dumps all the cars of a train. The winding axle terminates at one end in a ratchet wheel. A dog engages this ratchet and the dog is locked in place by a keeper. The dog and keeper are each pivoted to the body of the car. To dump the car, the keeper is knocked up so as to disengage the dog from the ratchet wheel. The details of this dump mechanism are shown in the drawing. One man is able to dump 15 cars in about 3 minutes.

Copper Range Man Car.—The car used for lowering and raising men at the mines of the Copper Range company, in the Michigan copper district, is shown in Fig. 160. It is designed for use in a shaft sunk at an inclination of 70° from the horizontal and is a two-deck car, there being room for 13 men on each deck. The sides of the car are covered with 1/2-in. mesh wire cloth, and the front is closed by gates, one for each deck, that slide in channel guides, a set of which is provided for each gate at the front of the car. The gates may be locked in position above the men's heads, or waist high by throwing the lever which pushes two rods out through holes in the channels provided for the purpose. The flooring of the upper deck is held in place by two pairs of channels, between which it slides, so that it may be removed when timbers are to be lowered. The lower parts of the gate guides are also cut away, so that the gates can readily be removed when the car is to be used for this purpose, and a bolt that hooks into eyes on each side of the car just below the upper deck, and which prevents spreading or bulging at the sides, can likewise be removed to make room for the timbers. The hood at the top of the car is hinged so it may be laid back when 30-ft. rails are to be lowered. In order to prevent accidents that might occur if one of the wheels should break while the skip is in motion, two pairs of skids are attached to the under part of the body, which, in the case of such breakage, would support the car upon the rail.

Copper Range Ore Skip.—The 6-ton skips used in the Copper Range mines in the Lake Superior copper country, present several novel details of construction, especially in the manner of attaching the draft lugs to the body of the skip, the system of oiling the wheels, and attaching the collars to the different lugs, bails and wheels. All hubs on which collars are used, are drilled for a split pin which holds the collars in place. The hubs are also cut with slots at 45° that receive the rivet pins of the collar.

In the drawings of the Champion skip given in Fig. 161 the details of the draft lug and the collar used with it are shown. Projecting from the inside

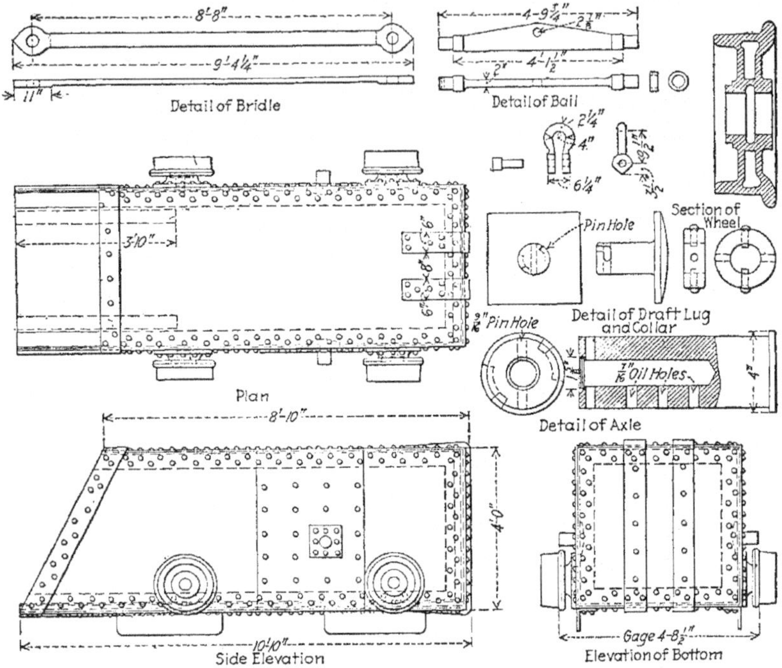

FIG. 161.—DETAILS OF THE SKIP USED AT THE CHAMPION MINE.

surface of the collar and at 90° from the hole drilled to receive the split pin, are two pins riveted into holes in the collar. In assembling the collar on the lug, the pins are slid into the notches to the bottom and then the collar is turned through an arc of 45°. This causes the pins to enter the grooves and when they are against the ends of the two grooves, the holes for the split pins in the collar are in line with the holes in the axle and the split pins may be forced through. In this way the collar is held at four points instead of at two, and thus held securely. All of the other collars are attached in a similar way.

The body of the skip is reinforced at the lugs by an extra plate. The bridle strap is bent so as to clear the front wheel. In order to keep the skip as low on the rails as possible, the wheels are mounted on hubs riveted to the sides instead of on axles. At the bottom there are two reinforcing straps, while on the under side and at the front end there are two straps that take the wear that comes on the skip's bottom when going over the roller of the dumps. By attaching the draft pins by lugs riveted to the body or the inside of the skip, the lugs are fastened much more securely than if riveted to the outside, as is the usual way.

The method of oiling the wheels is shown in the drawing of the axle. The ends are hollow and three holes in the walls let the grease out to the wheels. The wheels are made with a receptacle in them that also becomes filled with grease and acts as a reservoir, there being four holes drilled through the brass bushing to connect with the reservoir. These grease cavities are closed by a screw plug that cannot be taken out without removing the collar from the axle. The grease is shot into the wheels by a plunger gun that screws into the axle. Formerly when the wheels of the skip were filled with oil and the oiling of the axles was done by oil feeding down on them from these reservoirs, the wheels had to be filled four times per day, while the grease well has to be filled only once in 48 hours.

The rope is attached around a thimble. This thimble is now made solid instead of having its sides cut away and only reinforcing ribs going across, as they were formerly made; that construction proved to be too weak. The manner of attaching the lugs and collars, as well as the methods of oiling without the necessity of taking off the collars, was devised by W. J. Richards, the master mechanic of the Copper Range Consolidated Co.

The Franklin 10-ton Skip.—There is but one hoisting compartment in the No. 1 amygdaloid shaft of the Franklin mine near Houghton, Mich., so in order to raise the desired tonnage, a skip of 223 cu. ft. or 10 1/2 tons' capacity is required. This skip is by far the largest in the Lake Superior copper country. The front wheels of the skip, as shown in Fig. 162 are carried by bracket axles which are bolted to the body of the skip, the clearance between the inside face of a wheel and the body being about 1 in. A reinforcing plate through which these bracket-axle bolts pass stiffens the body of the skip at that point. The body is also reinforced at the draft lugs. Because of the small clearance of the front wheels it is necessary to bend the bridle strap so that it will clear the wheels. At that part of the bottom of the skip on which the ore from the loading pocket falls when the skip is being filled a wooden lining is used which extends a little below the reinforcing plate for the bracket axles. The rest of the bottom is reinforced by steel channels, every other one of which is inverted; all are riveted together as shown in the section A-A in the illustration. The skip is equipped with skids that will take the weight in case a wheel breaks, and also with guide shoes that run over a concrete stringer in the shaft and which will guide the skip in the event of its jumping from the rails. The rear axle is pivoted and is carried in sleeves which allow a play of 1 1/2 in. The front and back wheels are alike and are made of manganese steel, while the increase in the tread necessary on the rear wheels for dumping the skip is obtained by a separate hub that is carried outside the main wheels and on the same axle. This hub is a cheap steel casting as it does not have to take much wear. At the rear are lugs for attaching below the skip a truck for lowering timbers. The skip weighs about 6 1/2 tons. The skip track is 5-ft. gage and is made of 80-lb. rails carried on concrete stringers of the Mohawk type.

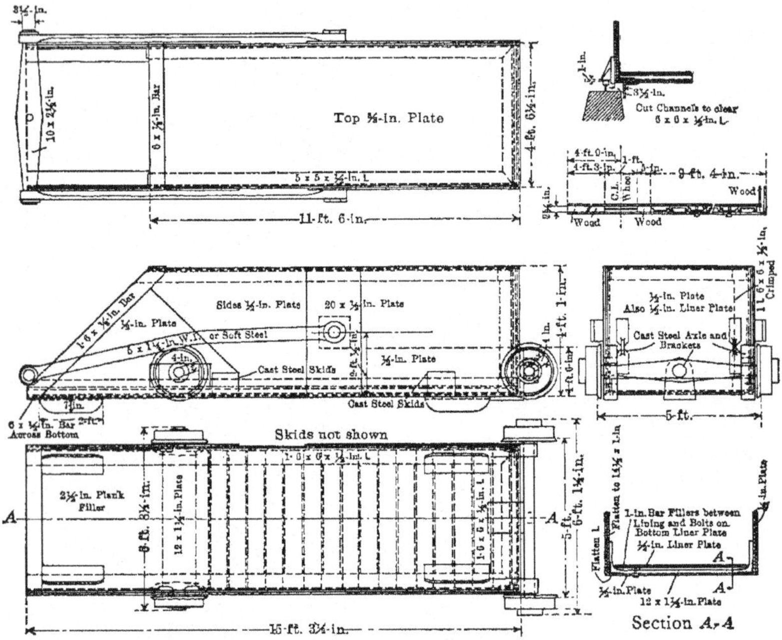

FIG. 162.—THE 10-TON SKIP FOR THE FRANKLIN MINE.

Skip and Dump Plate for Vertical Shaft (By Lee L. Wilcox).—A modification of the De Beer's type of skip for a vertical shaft, which is an improvement over the old type, is shown in Fig. 163. The top of the old type of De Beer's skip was square and trouble was experienced from ore falling back into the shaft. This not only gave the shaft an untidy appearance, but in a few instances the men working near the shaft were seriously injured by falling pieces of ore. Attempts were made to improve the skip; in one instance a lip was riveted to the bottom plate; in another a filling piece was attached to the dump plate, thus throwing the lip up farther from the edge of the shaft. These changes proved to be only partially successful in stopping the ore from falling back down the shaft, but they introduced other faults which were nearly as bad. The lip would spread the discharge of ore so that it would lie on the dump angles, causing the skip to stick as it descended; and the filling piece decreased the inclination of the skip when dumped so that it did not clear itself readily.

When the new Pettit headframe was built it was decided to make some changes in the design of the skip. It was made longer in proportion to the base than is customary; the dimensions being, base 3×4 ft. and length 6 ft. on one side and 5 ft. on the other. The base dimensions are unusual, because of local conditions. Under ordinary conditions it would have been made nearly

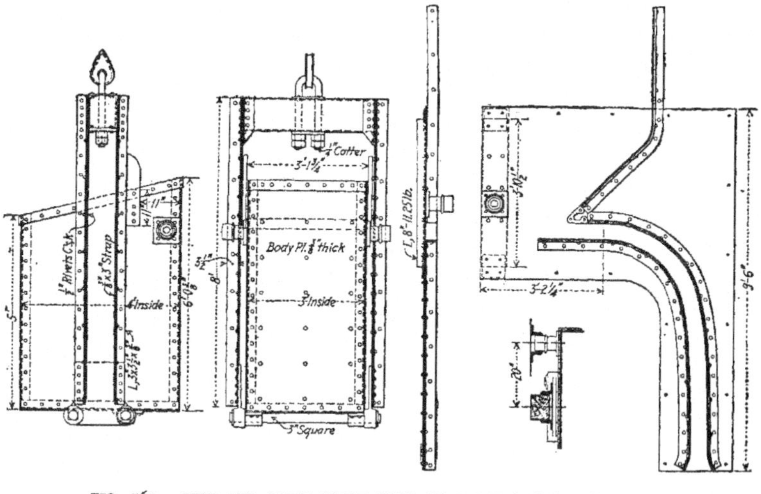

FIG. 163.—SKIP AND DUMP PLATE USED IN A MINNESOTA IRON MINE.

square. The additional length, however, throws the lip considerably farther away from the shaft and all the ore is discharged well back from the edge of the shaft. By incorporating the extended lip in the body of the skip itself, as shown in the illustration, the trouble caused by the ore gathering on the dump plates was entirely overcome. The new skip was satisfactory from the start. A few changes were made in the dump plates which are worthy of mention. The dump angles were shod with a 3/8-in. strap; as this strap is worn it can readily be replaced without removing the dump angle itself, thus making the repairs much easier and simpler. The plate was reinforced behind the dump roller by a channel. This prevented the bending of the plate at this point which on the larger skips is troublesome.

Automatic Skip for Inclined Shafts.—In Fig. 164 are shown the details of a 48-cu. ft., open-top skip used by the Salt Lake Copper Co. at its mine at Tecoma, Nev., in a shaft inclined at a low angle from the horizontal. The design presents several novel features, such as a curved bottom at the front for directing the ore when the skip is in discharging position and attachment of the bail at the lower end or back of the body of the skip.

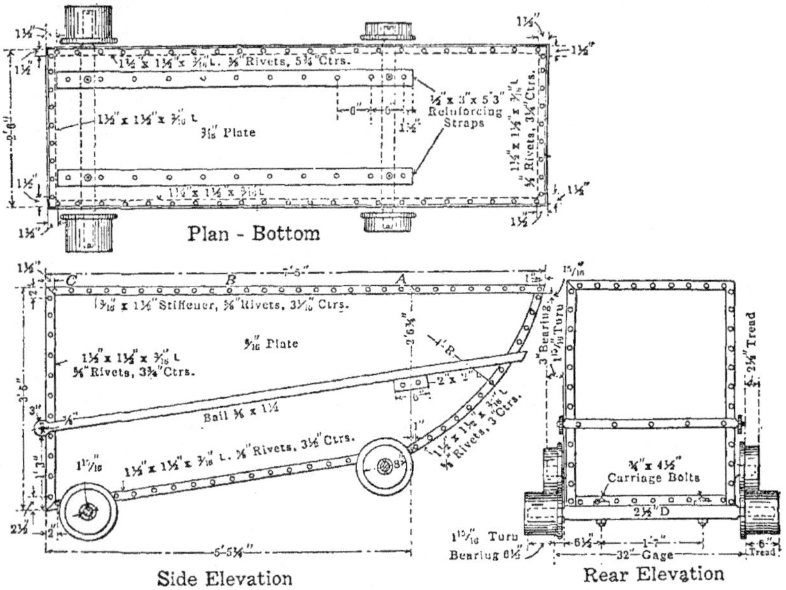

FIG. 164.—A SKIP FOR FLAT-DIPPING SHAFT.

Dumping Skip for Winze (By K. Baumgarten).—The automatic-dumping skip shown in Fig. 165 was installed by the Black Mountain Mining Co. in the vertical winze in the Cerro Prieto mine, 40 miles southeast of Magdalena, Sonora, Mex. The skip was designed for the purposes of sinking. The total length of the shoes was 20 ft., which admitted of lowering the bucket 16 ft. below the guides. The hoisting compartments were 4 ft. 6 in. by 5 ft. in the clear; the guides were 5 1/2-in. surfaced Oregon pine. A 1-ton skip was small for the size of compartment, which was larger than usual; the problem was to arrange a bucket of convenient height for shoveling and at the same time to provide sufficient height, such that when the bucket was turning over in the dumping guides, it would not begin to empty its load until the lip was clear of the edge of the winze. The capacity of the bucket was 23.5 cu. ft. or about 1.1 tons. The shipping weight was 1732 lb. After arrival at the mine, the springs were added, the "fingers" taken off and increased to 12 in. length, and a false bottom of 2-in. plank put in, thus increasing the weight by about 60 lb.—a total of about 1800 lb. The hoisting equipment was a 50-h.p. three-phase induction motor, 220 volts, geared to a drum carrying about 500 ft. of 3/4-in. wire rope. The load designed for this set was 3500 lb. but the hoist was operated to a depth of 400 ft. under a load of about 4000 lb., or, roughly, 15% overload, exclusive of rope weight. The cost of the skip was $195, to which may be added $18 for freight and $3 for duty, a total of $216 at the mine. No originality is claimed for the design, except that the wearing plates within the

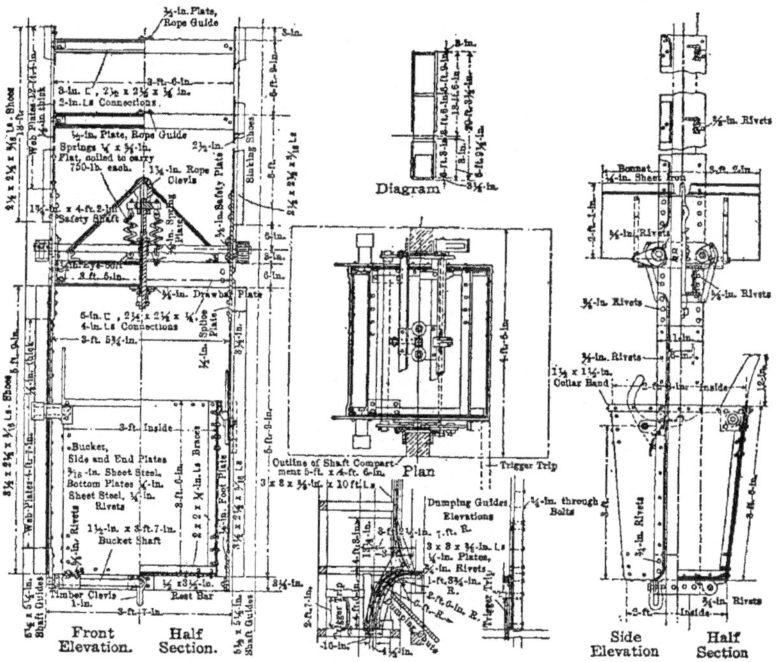

FIG. 165.—AUTOMATIC DUMPING SKIP FOR WINZE.

shoes are omitted, and to which may be attributed the smooth running, as there are no projections between the shoes and the guides. In six months' operation no repairs were necessary. The dumping guides were made at the mine, but proved to cost only slightly less than the skip. It would have been cheaper to have had them made at an outside machine shop, where the equipment for such work was at hand.

A Timber Skip.—At the South Eureka mine, near Sutter creek, Calif., a special skip is used for lowering timbers. The shaft is inclined at angles varying up to about 70°. The timber skip is hung below the rock skip from a ring on the bottom of the latter. It is made from an old skip by taking out the top plate and using an extra long bail, hung from the axles of the front wheels. By having the top of the skip open, timbers can be taken out by simply swinging them down, thus obviating the necessity of raising and lowering the skip for jacking. The extra length of bail enables long timbers to be handled.

Counterbalance for Skips.—The counterweight used at one of the iron mines of the Cleveland-Cliffs Co., at Ishpeming, Mich., consists simply of one 8-ft. section of 14-in. cast-iron pipe with flanges, mounted in a frame made of plate and angle iron. The two side pieces are 1/2×16 in.×10 ft., the upper and lower ends being fastened together with 1×10-in. plates. The runners are made of 3×4-in. angle iron spaced 9 in. apart. These run on 8-in. wooden guides in the shaft. The runners are also lined with 1/4-in. plates both on the angle iron and on the 1/16-in. side plate. This lining is easily removed in case of excessive wear and repairs can be easily made. A wooden block is placed in the bottom to act as a cushion and also add to the strength. The cast-iron

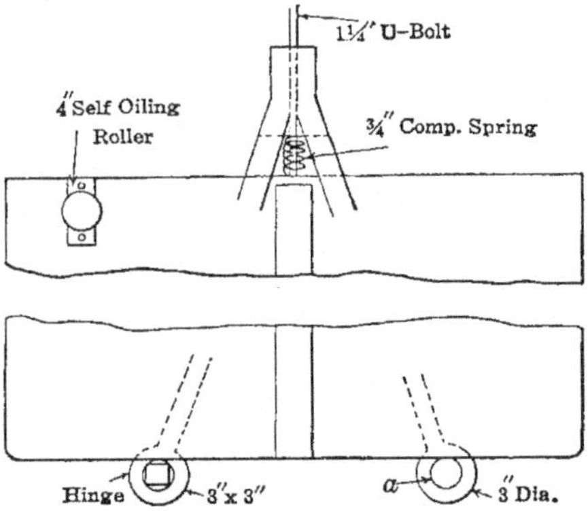

FIG. 166.—IMPROVED SKIP AT ADAMS IRON MINE.

pipe being hollow gives ample room to add any amount of scrap iron in order to give it the desired weight. Another type of counterbalance used by the same company consists of a piece of solid steel shafting about 10 in. in diameter which runs inside of a 12-in. pipe. The pipe thus takes place of a guide. Other counterbalances used at various mines consist of a steel frame work in which large pieces of cast iron are mounted and may be removed or added to, as desired, in the same manner as on elevators in buildings.

Skip Improvements.—The new skips that are being made for the Adams mine, near Virginia, Minn., are equipped with 3/4-in. compression springs beneath the crosshead to lessen the shock on the cable when starting to lift the skip full of ore. The use of springs on skips is not common although they have been used on cages. In the old skips two 3×3-in. iron bars were used under the skip to support the load. The skip is hinged to one of these bars for dumping purposes, and when vertical simply rests on the second one. These two bars are about 18 in. apart. Dirt will accumulate on the top of *A*, Fig. 166, if square, and prevent the skip from occupying its true position; hence the use of round bars. The skip being 5 ft. high, 1/2 in. of dirt or ice will throw the top of the skip 1 2/3 in. out of plumb.

MINE CAGES

A Three-deck Man-cage.—The hoisting of men consumes a large amount of time and hence reduces the amount of ore that may be hoisted daily. Witherbee, Sherman & Co. recently constructed a three-deck cage for hoisting men from the mine. The capacity is 30 men per trip. The framework consists of channel and angle iron, while the sides are inclosed with 3/8-in. wire screen, 2-in. mesh. The front of the cage is inclosed by means of sliding screen doors. The cage is to be operated in an inclined shaft and for this reason is mounted upon wheels. It is built so that the floor of the cage is horizontal, while the sides conform to the slope of the shaft. When the cage is not in use it will be removed from the shaft by means of a block and pulley which operates on an I-beam track. The cage will thus be carried to one side of the shaft house and entirely out of the way. The time required for making the change from the ore skip to the man skip is small. The present skip will hoist only 10 men so that with this new cage the men will be hoisted in one-third of the usual time.

Hiawatha Mine Cage (By H. L. Botsford).—In Figs. 167 and 168 are shown a type of cage and safety catch much used in the Lake Superior iron districts. The drawbar *A* is free to slide through the crosshead or supporting frame, until the cage is carried by the plate and jamb nuts *B*. Chain connections between the drawbar and the cam shafts cause the latter to rotate when the drawbar is raised. The springs *E* are of the helical type and are placed concentrically around the cam shafts, one to each shaft. The springs are fastened to the side of the cage and to collars *F* keyed to the shafts. Any rotation of the shafts produces a torsional stress in the springs. Should the hoisting rope break, the cam shafts are turned into such a position that the cams are brought into contact with the guides and cut into them sufficiently to stop any downward motion. This cage is also provided with two bolster springs *G* and *H*, one within the other and both concentric around the drawbar. Their purpose is to lessen the strains in the hoisting cable due to a too sudden starting or stopping of the cage, and also to draw down the drawbar in case of the failure of the hoisting rope, thus permitting the full force of the cam-shaft springs *E* to be expended in forcing the cams into contact with the guides. There are several holes in the chain pulleys J, and the spring collars F for the fastening of the chains K and the springs E. This permits of careful adjustment to secure the proper tension in the springs E. The cage is substantially a type made by the Lake Shore

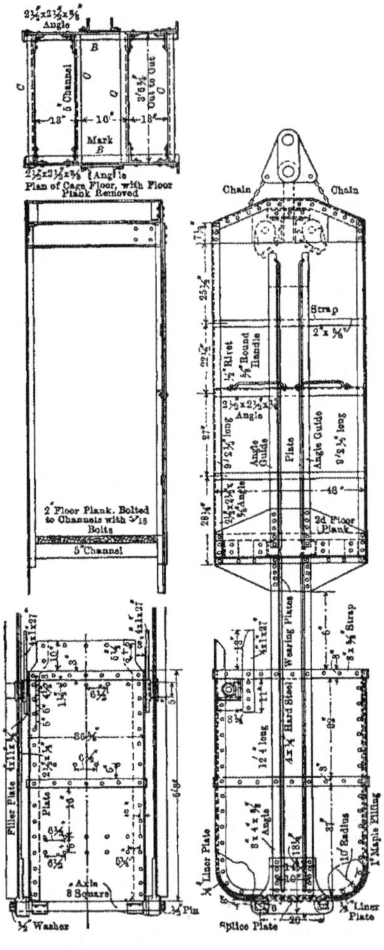

FIG. 167.—HIAWATHA CAGE WITH SKIP ATTACHED.

Engine Works of Marquette, Mich., and is shown with a Kimberley skip attached below it, for hoisting ore.

A Light Mine Cage (By H. L. Botsford).—A type of cage used in the West is shown in Fig. 169. This cage is of light construction, and is used for hoisting

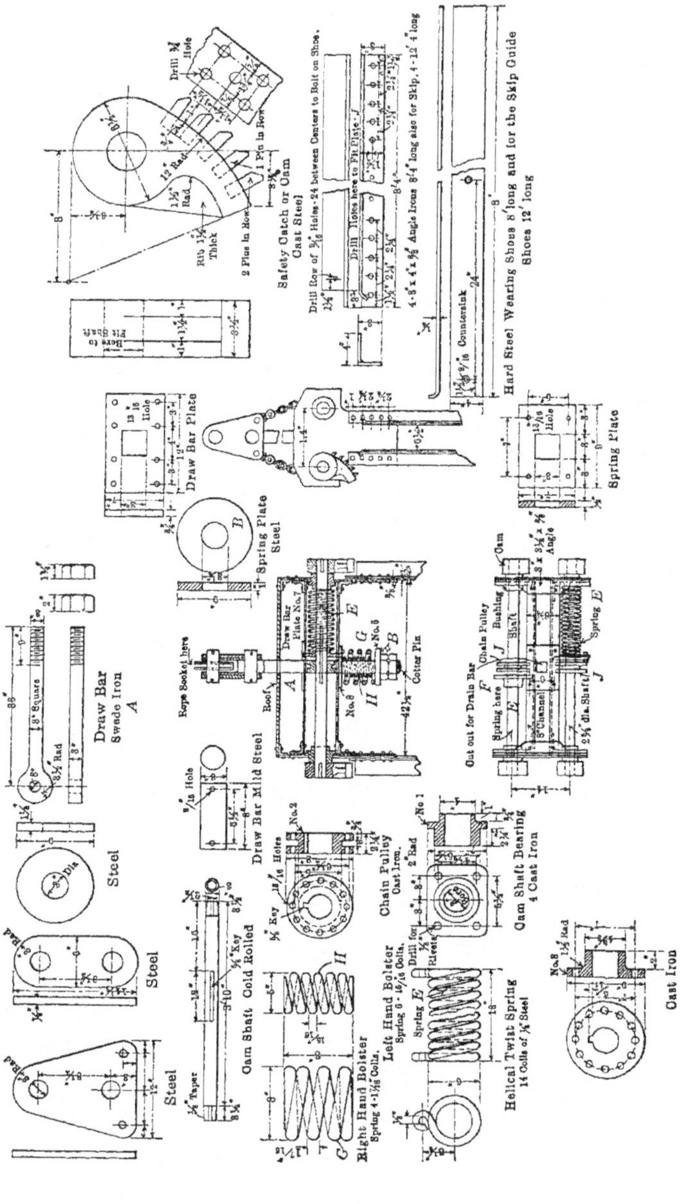

FIG. 168.—SAFETY CATCH USED ON THE HIAWATHA MINE CAGE.

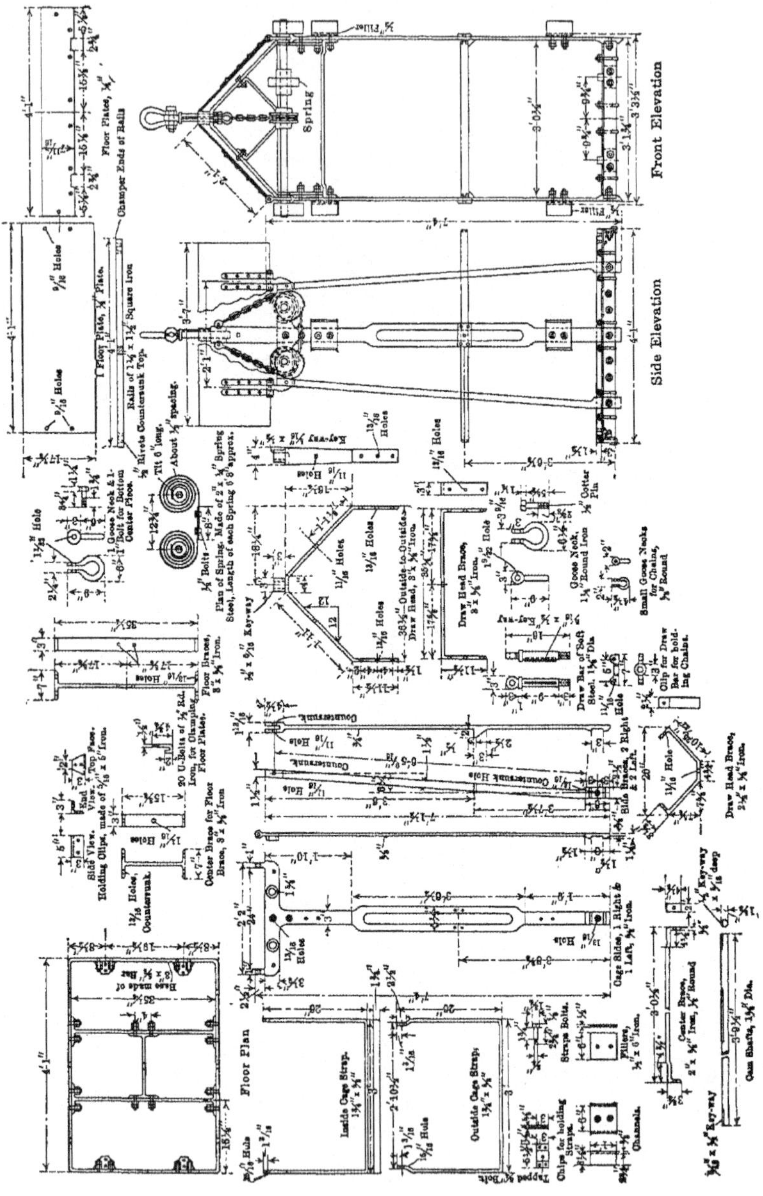

FIG. 169.—STRUCTURAL DETAILS OF A LIGHT MINE CAGE.

small loads. The springs are of the spiral type, two in number. The inner ends of the spirals are fastened to collars keyed to the cam shafts, and the outer ends are bolted together. Unlike the cage described in the preceding article, the springs are not attached to the frame of the cage. The cams are circular in outline, with grooved surfaces. They are mounted eccentrically on the cam shafts, which gives them their gripping effect upon the guides. The bonnets of these cages are usually attached by hinges so that they may conveniently be swung up out of the way when long timbers are being lowered. It will be noted that the guide shoes are short, whereas in the other cage the guide shoes were nearly as long as the cage. For a cage of this type which is intended especially for raising small loads it is not necessary that the guide shoes be long. The two sets of light short shoes should be set as far apart vertically as can be conveniently done as the greater the distance between them, the more will they tend to steady the cage during hoisting.

SPECIAL CARRIERS

Harness for Lowering Mule Down a Shaft (By W. F. Boericke).—Where there is no incline by which mules may be taken underground, the animals must be lowered either in a specially constructed mule cage, or swung down by means of a harness similar to the one here described and illustrated in Fig. 170. Wherever conditions are right, the preference should be given to the harness. The mule cage is heavy, cumbersome, difficult to get the mules into, and its use means a loss of time in getting the cage in and out of the shaft. To swing a mule with a harness is by no means as formidable a job as it appears at first. The harness itself is a simple affair, and can be made at the mine, out of pieces of old canvas or rubber belting, securely riveted together. The dimensions given in the sketch are adapted for a 900-lb. mule. The harness, when fitted on the mule, should be fairly snug, but the straps should not be buckled too tightly. If the mule is accustomed to harness, he will allow it to be put on without trouble. The parts lettered *T T* should come a trifle below his sides. The part *S* should be low enough to insure that the mule will be seated firmly when he is swung into the shaft. The head and neck, of course, come through the space *A*. When all is ready, the mule is led to the collar of the shaft, and a heavy chain attached to the hook, and made fast to the center of the cage. Planks should be laid across the shaft, so that as the mule is raised, his feet will not catch in the timbers. The signal is given to hoist slowly, and the astonished animal is swung out over the shaft, bearing a marked similarity to a dog begging. The planking is then removed, and the cage slowly lowered. If there are bad places in the shaft, it may be well to tie the feet together, but this makes a delay at the bottom, and is usually unnecessary. When the mule is near the bottom, a rope is tied to the back of the harness, and the animal drawn to one side, so that he lands on four feet. In an actual case, from the time the mule was swung into the shaft, until he was led to the stables at the bottom, was 6 minutes, the shaft being 150 ft. deep.

Automatically Discharging Bailers (By W. H. Storms).—Many mine managers prefer bailing to pumping water from mines, and undoubtedly there are conditions where bailing is less expensive than pumping. Fig. 171 shows two types of self-dumping bailers; one a cylindrical valve bucket, the other a skip fitted with a hinged valve which opens to admit the water when the skip

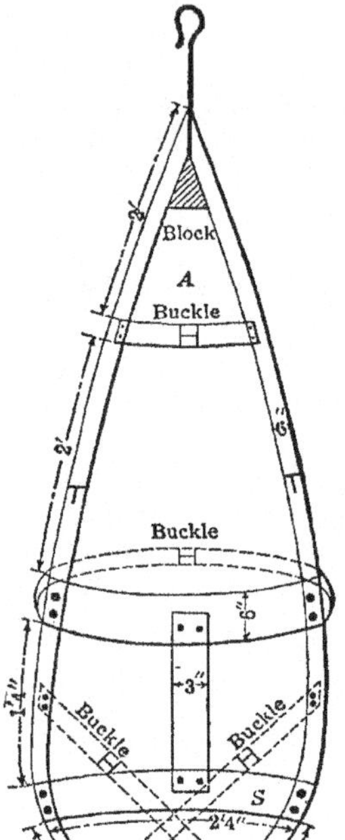

FIG. 170.—HARNESS FOR LOWERING MULES.

sinks in the sump. Both are arranged to operate in inclined shafts. The automatic-discharge features will be appreciated by those who have had no experience with bailers of this kind.

The right-hand illustration shows the steel barrel on the skids *S* and also its position, by means of dotted lines, when discharging. This is accomplished by cutting out a section *A* of the skids upon which the bucket slides. This section is supported on a pivot *B*. A counterweight *W* holds the section in place when the bucket is below, or even when resting upon it when empty, but when it is drawn up out of the shaft filled with water the excess of weight of the bucket and the water below the pivot, as compared with that above it, causes the loose section of the skids to swing backward into the vertical position indicated by the dotted lines. The engineer then lowers the bucket without leaving his station at the engine, until the valve spindle *V* rests upon the floor, which forces the valve upward, thus allowing the water to escape and flow away. The engineer then raises the bucket slowly and the counterweight *W* pulls the skids, with the empty bucket upon them, back into position and the bucket is lowered again into the shaft.

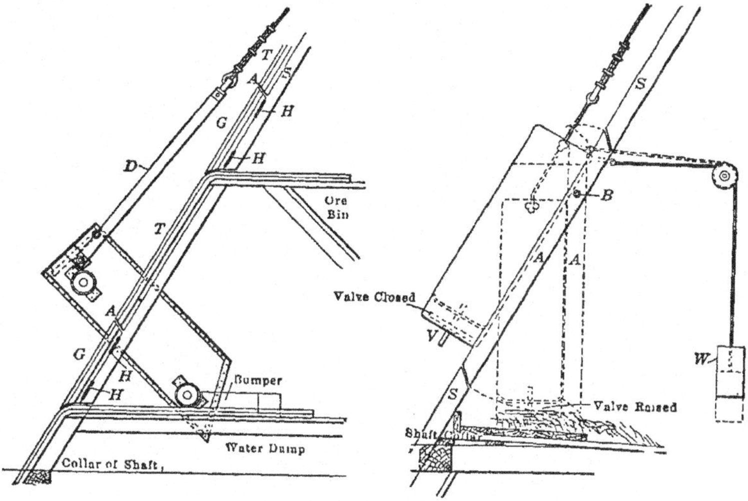

FIG. 171.—AUTOMATIC DISCHARGING BAILERS.

The illustration to the left shows an ordinary water-skip with a hinged valve in the bottom. The track, generally of T-rail, *T*, is spiked to stringers *S* in the headframe, and dumps provided for both ore and water by cutting out sections of the stringers and track and making gates *G G* as shown. The gates swing on the hinges *H H*, that for the water-dump being about 3 ft. above the collar of the shaft. When a skip is to be dumped the proper gate is opened and upon the arrival of the skip at this point the forward wheels follow the curved rails and run out upon the horizontal track, as shown in the sketch. The engineer continuing to hoist slowly, the lower end of the skip is lifted from the main rails and its contents dumped. There is one thing to be carefully avoided in operating this device. When the skip is approaching the dumping place the hoisting speed must be slackened to a moderate rate or the forward wheels are liable to collide with the upward extension of the track at *A*, thus doing serious damage to the headframe, or parting the cable. The bumper keeps the wheels from running so far forward that the skip will not automatically descend the shaft again when empty. By having a movable bumper, however, the water-skip can be run in on that dump and secured there by dropping a piece of light timber between the rear wheels and the skids. The cable can then be detached and made fast to the ore skip. In this way either ore or water may be hoisted in the same shaft compartment, by using two suitable skips.

A Two-ton Water Car (By Guy C. Stoltz).—The type of water car used at Mineville, N. Y., to unwater Mine 21 is shown in Fig. 172. Two cars of 528 gallons' capacity were hoisted in balance an average distance of 450 ft. on a 60°

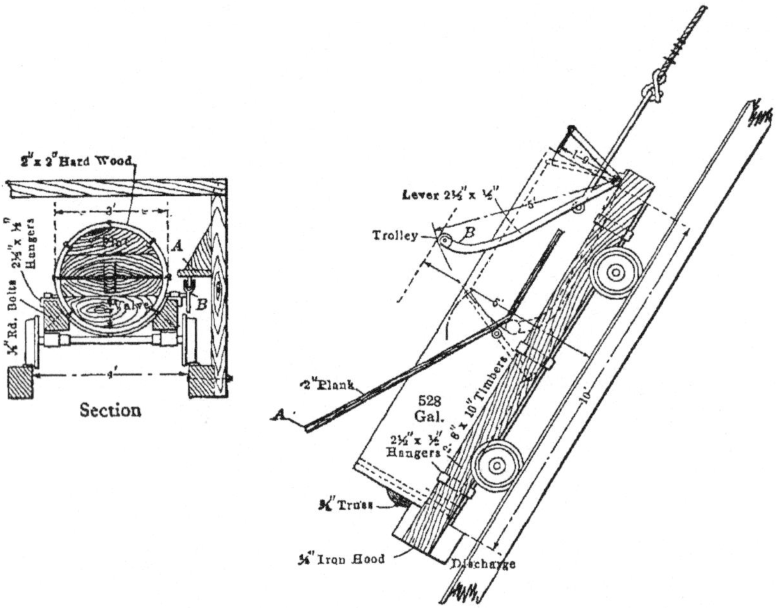

FIG. 172.—WATER SKIP USED AT MINEVILLE, N. Y.

double-track incline. They were automatically filled in the mine and automatically discharged at the surface into a series of wooden troughs. The greatest number of cars hoisted during a 10-hour shift was 812, while 600 was the average rate per shift. The hoisting capacity at maximum speed was equal to a pump fitted with a 10-in. discharge delivering 715 gallons per minute. The average quantity hoisted was 528 gallons per minute. The water tank was made from a 10-ft. length of 3-ft. stack, the lower end headed with 2-in. pine and held in place by a 2 × 2-in. hardwood ring which was bolted to the circumference of the stack at the bottom. The top was not headed. The bottom timber head was reinforced at its center by a truss of 3/4-in. iron bearing on a 4 × 4 × 12-in.

wooden strut. The tank, which was attached to a set of wooden stringers by flat iron hangers, was mounted on the ordinary skip axles. The lower 10-in. segment of the tank door was hinged and upon lowering the car into the water the pressure would automatically open the segment and allow the water to enter. As the car was hoisted above the level of the water the door would close due to the hydraulic pressure of the contents. The door was automatically opened at the dumping point on the surface by the operation of a system of lever arms. These were attached to the door and so pivoted that a plank guide *A* would intercept the trolley arm *B* and lower it to such a position that a lateral movement of 8 in. toward the top of the car would be delivered to the gate. Upon lowering the car the gate, which was heavier than the reciprocating system of levers, would close. Back rails should be provided to prevent the car from leaving its track as it enters the water.

Scraper for Cleaning Stopes.—The stopes of the Calumet & Hecla mine dip at about 38°. The empty stopes are not filled and after the stoping is finished,

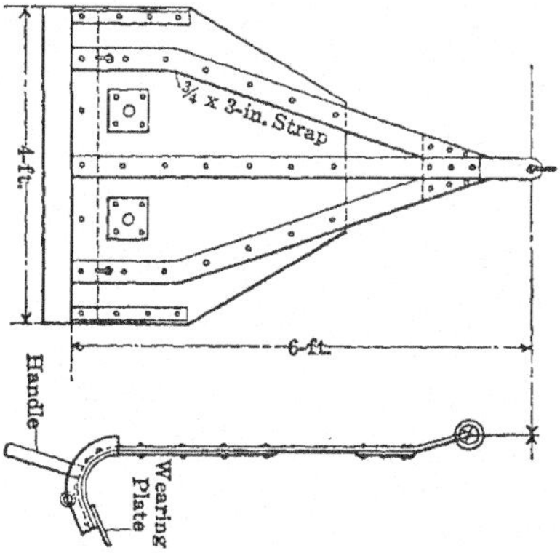

FIG. 173.—A LAKE SUPERIOR STOPE-FLOOR SCRAPER.

a considerable quantity of ore is left on the floor, especially if the floor is rough. This ore, when there is not much to be handled, is worked down to the level by hand. When much ore has accumulated on the floor, it pays to use the scraper, the construction of which is shown in Fig. 173. A sprag is set in the top of the stope, and a small air hoist is installed at the bottom. The rope that goes from the hoist to the rear end of the scraper passes through a pulley carried by the sprag, while another rope from the hoist goes directly to the front ring of the scraper, so that the air hoist, besides pulling the scraper with its load of ore in front of it to the bottom of the stope, also pulls it back again. The scraper works well when the foot wall is not too rough and practically no hand shoveling of the ore is necessary, but if the ore lies in pot holes in the foot wall, it must be shoveled out to a smoother part of the foot wall, where the scraper can get at it.

This scraper was designed by Capt. Samuel Richards, of the Calumet & Hecla company, and has now been in use for several years, both in the amygdaloid and in the conglomerate stopes. The main plate of the scraper is 5/16 in. thick, and is strengthened by means of three 3×3/4-in. iron straps that come together near the front, so as to distribute the load to the haulage strap. A ring is provided for fastening the rope from the hoist to the front end, while at the rear is a chain that is fastened to two eyebolts that go through the two outside straps. To this the return rope is fastened. There are also riveted to the rear of the scraper two plates carrying two handles. By means of these handles the men are able to steer the scraper, and by pulling up or shoving down on them regulate the quantity of ore that the scraper takes on its trip down the stope. In case the scraper has a tendency to ride over the ore, owing to the compact way in which it lies on the foot, car wheels can be hung on these handles, and the scraper made heavy enough to dig into the pile of ore.

A Scheme for Transporting Lumber (By W. F. Du Bois).—A contrivance that was very useful to me in transporting lumber is illustrated in Fig. 174. The

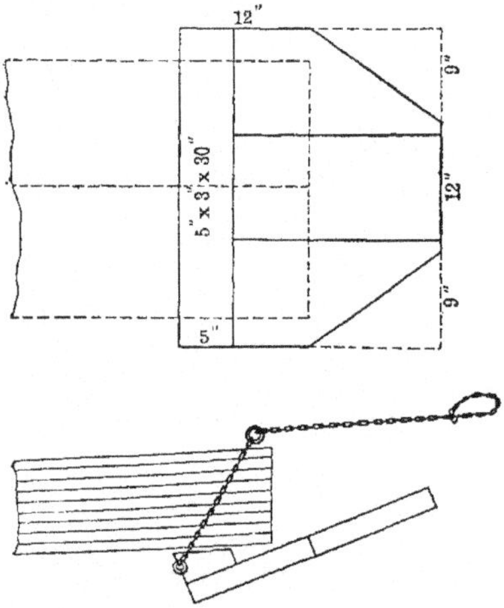

FIG. 174.—A LUMBER LIZARD.

wagon road ended 3/4 of a mile from the mine where I wished to put up an office and mine buildings. The slope from the end of the road to the mine was between 30 and 40°, so that it was impossible to go down with a wagon. The "lizard" used is 2 1/2 ft. wide, 2 ft. long and made of 2-in. plank. On the top is a piece of pine 3×5 in., beveled so that it is 3 in. thick on the back and 1 1/2 in. in front and spiked to the 2-in. plank with 20d. nails. The first planks or boards to be hauled are spiked on top of the 3×5-in. piece with 20d. nails and the successive layers of boards nailed to each lower layer. After 18 or 20 boards,

18 ft. long, are securely spiked, an ordinary 12-ft. log chain is put around the boards, back of the 3×5-in. plank. The end of the chain is fastened to a singletree by a grabhook.

One horse could easily pull the load down the back of the ridge to the mine. On arriving at the mine the chain was thrown off and each board pried loose. The chain and singletree were hung on the harness and a boy carried the lizard as he rode the horse to the lumber pile after another load. The lizard lasted 12 loads. Four trips a day were made.

A Wagon Oil Tank (Chester Steinem).—In view of the increasing use of crude oil as a fuel and for combustion in oil engines, a description of a tank

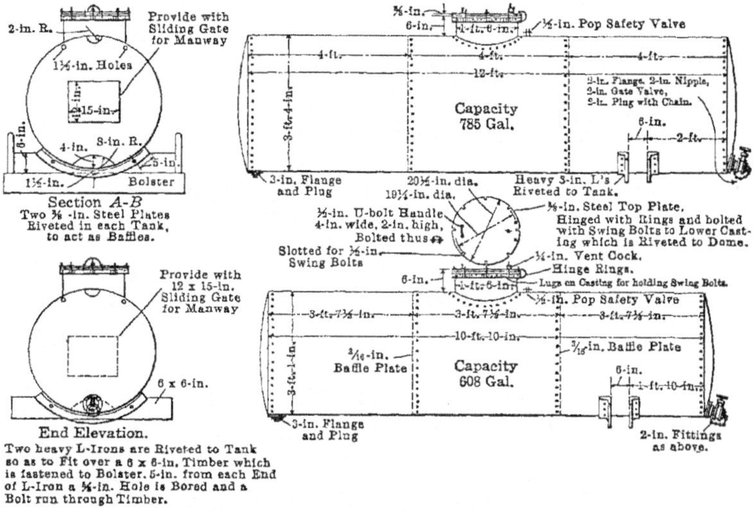

FIG. 175.—TANK FOR WAGONS FOR HAULING FUEL OIL.

designed specially for transporting crude oil is of interest. In the Mogollon district, oil has displaced wood as fuel for generating power, although wood is plentiful. It commends itself as being cheaper for many purposes, and to supply outlying districts a wagon-tank is often necessary. D. Ford McCormick designed the tank shown in Fig. 175. It is giving good service on a 99-mile haul over rough mountain roads.

UNLOADERS AND DUMPING DEVICES

An Automatic Bucket-tripping Device.—An ingenious bucket-tripping device especially adaptable for use in developing prospects is shown in Fig. 176. The bucket is made with lugs, which travel on skids, set below its center of gravity. When, however, these lugs reach the notch *A* the bucket is held, and as soon as the hoisting cable is relaxed it swings about the lugs as an axis and dumps in the bin or on a grizzly as the case may be. The bar *B* prevents the bucket dropping over too far. To lower, the bucket is hoisted until the lugs clear the end of the trip *C*, then it is lowered. The trip swings down and being arrested at *D* by a projecting 1/2-in. angle, forms a track over the notch *A* into which the lugs engaged in hoisting. The trip is heavier on the straight end, so immediately the lugs pass off it swings back into position. This simple device is valuable for small mines or prospects where the shaft is inclined at an angle less than about 70°, as it saves the labor of the man who usually has to swing and dump the bucket or attach the dumping line to the bottom of the bucket.

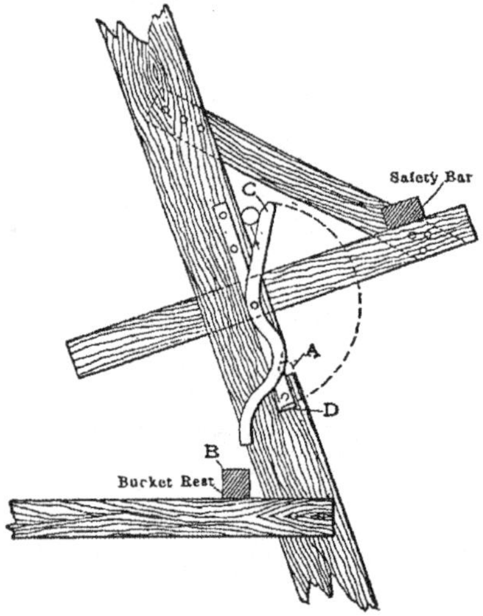

FIG. 176.—AUTOMATIC BUCKET TRIP.

Safety Dump for Sinking Bucket.—The accompanying sketch, Fig. 177, shows a satisfactory and novel method of automatically dumping a sinking bucket. The scheme was used in sinking from the 3150 to the 3300 level in the Kennedy mine, Jackson, Amador county, Calif. Sinking was conducted in the third, or pipe and manway, shaft compartment so that hoisting was not interrupted. At the upper level station a wooden wing door *W* is hinged so that when released after the bucket has been raised above the level, it swings from a nearly vertical position to one across the shaft compartment as is shown in the drawing. In this position it completely closes the shaft. A second door *X* built of sheet iron is hinged at its lower edge to the wing door and fastened to it by a chain *C* at its upper edge. There is an iron lug *L* riveted on the upper side of the iron door. From the bottom of the bucket is rigidly suspended, by four iron straps, a circular hoop *R* of 1-in. drill steel. When the bucket is raised above the level, the same lever that drops the wing door throws out the dogs or chairs *D* which hold the crosshead while the bucket is dumping. On releasing the bucket the iron hoop engages the lug and the weight of the loaded bucket swings down the iron door into the position shown by the dotted lines. The chain keeps this

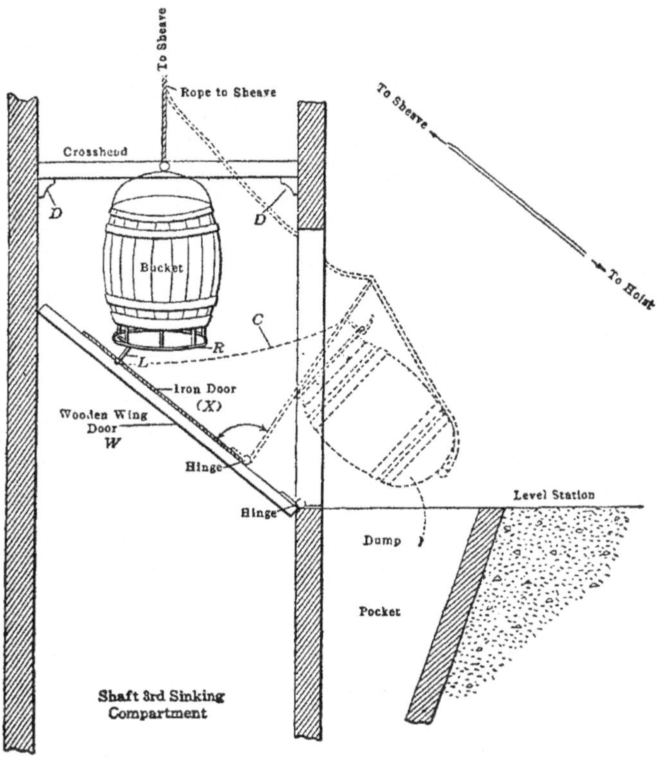

FIG. 177.—ARRANGEMENT USED AT KENNEDY MINE FOR DUMPING SINKING BUCKET.

flap door from swinging beyond the position indicated. The bucket is then hoisted, this swinging the iron door back against the wooden wing door. The engineer then pulls back the wing door and at the same time releases the dogs holding the crosshead. The bucket may then be lowered for another load. The operation of dumping is automatic save for the swinging of the wing door. However, the man operating the hoist attends to this without leaving his position at the engine. The men at work in the shaft are absolutely protected from falling rocks while the bucket is being dumped.

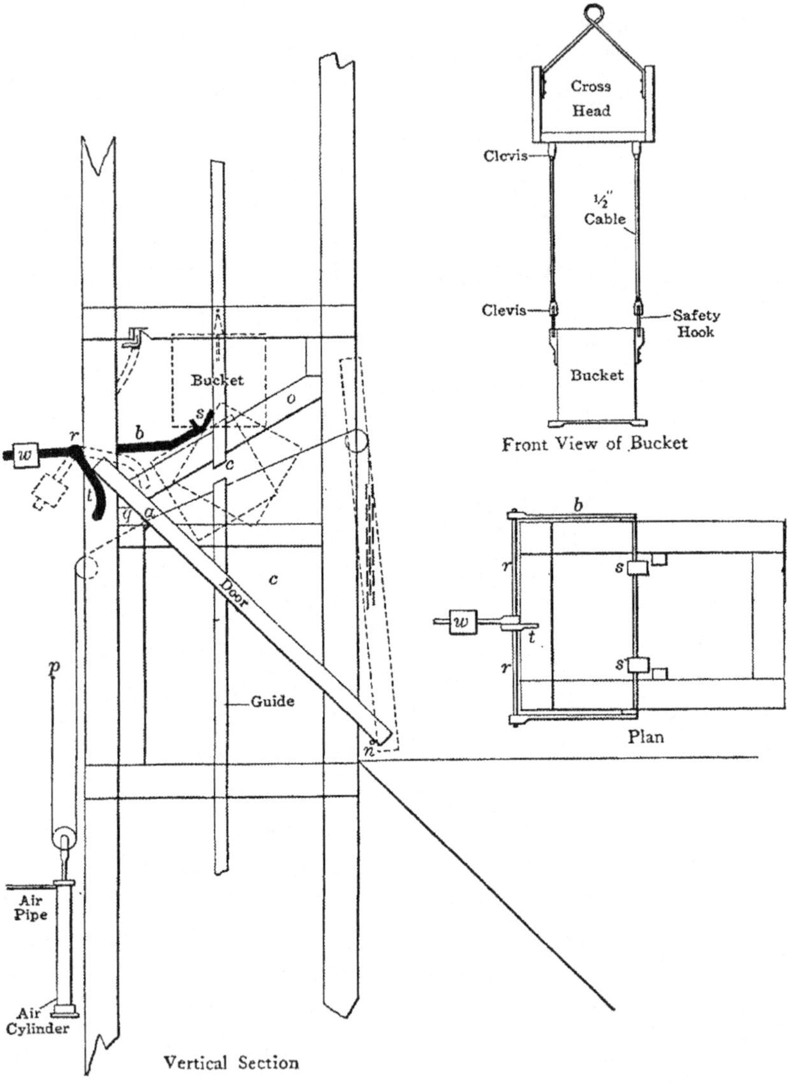

FIG. 178.—SELF-DUMPING BUCKET USED AT BULLY HILL MINE.

Self-dumping Bucket for Winze (By Lawrence May).—The self-dumping bucket arrangement shown in Fig. 178., was designed to meet peculiar conditions for handling ore underground; large capacity was also required. As the shaft is a winze shaft and in extremely hard rock, work above the station level is expensive. Accordingly, the cost of a skip installation put it out of consideration, and the arrangement of this bucket on a crosshead, with platform for men and timber was installed. The bucket is rectangular, 24 in. wide, 30 in. long, and 30 in. deep. A lug projects 3 in. from the side and is set behind the center. Two "ears" on the upper rim of the bucket are drilled for safety hooks hung from the crosshead with flexible steel cables. The door upon which the bucket dumps its contents is mounted on the shaft *N;* at *A* a heavy piece of angle iron on the under side of the door projects beyond the shaft timber, and to it are attached the ropes which operate the door. To allow this angle iron to pass, the guide is cut at *C*. The dumping mechanism consists of the two arms *B*, to which is bolted a heavy iron strip carrying two stirrups *S*, the trigger arm *T*, and the weight arm *W*, all firmly keyed on the shaft *R*. In operation, the bucket is hoisted far enough to clear the door. The engineer then throws a three-way valve, admitting compressed air to the cylinder shown. The piston is forced down, pulling the door by the arm *A*, to its closed position on the bumper *Q*. As the door descends, it strikes the trigger *T*, throwing the dumping mechanism into the position shown by full lines. The bar holding the stirrups rests on the timbers *O*. The whole gear is now held in position by air in the cylinder. The engineer lowers the bucket, the lugs drop into stirrups *S*, and the bucket dumps its contents on the door by turning about the lugs as an axis. After dumping, the bucket is raised to clear the door. The engineer releases the air in the air cylinder. The chain counterweight pulling on arm *A*, draws the door into its clearance position. As the door rises, the weight *W* causes the dumping mechanism to assume the position shown by dotted lines and the shaft is clear for lowering the bucket. The bucket drops on a low truck, is quickly unhooked and run away. A loaded bucket is run under the crosshead, hooked on, and is ready for hoisting. A large weight is needed to start the door up, but less as it ascends. A solid weight heavy enough to start the door would pull it too severely on the finish. Therefore a chain is used, and after the door starts and moves a few inches, the end of the chain rests on a platform. As the door ascends, weight is taken off the pulling rope, by the chain piling up on the platform. Thus the door is almost perfectly counterbalanced in any position and operates smoothly.

Automatic Bucket Dump.—The automatic-dump apparatus shown in Fig. 179, is the invention of R. H. Pascoe, former superintendent of the Burke mines of the Federal Mining & Smelting Co., Cœur d'Alene district, Ida. The main features of construction and operation are clearly indicated by the engraving. The laterally extending trunnions *A* on the bucket (which is suspended by chains from a crosshead) engage in the trunnion recess *B* on the skids when the upper end of the bucket comes in contact with the deflector opposite to the skids, thus causing the bucket to capsize and discharge its contents into the chute or other receptacle. In the engraving the deflector is shown as being composed of wooden pieces faced with iron; 3-in. angle irons are used for this purpose in the Morning mine of the Federal company. The crosshead is preferably constructed of steel I-beams or angle irons, properly braced, with an eyebolt for suspension from the cable. The bucket is suspended from the crosshead by a pair of eyebolts connected with chains of sufficient length to allow easy dump-

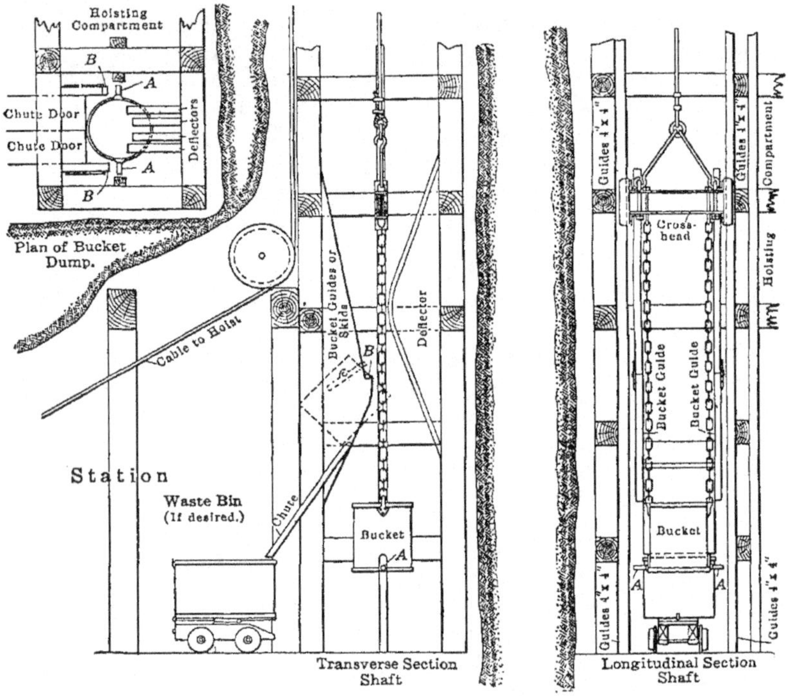

FIG. 179.—AUTOMATIC DUMPING BUCKET.

ing. The crosshead runs between a single pair of timber guides secured in proper alignment. If desired a safety clutch can be applied to the crosshead.

This apparatus has been thoroughly tested under actual working conditions in the mines of the Federal and other companies operating in the Cœur d'Alenes. By its use a 1-ton bucket can be hoisted, dumped, and delivered back to the bottom of 100-ft. shaft in 90 seconds, this having been proved in the actual work of sinking two three-compartment shafts 300 ft. each. This work also demonstrated a 1-ton bucket would more than take care of the rock broken, in addition to hoisting the men and supplies. One great advantage of using this apparatus in shaft sinking lies in the safety afforded the men. No rock falls back into the shaft, and the quick, positive action and freedom from sway and sideplay reduces confusion when ready to blast. As the bucket rides as steadily as a skip, it is unnecessary to line a shaft in good ground. In sinking two- or three-compartment shafts, the chain on one side can be unhooked and the bucket taken to any part of the shaft for loading, thus enabling this operation to be performed readily and conveniently. After loading, the bucket is pulled back into position by the hoist and the loose chain hooked on, when the load is ready for hoisting. Owing to vibration being taken up by the chains, a bucket fitted with a valve makes an ideal bailer. Dumping can be effected at any level at or below the hoist station, making it unnecessary to move the hoist until the limit of capacity of the engine is reached, which cannot be done with an ordinary bucket without the services and extra expense of a topman at the dumping point. If the shaft has more than one compartment this apparatus is best installed in the pipe compartment.

Method of Handling Sinking Buckets (By W. B. Baggaley).—The method of removing buckets from the cage at one of the newer Lake Superior copper properties, which is still in the development stage is shown in Fig. 180. A vertical, five-compartment shaft is being sunk to an ultimate depth of 4000 ft. Two compartments are being used for hoisting rock and water and lowering timber, etc. A floor plan and elevation of part of the shaft house are shown in the illustration.

On nearing the surface the cage slows down sufficiently to pick up the shaft cover, and then proceeds to an elevation of 15 ft. above the floor of the shaft house. Chairs are used in the shaft to prevent the cage descending. The floor car, operated by the compressed-air cylinder shown, and having the bucket car upon it, is run over the shaft. The cage is not attached to the hoisting rope, but rests upon a lug, while the bucket hangs from chains which are permanently fastened to the end of the rope itself. Consequently, when the bucket is lowered the cage remains on the chairs. The full bucket is then lowered to the car below, which has a mounted and pivoted frame into which the bucket fits. The floor car is then pulled back to the normal position, and as it stops suddenly, sufficient impetus is given the car to assist materially the trammers, who push it to the dump. It will be noted that the tram car is of such dimensions that it could not drop down the shaft if any accident should occur.

An empty bucket is now placed on the rope, and hoisted until the lug lifts the cage from the chairs. The latter are then drawn back, allowing the cage to be lowered slowly until the shaft cover comes to rest on the first set of stringers at the level of the floor. The shaft cover has a hole in the center through which the hoisting rope runs. The cage is necessary to act as a guide for the bucket, which contains about 2 tons of rock, and which is hoisted at the rate of 3000 ft. per minute.

This arrangement, which keeps the shaft covered at all times, not only prevents anyone falling in at the surface, but also safeguards the miners working below, from having anything dropped on them when the cage is being loaded with drill steel, etc. The loading of shaft timbers which are swung under the cage, is greatly facilitated. Also the substitution of a bailer in place of the cage

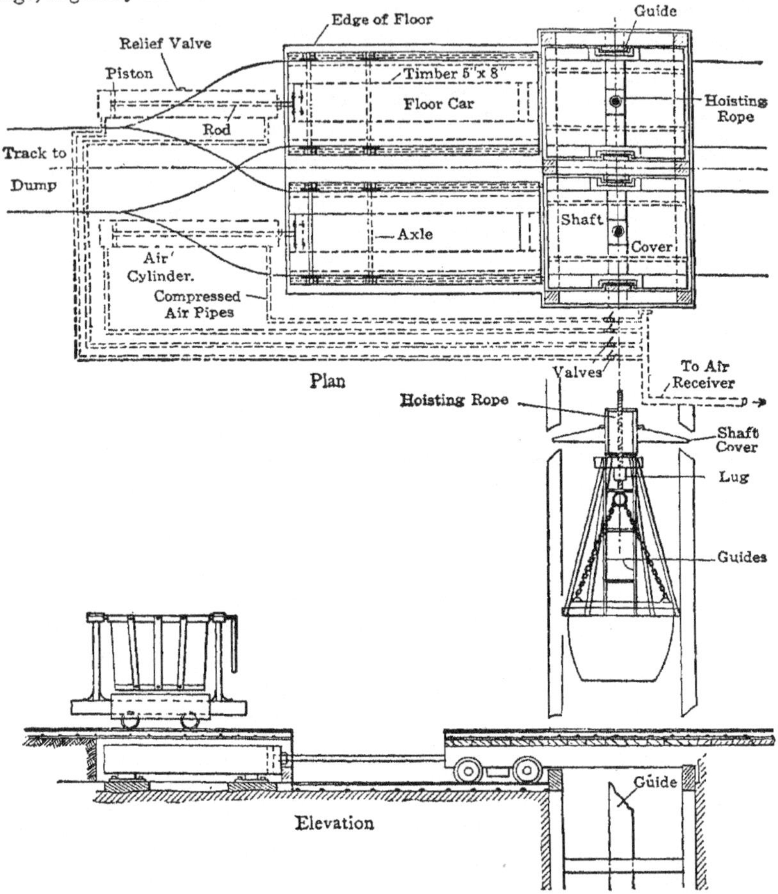

FIG. 180.—SURFACE ARRANGEMENT FOR HANDLING SINKING BUCKETS BY COMPRESSED AIR.

is made easier and safer. The services of two men are saved underground, by having the cage rest upon the lug, as the latter comes to rest on chairs at the lowest level and the bucket continues to the bottom of the shaft. Ordinarily, an auxiliary hoist is installed at the lowest level, which, being operated by compressed air, is much less efficient.

Types of Skip Dumps in New York Iron Mines (By Guy C. Stoltz).—The drawings presented in Figs. 181*a* and 181*b* illustrate several methods employed for dumping skips in the head-frame storage bins at the magnetite mines of northern New York. The car dump, shown in Fig. 1, has been used at the Forest of Dean mine in Orange county. Here the regular 2-ton mine

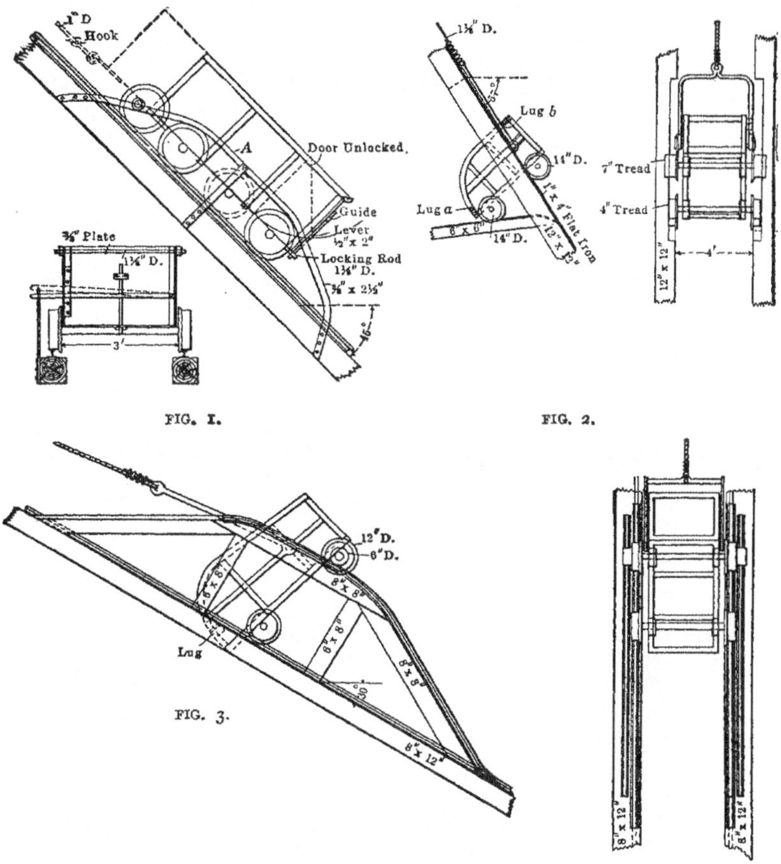

FIG. 181 a.—TYPES OF SKIP DUMPS IN NEW YORK IRON MINES.

cars, equipped with an end-locking door, are hoisted about 1800 ft. up a 23° slope, to a point near the surface where the inclination is increased to 45°. At the storage bin the lever which actuates the locking device on the door is mechanically raised on encountering a flat iron truss *A* bolted to the rail stringer, the ore being discharged from the rear of the car. The car is then lowered to the hoisting level, the door locked, the cable disengaged and the car is trammed to the stope to be filled. The cable provided with a locking hook is then attached to a loaded car which is in waiting at the foot of the incline.

The usual method of dumping adopted on incline hoists at Lyon mountain, Arnold hill and Mineville is shown in Fig. 2. At Mine 21, Mineville, the 2-ton skip cars are hoisted in balance on a 57° slope, 800 ft. from the lowest level, and are dumped above the grizzlies by allowing the front wheels of 4-in. tread to leave the main stringers and take an almost horizontal course on a set of auxiliary

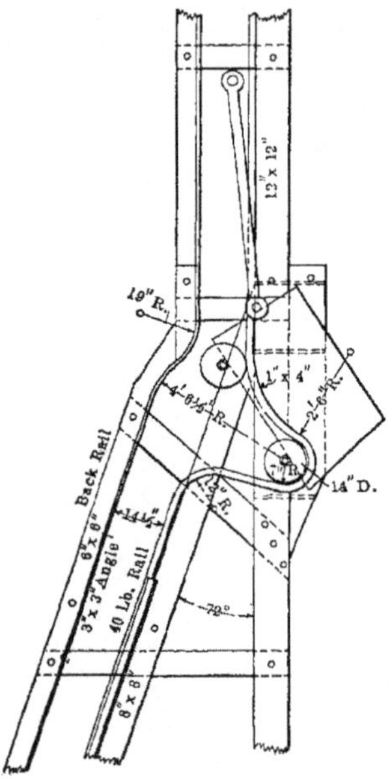

FIG. 4.

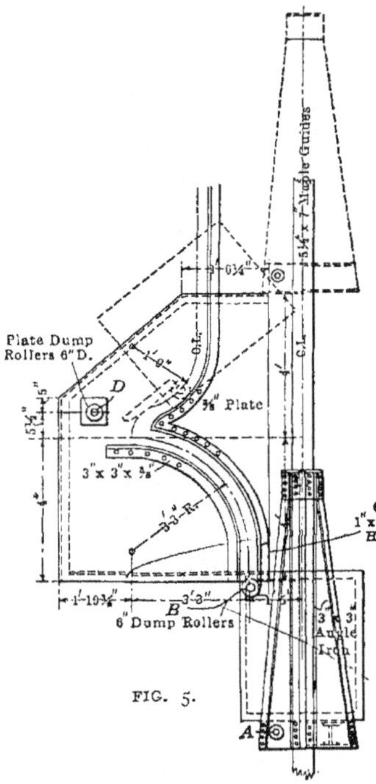

FIG. 5.

FIG. 181 b.—TYPES OF SKIP DUMPS IN NEW YORK IRON MINES.

stringers, while the rear wheels of 7-in. tread run over the dumping point on the inclination of the main stringers. No guides are used on the hoist and the front wheels of the skip are prevented from lifting off the rails in hoisting by having the bail press against lugs riveted on each side of the skip.

The wheels are stationary on the axle, the latter turning in brass-lined journals. This method of stationary wheels is to be recommended since the wheels cannot become loose or wabbly by wearing of wheel hubs or axles and consequently reduces to a minimum the chances of skips jumping the tracks. In case of an overwind through the dump the skip is hoisted in its dumping position of 45°, since lugs on the skip at the back bear against the bail as the skip assumes the angle of discharge.

The skip used at the Cheever mine, near Port Henry, N. Y., is shown in Fig. 3. The 4-ton skip cars are dumped by having the front wheels of 4-in. tread keep to the 30° inclination of the main stringers, while the rear wheels (which are cast with the usual 4-in. tread and diameter of the front wheel together with smaller wheels of like tread and smaller diameter) by means of the extended tread pass over a timber truss which causes the skip to assume the angle of discharge. In case of an overwind the upper members of the dumping truss provide for this and are inclined toward the inclination of the main stringers, so the skip on descending would readily stay to the stringers and lower over the truss and down the shaft.

The skip-dumping arrangement, in Fig. 4, used by Witherbee, Sherman & Co., at the Harmony A and B, Smith, Joker and Bonanza mines, in the Mineville district, is well designed and differs much from the usual types of dump. The fore and rear wheels have the same dimensions and are so guided at the dump that the tilting of the skip is positive, and in case of an overwind the back track guides will enable the skip to right itself quickly. At the Joker steel headframe 4-ton skips are dumped, while the same method is applied at all of the hoistways, vertical and inclined.

The dump used by the Port Henry Iron Ore Co. in the steel headframe at the Clonan shaft is shown in Fig. 5. The 4-ton Kimberley skips are turned on the axle *A* by means of the rollers *B* taking a course away from the vertical guides as described by the angle-iron guides until the horns *C* intercept the rollers *D*, whereupon the skip rollers *B* are elevated to the upper angle-iron guides. The skip overwinds at approximately the angle of discharge.

Throughout the district, the hoisting cables are plow steel, 1 in. to 1 1/8 in. in diameter, and are generally attached by passing through a clevis on the bail of the skip, being lapped 3 ft. and held by three to six grips. Iron head sheaves are 5 to 8 ft. in diameter. Eight to 14 ft. are allowed for overwinding. The usual gage of skipways is 4 feet.

[The arrangement illustrated in Fig. 5 of Mr. Stoltz's article is that of the ordinary Kimberley dump. At the Kennedy mine, at Jackson, Amador county, Calif., an improvement of this idea is used by which the necessity for lugs on the skips is obviated. In place of having the "plate-dump roller," indicated at *D*, which engages the horns *C* on the skip and thus makes it possible to further elevate the bottom of the skip, two hooks of flat iron, spaced about 2 ft. apart, are hung from the headframe so as to engage the lip of the skip as it is tilted by the curved tracks. These hooks support the skip while the bottom is being raised, just as does the bar on which the horns catch in the skip

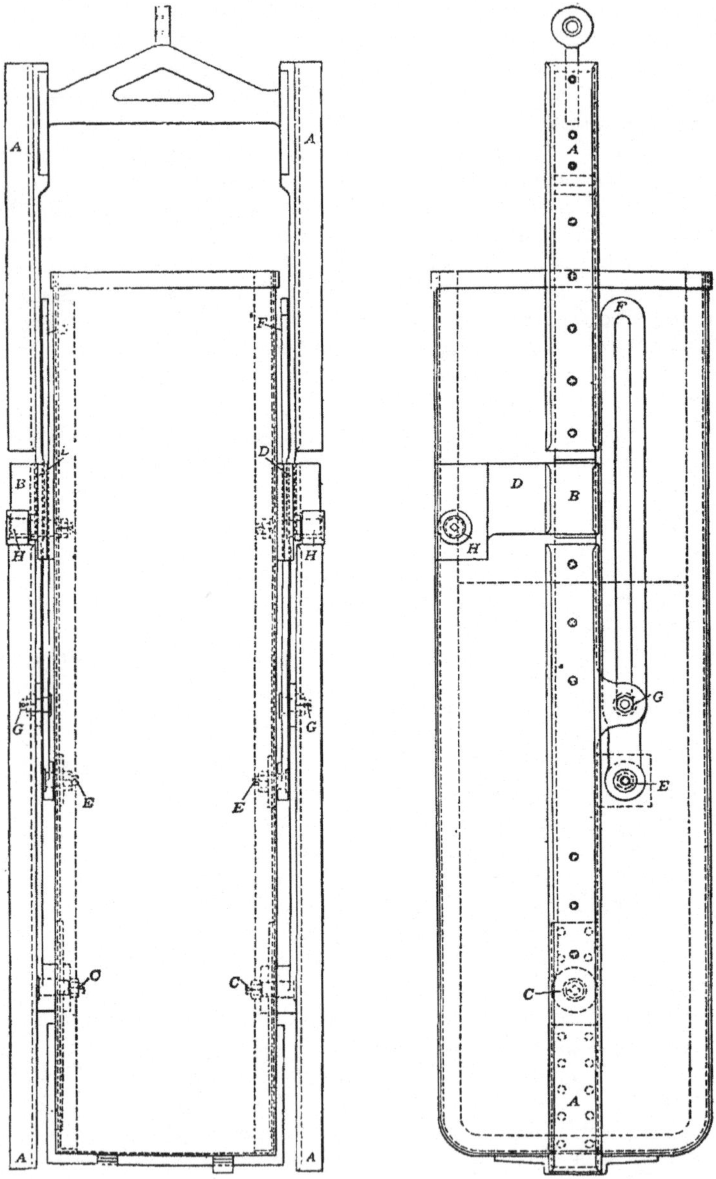

FIG. 182.—SELF-DUMPING SKIP USED AT ORIGINAL MINE, BUTTE, MONT.

used at Port Henry. As the hooks are free to swing there is less jar on the skip when they are used than is usual with the ordinary device.—EDITOR.]

The Original Consolidated Self-dumping Skip.—A self-dumping skip of novel and satisfactory design has been worked out and used for some time at the Original Consolidated mine, Butte, Mont. In fact, so satisfactory has the operation of these skips been at the Clark mines, that they have been copied by the Amalgamated company, and more recently by the Miami Copper Co., at Globe, Ariz. The particular feature of the design of the skip is the dump mechanism which is simple and self-contained. The skip is durable and also of convenient proportions.

The details of the Original skip are shown in Fig. 182. As will be noticed, the guide shoe extends over the full length of the skip bail. This is one of the most important features of the design, for the usefulness of any hoisting conveyance is limited the life of the shoes. By distributing the wear over a large surface the life of these is prolonged materially. This is especially necessary with the fast hoisting practised at Butte.

The portion of the shoe marked *B* is not connected to the bail, as will be explained. The skip is hung from the bail on the axles *C*. The portion of the guide shoes *B* is riveted to plates *D* on the skip body, instead of being fixed to the bale. The support afforded by this part of the shoe serves to keep the skip in a vertical position in the shaft. Slotted steel bars *F* are bolted to the skip body and rotate about *E* as pivots. The skip bail has a projection which extends over the lower end of the slotted bars and the bolts *G* pass through the slots and the extension of the skip bail. In dumping these bolts slide up and down the slots.

As usual with self-dumping skips, there is a roller *H* fastened on either side of the skip at a point near its top and close to its front edge. These rollers engage, at the discharge point, in tracks curved up and away from the shaft so that as the skip is hoisted it is tilted forward about the axles *C* and discharges its contents. As the skip swings down, the bolts *G* slide up the slotted bars *F* which swing between the skip and its bail. The guides in the shaft (or headframe) must, however, be cut away at the point of discharge so as to allow the portion of the shoe *B*, that is fixed to the skip to swing free. The skip turns over until the bolts reach the end of the slotted bars. On lowering the skip, the rollers run back down the curved guide track and the skip body is again swung into a vertical position. The governing action of the slotted bars is more positive and quicker than that of the arrangement of projecting lugs to engage a crossbar, ordinarily used on self-dumping skips. The top of the curved guides for the rollers is cut away at the point reached by the rollers when the skip is tilted as far as the slotted dump bars permit. This allows the dumped skip to be raised farther so that slight overwinding does no damage.

The skips at the Original mine are 11 ft. 4 in. deep, 3 ft. 5 in. from back to front and 3 ft. 4 in. wide. The total length of the guide shoes is 14 ft. 2 1/2 in.,

1 ft. 2 1/2 in. being cut away from the bail at a point 5 ft. below the top. The diverting rollers are centered at a point 8 ft. 1/2 in. from the bottom of the skip and the bail hung 2 ft. 1 1/2 in. above the bottom. The slotted bars are centered 2 ft. 8 in. above the skip bottom. The skip is constructed of 5/16-in. steel.

Skip Changing Device at Leonard No. 2 Shaft.—A device by which skips and man cages are interchanged in 2 1/2 minutes is used on the steel headframe

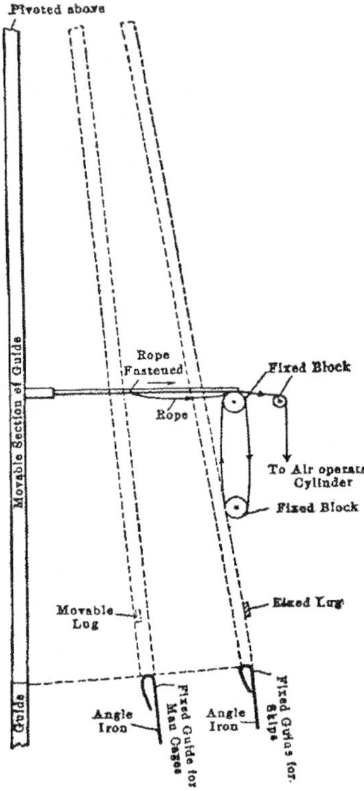

FIG. 183.—ARRANGEMENT FOR INTERCHANGING SKIPS AND CAGES ON LEONARD NO. 2 HEADFRAME, BUTTE.

at the No. 2 shaft of the Leonard mine at Butte, Mont. As shown in the diagrammatic sketch, Figs. 183, a portion of the guide in the shaft is cut away and pivoted at the top so that the lower end may be swung over to either of two other fixed guides on which cages and skips are supported when not in use. The movement of the loose section of the runners is effected by a compressed-air cylinder.

A rope is fastened at one end to a rod that is connected to the guide, passed over blocks, as shown in the sketch, and the other end fixed to the piston rod of the air cylinder. Movable lugs operated by hand levers are so arranged as to engage the moving section of the guide at the proper point so that it will form a continuation of a fixed guide for the man cages. By throwing these lugs back the guide may swing over against fixed lugs so as to form a continuation of a fixed guide for skips. These two fixed guides are made of angle iron. When it is desired to take off the skip and use man cages, the skip is hoisted above the lower end of the section of movable guide, this section is swung out against the fixed lugs, the skip lowered on to the fixed angle-iron guide and the hoisting cable disconnected. The section of the guide is then swung down to engage the movable lugs, which are thrown out, and the hoisting cable fastened to the man cage which is then hoisted far enough to allow the section of guide to be swung back into the vertical position. In interchanging skips and cages it is necessary to fasten the hoisting cable with ropes before it is removed from the skip or cage. The entire operation of changing from skip to cages or *vice versa* is quickly performed by a man stationed at the platform in the headframe where are situated the levers for operating the air cylinders that control the swing of the guide and for throwing out the lugs to engage the guide.

Crane for Changing Skips.—At mines where hoisting is done in skips it is advisable, and sometimes necessary, to have a special cage for lowering and hoisting men. At the Burra Burra mine of the Tennessee Copper Co., Ducktown, Tenn., cages are interchanged and made ready for lowering in from 30 to 60 seconds. A similar arrangement was successfully used for a number of years at the famous Gwin mine on the Mother Lode of California. The changing device consists of two cranes at the shaft mouth, one to handle each skip. The cranes are made up of iron-pipe verticals with channel-iron swinging arms stayed from the top of the uprights. The swinging arms carry small traveling crawls to which are attached the chains by which the skips are lifted. The cranes are connected across the top and are also stayed in several directions with wire cables. The chains can be attached to the skips by slipping the hooks at their ends into eyes which are provided on both the man and ore skips.

The operation of changing the skips is as follows: When a skip is raised to the surface the chains are attached to it, the cable secured (as will later be explained), the thimble removed and the skip then swung out of the way by a rope attached to the arm of the crane; the other skip is then swung in from the other crane, the cable attached to it, chains knocked loose, cable made free, and it is ready to lower. A small winch is provided on the headframe above the shaft's mouth. When skips are to be changed, a couple of men must stand ready at this winch in order to fasten the cable so that there will be no danger of its being pulled over the sheaves. A hitch is thrown about the cable with a piece of rope secured to and drawn taut by the winch. As stated, the whole operation of changing skips occupies only about a minute, and from the added security obtained in raising and lowering employees, the small cost of installation is certainly warranted.

Self Acting Tipple.—In Figs 184 and 185 are shown the details of a satisfactory self-acting tipple which was developed in the Birmingham district and is used by the Sulphur Mining & R. R. Co. at its mine near Mineral, Louisa county, Va. The loaded ore car is run by gravity on the tipple where the curved

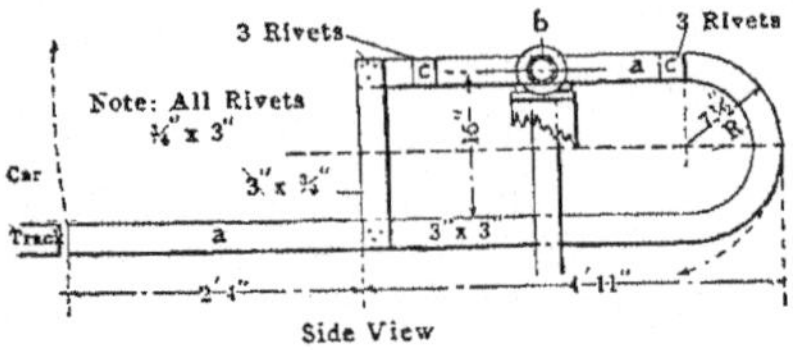

FIG. 184.—ARRANGEMENT OF TRACK FOR TIPPLE.

3×3-in. irons *a* engage the wheels. The tipple is so designed that the weight of the loaded car is sufficient to cause it to rotate about the axis *b*, and hence dump the car. It is also heavy enough to swing back into position after dumping with sufficient force to run the car off the tipple and back on the track. A wheel and band brake may be used on axis *b* at *A*, but this is not necessary. This form of tipple is simple in operation, easily constructed and gives good service.

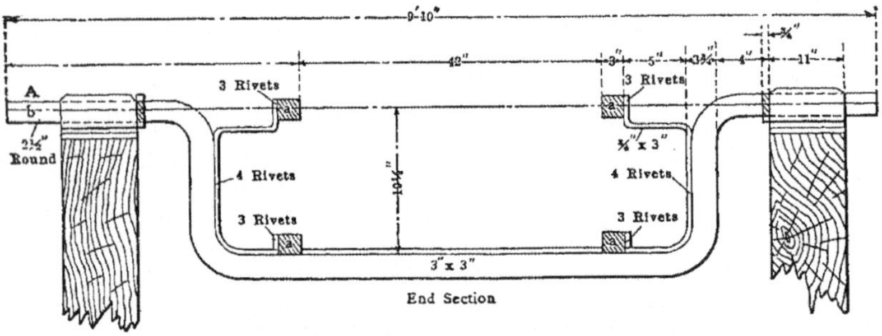

FIG. 185.—AUTOMATIC TIPPLE WITH SUPPORT.

The Republic and Tennessee companies, in their newer slopes on Red Mountain, Ala., hoist in 10-ton skips which, in headings are filled from 2-ton cars, dumped by tipples directly into the skips. The details of the tipples used in headings by the Tennessee Coal, Iron & R. R. Co., are shown in Fig. 186. The cars are strong and simple in construction. They are made of 3/8-in. steel, braced and reinforced as shown. By having the body of the car low and long, 2 ft. 4 in. × 6 ft. 6 in., several advantages are gained. In the first place the cars are easy to load, which is an important consideration. This construction, on account of the low center of gravity, also adds stability and reduces the liability of the car to overturn or jump the track; at the same time the long body does not make tipping excessively difficult. The low body also allows the dump end of the car to be hung 6 in. above the side, thus giving an opening sufficiently wide to allow the ore an easy passage from the car. The illustration shows the car in the tipple in position of dumping the extra clearance gained by having the movable end hung above the sides of the car being brought out by the dotted lines.

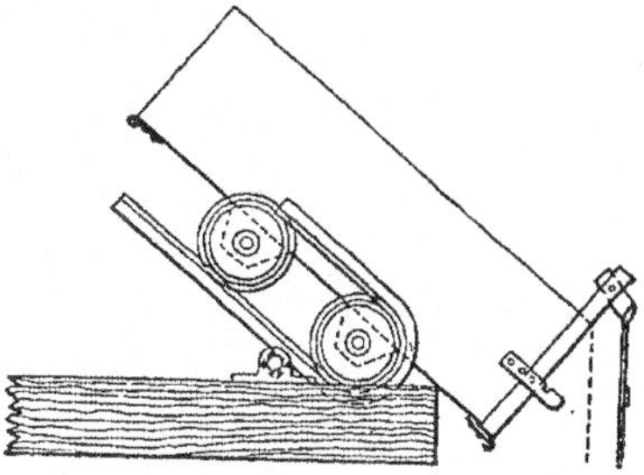

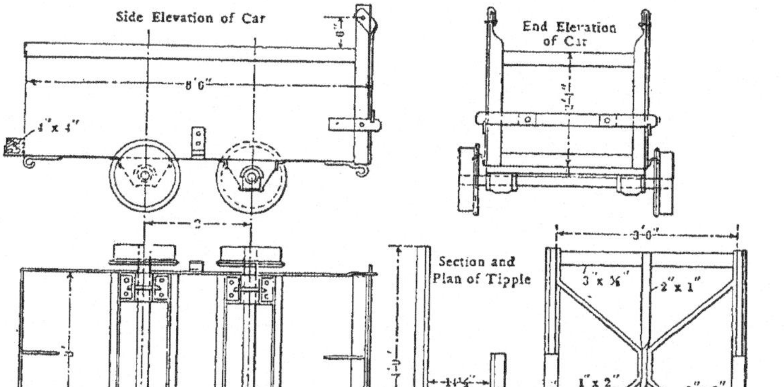

FIG. 186.—CONSTRUCTIONAL DETAILS OF CARS AND TIPPLES.

The tipple is of simple construction, being built of pieces of 40-lb. track rail curved back, so as to grip the wheels of the car, and cross-braced with a 3×5/8-in. iron at the track end, and a 2 1/2×5/8-in. iron on the curve, 7 in. above its base. A lateral brace, 2×1 in., connects these cross-braces and it is twisted at the center so that it may be riveted to pieces of similar material, with the larger dimension placed vertically. A piece of 2 1/2×2 1/2-in. iron is used for the axle, 8 in. being allowed at either end for bearing.

Tram Car Tipple (By Guy C. Stoltz).—A tram-car tipple used by the Cheever Iron Ore Co., and Witherbee, Sherman & Co., in the Mineville, N. Y., iron district, is similar in principle to the one described above. A distinctive feature in this tipple, however, is the bumping-block arrangement which is attached to the frame proper as shown in Fig. 187. The tipple is made of 8×8-in. timbers, securely tied together and operating on a frame of 8×8-in. timbers. As the car assumes a position of 45° in discharging its contents, the back of the

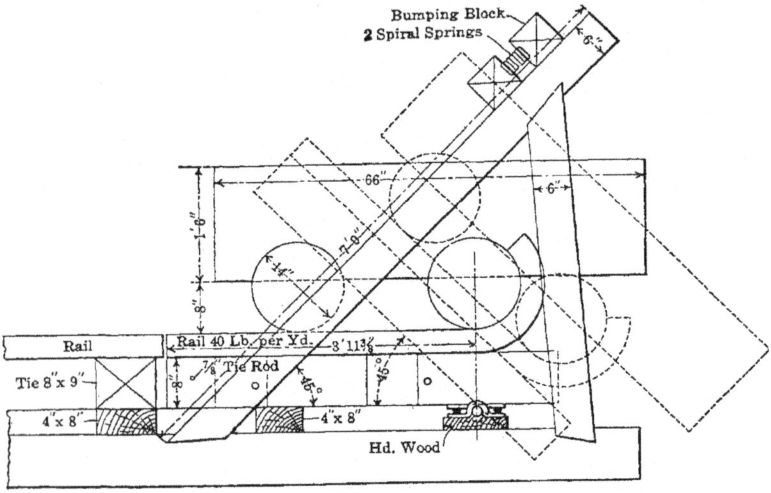

FIG. 187.—TIPPLE USED BY WITHERBEE, SHERMAN & CO., MINEVILLE, N. Y.

car strikes against a bumping block, which has behind it, two spiral springs which contract 3 in. before the block will press against a secondary bumper. This device lessens the shock to the tipple and frame in dumping and makes a saving on tram-car repair bills. The axle is in such a position with respect to the center of gravity of the car and tipple, that a tram car will automatically dump when pushed on the tipple at the average rate a man walks.

Revolving Tipple.—The revolving tipple for mine cars has the advantage that it simplifies the construction of the car. It saves a large number of castings and no hinges are necessary. The car can be made much lighter on this account—simply a sheet-iron box mounted upon wheels. The tipple may be made any size, but a convenient one for a car that one man can handle is 50 in. in diameter and 6 ft. long. The two rings are made of 3/4×4-in. steel bars, held together by 3-in. angle irons at the top which engage the top of the car. The ends of the angle iron are split and bent back on the ring and fastened.

Rails are fastened in the bottom of the ring as shown in Fig. 188. The track is 24-in. gage and the entire frame is mounted upon four ordinary car wheels, two at each end. With the loaded cars the center of gravity is above the center of the circle and the car is practically self dumping. When the car has been

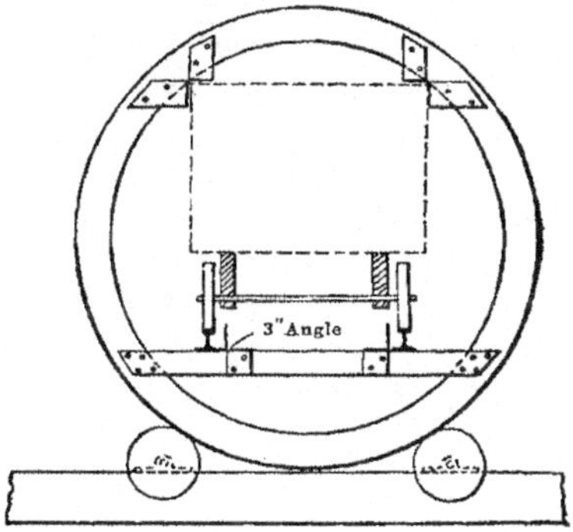

FIG. 188.—REVOLVING TIPPLE FOR ORE CARS.

emptied, the center of gravity is below the center of the circle and the car again returns to its original position.

Automatic Trip for Ore Cars.—The automatic tripping device shown in Fig. 189 is used in several mines at Iron Mountain, Mich. By means of this

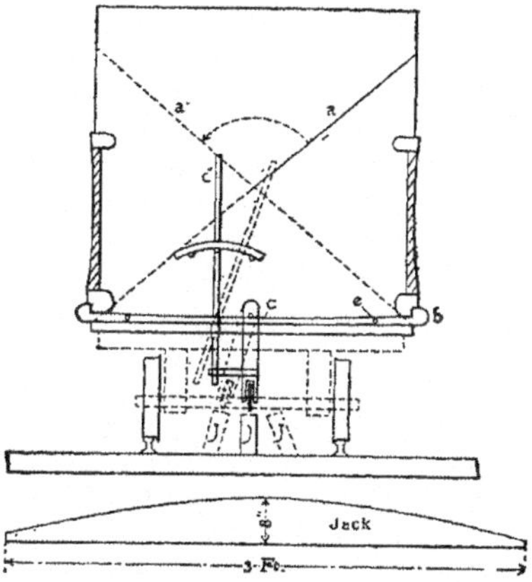

FIG. 189.—CAR WITH AUTOMATIC TRIP.

device it is possible to dump one car on any one of three piles. This is accomplished by a lever *d* in front of the car, which may be set in three different positions. Beneath the car is a small wheel which strikes a jack *j*, set between the rails at any required position on the stockpile. As the wheel passes over this jack it is lifted and the catches *b*, which hold the side doors, are released and the doors opened. The arm of the catch *b* is slotted at *c* to give free movement when the wheel encounters the jack. Similar catches are operated on the other end of the car by a connecting rod through *e*. One jack is set in the center of the track and the wheel strikes this when the lever is in a vertical position. The other two jacks are set a little off center and are encountered by the wheel when the lever is set at one side. The car tender at the skips sets this lever for a certain stockpile before the car leaves the shaft. During the winter months, 100,000 tons of ore are handled per month in this way from a single shaft. At the Pewabic mine a similar car is used for sending ore to the various bins, but instead of the automatic trip being changeable the side-dump car has a movable bottom *a*, extending from the center to the top so as to deliver the ore on either side of the car, but not on both sides at once.

X

SAFETY APPLIANCES

Hooks and Thimbles—Crossheads and Safety Catches—Landing Chairs, Shaft Gates and Other Appliances—Notes on Mine Track and Switches.

HOOKS AND THIMBLES

Hoisting-bucket Hooks.—The greater part of the ore produced by the Joplin mines is raised to the surface in buckets of 800 to 850 lb. capacity. These buckets are raised at such a speed that a round trip from a depth of 250 ft. is made in 35 seconds. To maintain hoisting at this speed, a hook must be used on the hoisting rope that will permit the hooking and unhooking of buckets with the utmost facility. Snap and "pig-tail" hooks, as shown in Fig.

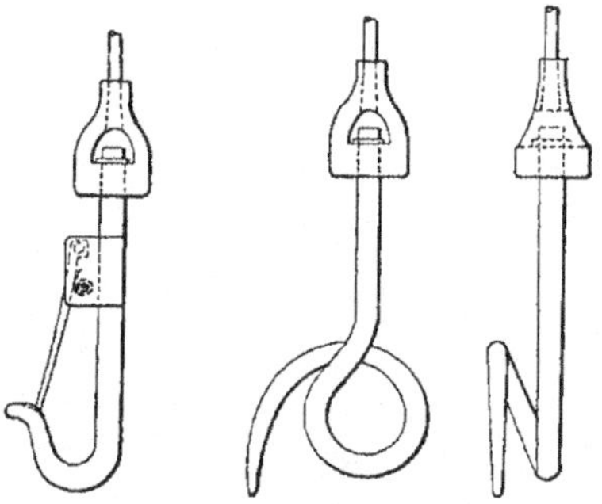

FIG. 190.—SNAP AND PIG-TAIL BUCKET HOOKS USED IN JOPLIN MINES.

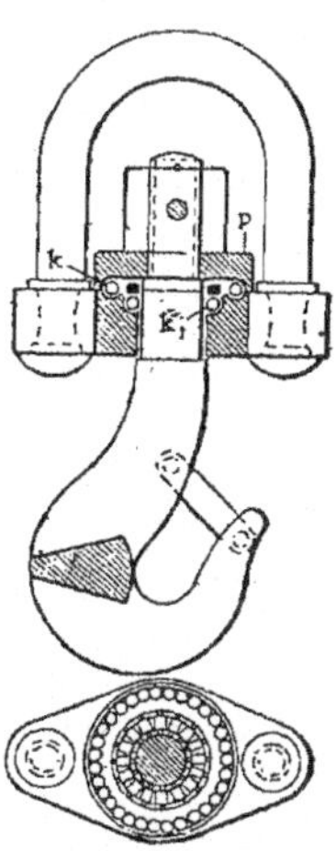

FIG. 191.—SWIVEL HOOK FOR HOISTING.

190, are used, the former being considered safer. The snap hook is made of 1 1/4-in. steel and the snap is held in a groove in the point of the hook by a coiled spring that is housed in the sheath welded to the shank of the hook. The pig-tail hook is made of 1 1/4-in. iron, bent in a spiral 4 in. in diameter and of

2-in. pitch, which permits easy entrance of the 1 1/4-in. bail of the bucket. Both the snap hook and the pig-tail hook are attached to the hoisting rope by a swivel. By constructing the hook with ball bearings in the swivel joint as shown at *K* in Fig. 191, twisting of the cable is not communicated to the bucket.

A Safety Hook for Hoists.—A patent was recently granted in Great Britain for the safety hook shown in Fig. 192, the operation of which is apparent. The opening of the hook is closed by the simple lever *A*, one end of which is attached to the sliding member *B* as shown. When the rope is slack the sliding member slips to its low position opening the hook by raising the end of the lever *A*. Upon taking up the slack of the hoisting rope, the hook opening is closed by the lever.

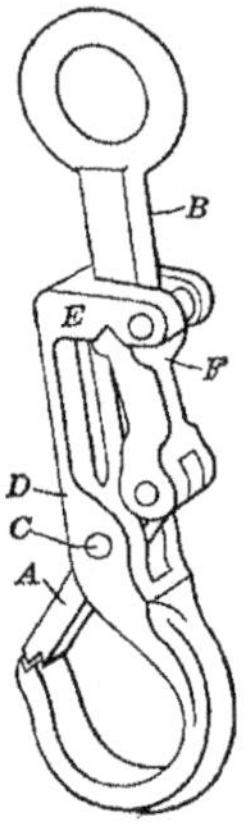

FIG. 192.—A SELF-LOCKING HOIST HOOK.

The pull of the rope does not come on the pin *C*, but upon the part *E* of the body of the hook *D*, a projection *F* of the sliding member engaging the part *E* of the body.

Safety Crane Hooks.—Three safety hooks in use in the works of the National Tube Co. are shown in Fig. 193. A safety hook having a lock or safety piece which closes the entrance to the hook when in place, and is held from opening by the sliding collar on the shank of the hook, is shown at *A*. When the safety locking device has closed the entrance and the collar has dropped into the notch, it is practically impossible for the sling to get away. If the latch is not in place and the hook made safe, the collar will not fall. When the collar has fallen into place, the sling cannot come out of the hook except by exerting a force sufficient to bend and displace the latch. At *B* is shown a form of toggle hook for use in lifting large plates. The sister hooks were formerly forged in one piece and the weight of the plate was depended upon to jam the jaws so tightly by friction that the plate was not likely to slip out and hurt anyone. In the modification and improvement shown in the cut, it will be apparent that the end of the flat link can be forged with the hole eccentric to the curve of the end, so that when the strain comes, the bite of the cam will be the stronger upon the plate the heavier the applied load. It is practically impossible for such a hook to slip off as long as the weight of the plate is borne upon the suspending chain. At *C* is shown a crane hook with a handle for guidance, as used in some of the western smelteries. By using the handle the operator is in no danger of having his hand caught while holding the hook in place until the chain to which it is attached has been drawn taut.

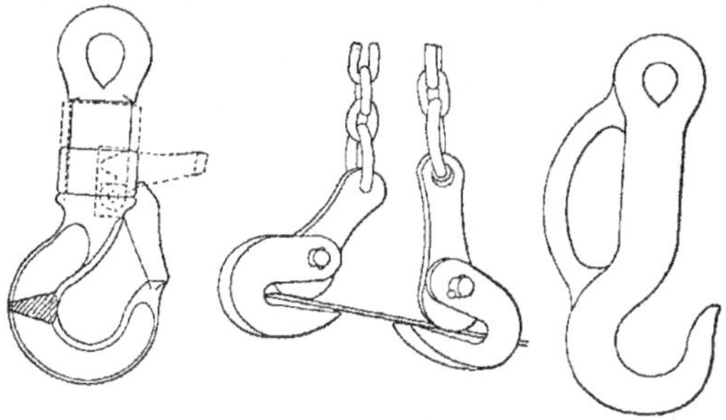

FIG. 193.—TYPES OF SAFETY CRANE HOOKS.

Details of a patented self-closing safety shackle hook for hoisting work are shown in Fig. 194. There have been many accidents due to the disengaging of ropes and chains from hooks and shackles, and this device aims at their prevention. As shown in Fig. 1, *A* represents the attaching member; *B* the hook member. The attaching portion is formed with an eye *C*. A pivot bolt *H* is passed through the attaching member and the hook member to secure the two parts together. This pivot bolt is arranged a little to one side of where the line of load falls on the hook. When there is a pull simultaneously on the eye and hook, it will force the hook tightly against the guard-finger *D*, and the harder the pull the more the hook member will be sustained by the pressure of the guard-finger against the end of the hook, thus preventing to a considerable degree any chance of straightening out the hook under a heavy load. From the lower end of the attaching member there is a heavy inward thrust against the lower end of the hook member when a load is applied, which together with lug *E* greatly strengthens the hook. It is claimed that a hook made in this manner, of good material, should be safe, handy and stronger than an ordinary open hook as the stress is transmitted to all parts; that it is useful for all kinds of hoisting work where ordinary hooks are not to be relied upon, and particularly where shackles are used and quick changes are required. In using shackles sometimes the pin is carelessly or incompletely inserted, sometimes resulting in serious accidents. This danger is obviated by the use of a hook of this kind because as long as the weight is on the hook it cannot open, but for still further security, a cotter pin can be inserted in the hole formed in the lug *I* if desired. This hook has been indorsed by the leading casualty companies and the American Museum of Safety, after a thorough investigation. It is the invention of F. F. Norden, and is being used and placed on the market by the Foundation Co. of New York.

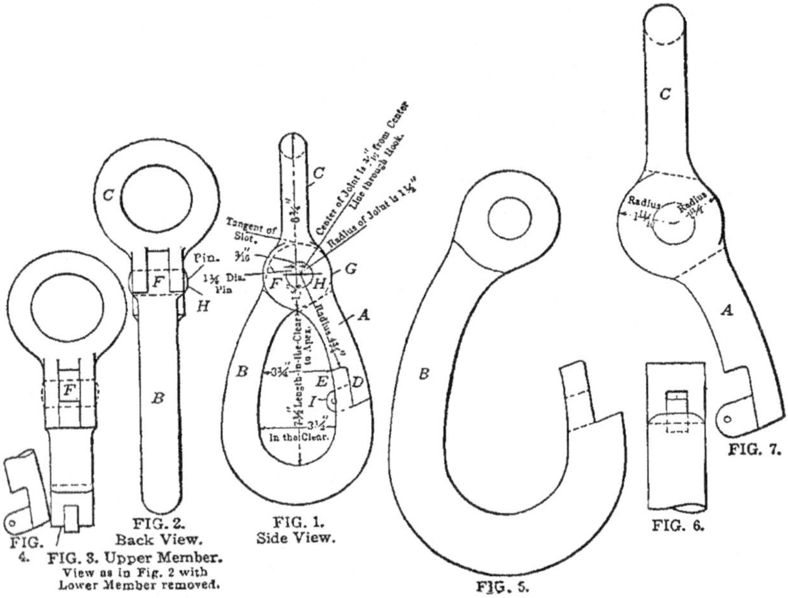

FIG. 194.—A NEW SAFETY CRANE HOOK.

Thimble for Hoisting Cable.—In hoisting the 10-ton skips used in the slopes of the Tennessee Coal, Iron & R. R. Co. on Red Mountain, Alabama, much difficulty was experienced in fixing the wire rope to the bails of the skips so that they would not pull out. The pull exerted by the hoisting engines is so great that ordinary clamps will not hold at all. This trouble has been overcome by using thimbles of the type shown in Fig. 195. The eye-piece *a*, through which the bail of the skip passes, is keyed to the main shank *b* of the thimble by two square lugs *d*. The clamps *e*, which are drawn down on the cable by the bolts *g*, hold the keys in place. The clamps *f* are similar to the larger ones and are bolted in the same fashion. The thimble is fitted to the skip bail by removing the larger clamps, thus releasing the eye-piece which is placed about the bale, the key lugs and clamps then being replaced. The hoisting cable passes under clamp *f*, under *e*, around *a* and then back under the opposite ends of *e* and *f*; *b* and *a* are grooved to receive the rope as shown in the sectional views. Finally the clamps are drawn down upon the cable by the bolts upon which lock nuts are used. Usually several ordinary clamps are used to hold the loose end of the rope to the taut end.

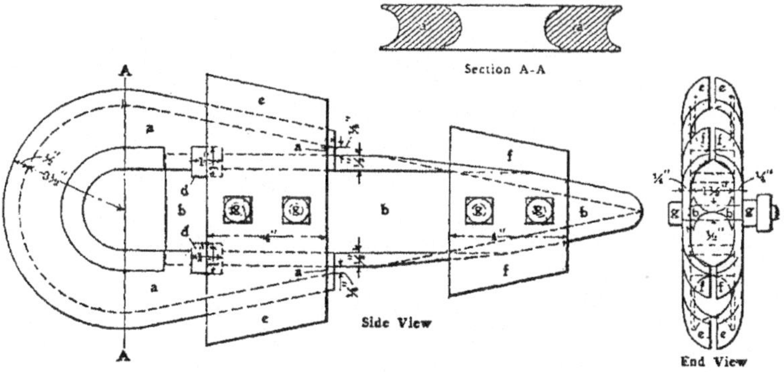

FIG. 195.—DETAILS OF HOISTING-CABLE THIMBLE.

CROSSHEADS AND SAFETY CATCHES

Safety Crossheads for Hoisting Buckets.—Because accidents have been caused by the falling of crossheads in shafts, the Ontario Bureau of Mines calls attention in its report for the year 1911 to the safety crossheads that have been designed and patented by Mr. Morin, master mechanic at the Nipissing and Mr. Sargeson, master mechanic at the Waldman mine. The object of the design of these crossheads is to prevent them from falling when they stick in the shaft. In the Sargeson crosshead, which is shown in Fig. 196, the attachment *A* is fastened to the crosshead at *C*. If the crosshead sticks, this arm automatically engages the clip *B*, attached to the cable, and so stops the bucket. In sinking operations the arm *A* is automatically tripped by the stop block *E*, allowing the bucket to descend to the bottom of the shaft. In Fig. 196, *A* shows the the attachment in normal position; *I*, the attachment tripped by the crosshead stop; *B*, the clip in normal position; *J*, the clip lowered through the tripped attachment; *C*, the daw pin; *D*, the cable; *E*, the crosshead stops; *F*, the guides; and *H*, the crosshead.

The same principle is followed in the design of the Morin crosshead. It is further equipped with an automatic safety device, which by the aid of springs enables dogs to grip the guides, thus preventing the crosshead from falling.

Safety Crossheads for Bucket Shaft.—A crosshead is in use at the Colby mine at Bessemer, Mich., which holds the bucket whenever the crosshead is stopped either accidentally or intentionally. C. E. Holley describes the device as follows: At a suitable distance above the bucket, the thimble shown in Fig. 197 is attached to the cable by U-bolts. The crosshead is made up of channels bolted to vertical timbers. Across the bottom channels is a yoke *Y* through which the cable passes, and this yoke rests on the top of the thimble, thus supporting the crosshead. The dogs *D* close in below the collar of the thimble preventing the bucket leaving the crosshead in case the crosshead should stick in the shaft while being lowered. When the crosshead strikes the stop at the bottom of the guides, the lever *L* is raised, raising the wedge *W*. This wedge forces the dogs apart and permits the thimble to pass down be-

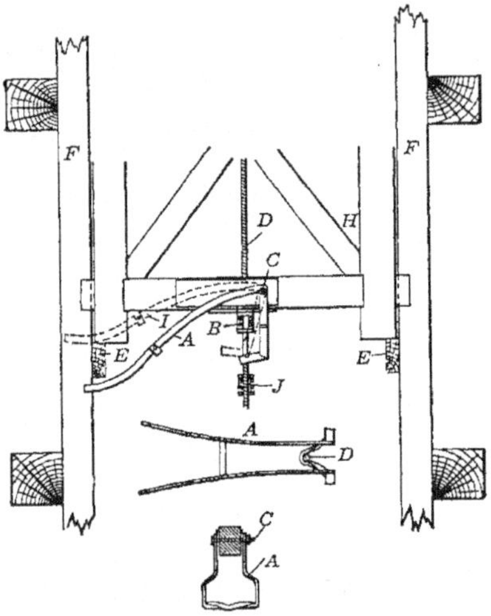

FIG. 196.—SAFETY CROSSHEAD USED AT A COBALT MINE.

tween them, lowering the bucket without the crosshead. The wedge is guided by an extension, which slides in a hole in the yoke.

The details of the crosshead used in sinking the Morning shaft, at the Federal Mining & Smelting Co.'s mine, near Mullan Ida., are shown in Fig. 198. The design is satisfactory, being the result of several experiments made by the company along this line. One great advantage of such a crosshead is in its lightness. One man can handle it when necessary. At the same time it is strong, sure in action and of simple construction. The main frame of the crosshead is built up of members constructed of two angle irons. The verticals are 2×2×1/4-in. angles, 6 ft. long and set 2 1/16 in. from each other. The guide shoes *A*, spaced 3 ft. 10 in. apart, are riveted to a filler plate and the angles. The shoes are 1/4 in. wider than the guides, which are 5×6 in. The top and bottom members of the frame are made up of two 2 1/2×2 1/2×5/16-in. angle irons. A 3/8-in.

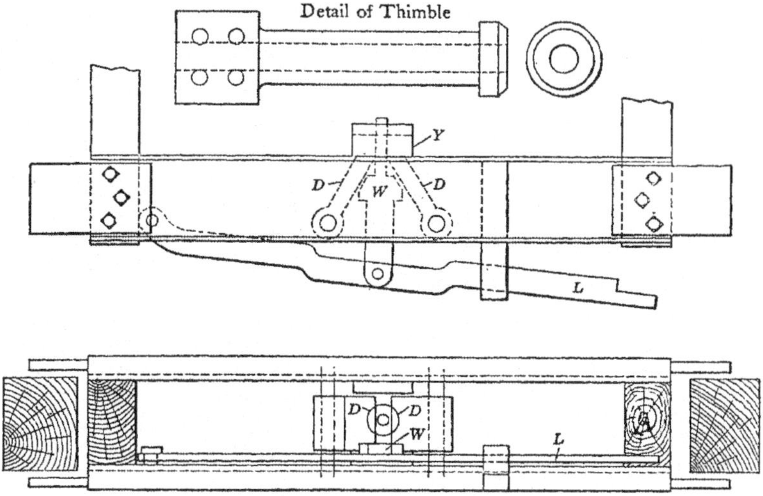

FIG. 197.—SAFETY CROSSHEAD FOR HOISTING BUCKETS.

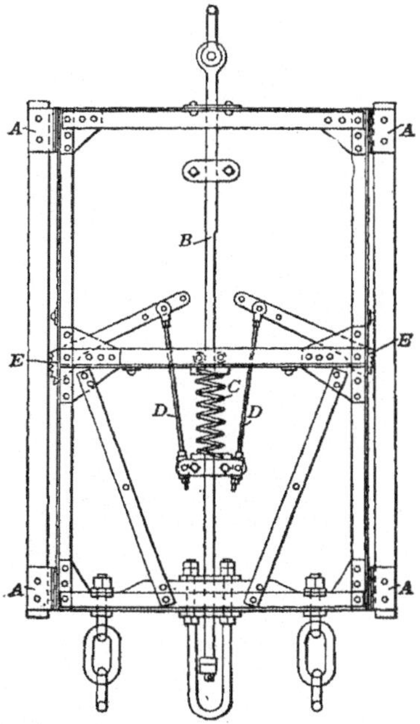

FIG. 198.—CROSSHEAD USED ON SINKING BUCKET IN MORNING MINE.

cover plate is used on the upper member. At a point slightly below its center the frame is cross-braced with two angle irons. In the lower half 1/4×2-in. diagonal braces are used to stiffen the frame. The center rod *B*, to which the hoisting cable is fastened, is made of 1 1/4-in. material. A spring *C*, 4 in. in diameter, made of 7/16-in. wire, 22 1/2 in. long, is coiled about the rod and bears up against a disk on the lower side of the central cross-brace of the frame. A clamp on the rod bears on the lower end of the spring. This clamp is connected by 1/2-in. rods *D* to the dogs *E*. There are three points at which the connecting rods may be joined to the dogs. The proper adjustment may thus be obtained. There are two other clamps on the central rod which prevent the spring from being compressed beyond its limit by coming up against the horizontal members of the frame. The bucket is suspended from the crosshead by two chains.

By using a crosshead with a sinking bucket, much time can be saved. However, the additional safety afforded the men while riding the bucket, should be sufficient consideration to warrant the necessary expense for the installation. The bucket being suspended by comparatively short chains from two points instead of one, the tendency to twist and swing while it is being lowered or hoisted is much reduced. This, of course, makes riding the bucket safer and enables hoisting to be carried on at a greater speed than is permissible without the use of a crosshead.

The Bryant Safety Crosshead.—In Fig. 199 is illustrated a safety crosshead for use in shaft sinking that has been patented by Thomas Bryant of Butte,

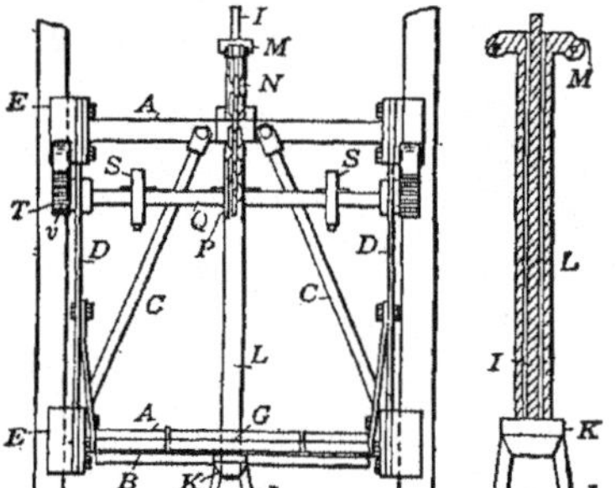

FIG. 199.—BRYANT SAFETY CROSSHEAD.

Mont. A crosshead of approximately this type was installed at the Original Consolidated mine in that district. The frame of the crosshead comprises, upper and lower cross-bars *A*, platform *B*, braces *C* and side arms *D*, which are provided with guide shoes *E*. The cross-bars are provided with the central passages *G*. The skip, cage, or bucket is secured to the rope or cable by a shackle *J*, having a head *K* which normally engages the bottom of the lower cross-bar. A tube or sheath *L* incloses the lower portion of the rope and is movably mounted on the cross-bars in the passages *G*. Normally this sheath rests upon and is supported by the head *K* of the shackle, but when the support is withdrawn, downward movement is limited by a collar *M*. The upper part of the sheath projects a short distance above the upper cross-bar and has chains *N* secured at the opposite sides of the collar *M*. Each chain is connected with a sheave *P* keyed to a shaft *Q*. One or more springs *S* are secured to each shaft so they will normally operate to wind the chains upon the sheaves. The dogs *T* are rigidly fastened to each shaft.

When the crosshead is operating in its usual manner it rests upon the shackle and moves up and down upon the guides with the rope. The sheath *L* is held in its raised position and pulling upon the chains *N* holds the dogs *T* in their nonoperative position against the tension of the springs *S*. When the crosshead strikes any obstruction or sticks accidentally while the cable is descending, the rope with the carrier attached will continue its downward movement independently of the crosshead. The shackle disengages from the lower end of the sheath which thereupon drops by gravity, and under the tension of the springs, into its lowermost position; simultaneously the springs rotate the shafts to bring the teeth on the dogs into engagement with the guides. The crosshead will thus be securely held until the bucket is raised again and the cable engages with and raises the tube, thereby rotating the shafts *Q* by means of the chains and returning the dogs to their nonoperative position.

By this construction the disasters caused by the falling of the crosshead upon the bucket, after being accidentally caught on the guides, are effectually prevented. At the same time the cable is always free to move independently of the crosshead and if the crosshead should be caught no delay is occasioned as the cable can be lowered as far as desired and then when raised again the crosshead will be automatically released. In case the traction cable breaks it is obvious that the safety means will operate automatically in the same manner as when the crosshead is accidentally caught and will prevent the crosshead from falling to the bottom of the shaft. The mechanism is especially adapted for use in shaft sinking in places where it is necessary to lower the bucket below the end of the guides, since a stop or other obstruction placed at the ends of the guides will cause the safety device to operate and hold the crosshead positively and independently while the bucket is lowered to the bottom.

An improvement of the Bryant crosshead, shown in Fig. 200, consists in a safety device for preventing the falling of the bucket in case of the breaking of the hoisting rope. In the illustration the bucket *A* is shown suspended below the cage *B* by the hoisting rope *C* which passes through a hollow shaft *D* set vertically in the cage. The end of the hoisting rope terminates in a shackle with collar *E*, the collar of which engages and holds the bottom of the cage so as to carry the cage; the point of support being at the bottom instead of the rope being attached at the top. A chain *F* is suspended from the hoisting-rope shackle, which carries the bucket as shown. This chain carries a circular disk *G*, which is dished so that the edge fits into and engages the hooked ends of the two bell-crank levers *H* which are pivotally supported from the lugs *I* dependent from the bottom of the cage.

To any part of the guides, at a point below which it is not desired that the cage should pass but below which the bucket is to be used, two lugs *J* are

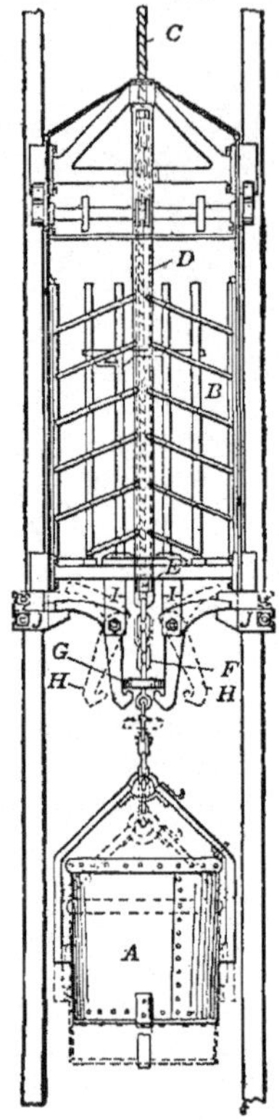

FIG. 200.—DEVICE FOR LOWERING BUCKET INDEPENDENTLY OF CAGE.

bolted. These engage the ends of the bell-crank levers *H* throwing their other ends out of engagement with the disk *G*. These lugs hold the cage from sinking but the bucket may be lowered, the hoisting rope passing freely through the hollow vertical shaft in the cage. The hooks and disk provide safety inasmuch as the bucket will not drop from the bottom of the cage in event of the breaking of the hoisting rope. When the cage is resting on the guide lugs, and the bucket is being lowered, the safety dogs at the top of the cage are released so as to bite into the wooden guides so that, in case the lugs should break, the cage will not fall. In hoisting the bucket the rope passes freely through the hollow shaft in the cage until the disk is at a point above the hooks of the levers H. When the collar E comes in contact with the bottom of the cage, the safety dogs are released and the cage raised together with the bucket whereupon the bell-crank levers drop into position so as to engage the disk. The device is patented.

Safety Catch for Cage (By H. L. Botsford).—Many devices have been designed to prevent accidents to mine cages from the breaking of the hoisting rope. They may all be divided into two general classes: safety stops, and safety brakes. The former aim to stop the cage instantly, by some positive means, such as the slipping of bolts into holes in the guide timbers. This class of safety catch would be satisfactory if the cage could always be stopped just at the instant the rope broke, on an ascending cage, and before the cage had acquired any appreciable downward velocity. Should the catches refuse to work until the cage had gained considerable momentum, or should the rope break while the cage was descending, and the speed of the cage should not be entirely checked before the hoisting rope failed, the stresses produced in the safety catches would be too severe for them to resist successfully. Even should the catches be able to stand the strain, there yet would be great danger of bodily harm to the occupants of the cage from the shock of sudden stoppage.

These serious objections to the safety stops are not so inherent in the second class of safety catches. Safety brakes are designed with the aim to arrest gradually the momentum of a descending cage, without danger to the occupants of the cage, or to the safety-catch mechanism. In one type of safety brake wedges are made to grip the guide timbers, when there is no tension in the hoisting rope. The motion of the descending cage forces them tighter and tighter against the guides until the cage is stopped. This type has not proved satisfactory. In a second type of safety brake, friction shoes are forced against the guides by the action of compressed carbonic acid gas against plungers. The gas is stored in cylinders under a pressure of several hundred pounds per square inch. This type has given good service at the mines of the Rand. It may be used with steel guides. In a third type of safety catch, or brake, much used in America, cams or "dogs" are made to engage the guides, one on each side of each guide. These cams have toothed or grooved surfaces, which cut into the guide timbers. They are quicker in their action than the friction blocks described above, although they may properly be classed with safety brakes because their action is to stop the cage by the resistance which they meet in deforming or cutting the guides. Obviously they can only be used with wooden guides.

This type of safety catch depends for its action upon springs which force the cams against the guides when the tension in the hoisting rope is released. While the weight of the cage is supported by the hoisting rope the cams are drawn back from the guides, and in no way interfere with the free running of the cage.

Much variation is shown in the design of cams, and in the choice of springs. The springs may be of the laminated, spiral, or the helical type. The laminated spring is but little used. In many mining districts one type of cage construction is used almost exclusively.

In Fig. 201 a cage long in use in the copper country of Lake Superior is shown. In this construction the cage is hung by two hanger forks *H* from the drawbar sleeve *G*, through which the drawbar *A* is free to slide. While the

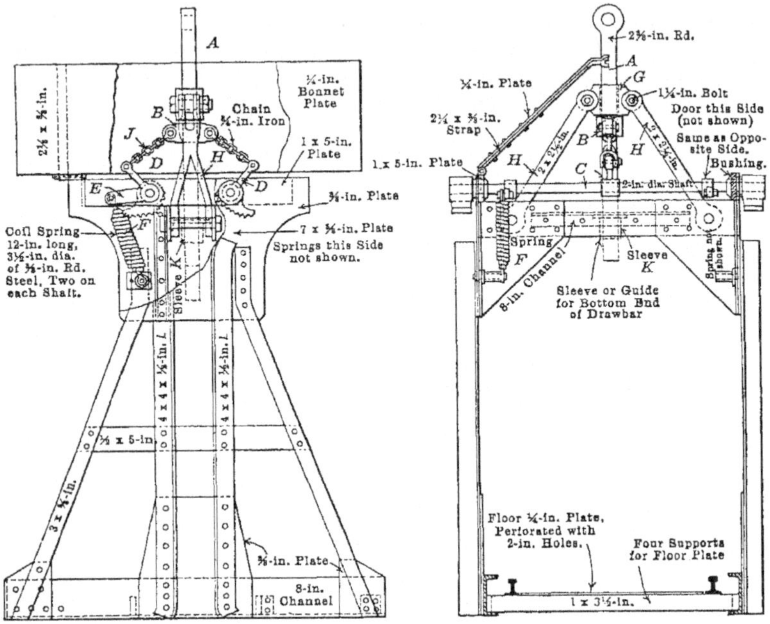

FIG. 201.—A LAKE SUPERIOR TYPE OF MINE CAGE WITH SAFETY CATCH.

cage is hanging on the hoisting rope the yoke *B* on the drawbar carries the weight of the cage. The springs *F* have one end fastened to the side plates of the cage and the other end to the lever *E*, keyed to the camshaft. The camshafts are connected to the yoke *B* by the lever arms *D* and the chains *J*. If the chains *J* are of the proper length, the cams will be turned back from the guides, when the drawbar is raised to its full height, that is, when the yoke *B* is in contact with the drawbar sleeve *G*. This is the position the drawbar assumes while the cage is suspended on the hoisting rope. The springs *F* are now under tension, and should the hoisting rope break, or the cage be rested upon landing chairs, these springs will rotate the camshafts sufficiently to bring the cams into contact with the shaft guides. Any downward movement of the cage will now force the cams into the guides, until they meet with sufficient resistance to support the weight of the cage. The lower end of the drawbar passes through the sleeve or guide *K* which prevents it from binding in the sleeve.

The design of the cams themselves is a very important detail in cage construction. They should be so shaped that when they once come in contact with the guide timbers, any downward motion of the cage must force them deeper and deeper into the guides. As to the relative merits of cams with notched or grooved faces, and those with projecting teeth, much may be said in favor of each. The sharpened teeth will be more certain to cut into the guides at the start, while the other type makes a stronger construction.

Regardless of the type of cage, the safety catches should be tested regularly at least once each month. This is conveniently done by hoisting the cage a few feet above the collar of the shaft, and suspending it there by a hemp rope fastened to a convenient timber in the headframe. The hoisting rope should now be slackened a few feet and the hemp rope severed. This gives the cage a free drop until the catches work, or the slack in the hoisting rope is taken up. With safety catches in good order they should not permit the cage to drop more than a few inches.

LANDING CHAIRS AND OTHER APPLIANCES

Skip Chairs at Argonaut Mine.—Few mines hoisting in skips use chairs in the shaft, but rather rely upon the engineer to stop near enough to the proper point for loading. In hoisting from depth the stretch or give of the hoisting cable is enough to cause considerable motion of the skip up and down the shaft, after stopping the hoist. Loading while the skip is in motion is bound to result in the spilling of a considerable amount of rock down the shaft. This is always dangerous and should be avoided. In the Argonaut mine at Jackson, Amador County, Calif., Superintendent Ralph S. Rainsford has devised a simple type of skip chair, which is shown in Fig. 202. At the loading station in the shaft there is a rod *A* connected by a shorter lever to the 4×6-in. post. This piece is capped at its end with iron and acts as a buffer upon which the skip rests during loading, as is shown by the dotted lines. There is a counterbalance *C*, made of 8×8 material, connected to the buffer; this serves to swing the chair between the shaft timbers and clear of the skip, when not in use. As shown in the side view, a spring connecting the buffer and shaft wall plate is also used to assist in swinging the chair back. Where it is possible to use a sufficiently heavy counterbalance, the spring is unnecessary. Skip chairs of this design are easily and cheaply constructed and placed in the shaft. Their use is certainly made worth while from the time saved in loading (not having to wait for the skip to come to rest) and the added security afforded men working at lower stations in the shaft.

Landing Chair for Skips in Inclines.—The usual form of chair or gate that is used in inclines for the skips to rest on consists, as shown in Fig. 203a, of a strong piece of timber bolted to the ends of two arms that are equipped with a counterweight to lift the gate out of the way as soon as the skip is lifted from the chair. These gates are carried by two posts in the incline, and from the end of the arm that carries the counterweight a chain goes to the station floor by means of which the chair is raised so as to catch the skip. The objection to this kind of chair is that when the skip is lowered heavily on the chair or gate, the arms are bent. It therefore takes considerable time to repair this form of gate. On this account a gate whereof the cross piece rests loosely in hooked carriers, which was devised by Capt. Samuel Richards, is used at the South Hecla shafts of the Calumet & Hecla company. This gate has a three-armed spider at one end and

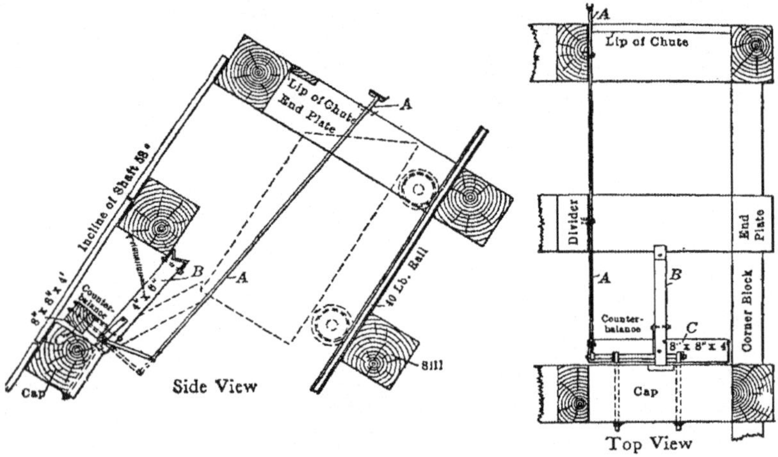

FIG. 202.—SKIP CHAIRS USED IN ARGONAUT MINE, JACKSON, CALIF.

a two-arm spider at the other. To the end of one of the three arms is connected by a hinged joint the counterweight that lifts the cross piece out of the way of the skip as soon as the skip is lifted off the gate. To the arm extending out in the opposite direction is connected by a hinged joint the hook in which is carried the cross piece that stops the skip, while to the third arm, which is at right angles to the other two, is connected the chain or rod that extends to the floor of the station, by means of which the gate is thrown in when desired. The details are shown in Fig. 203b. Owing to the fact that the hooks that carry the cross piece are loosely connected to the arms, the cross piece as it moves up or down slides along the top side of the two posts that carry the spider. Consequently, in case a skip comes down heavily on the chair or gate, only the cross piece can be broken as it rests directly against the two posts. A new gate timber can be put in quickly to replace a broken one. Chairs of this type have been in use in the South Hecla shafts for some time, and now this type of gate is to become standard for all the inclines of the Calumet & Hecla company.

Emergency Chairs on Headframe.—The weak point of many safety devices is that they depend upon springs for their action when the cable strain is released. In mine usage joints and bearings rust and the springs deteriorate, so that when the accident finally happens the mechanism may refuse to work.

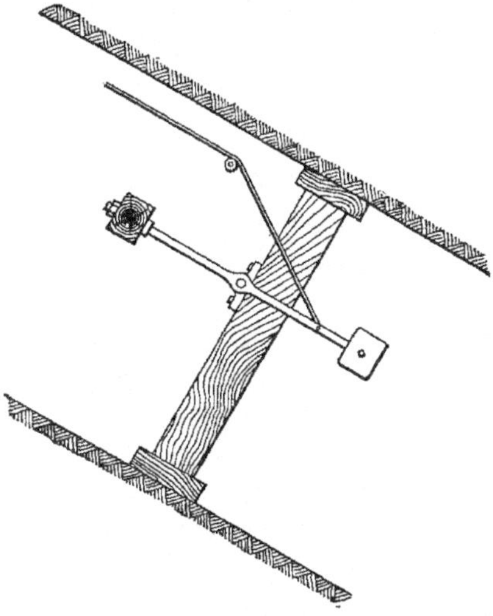

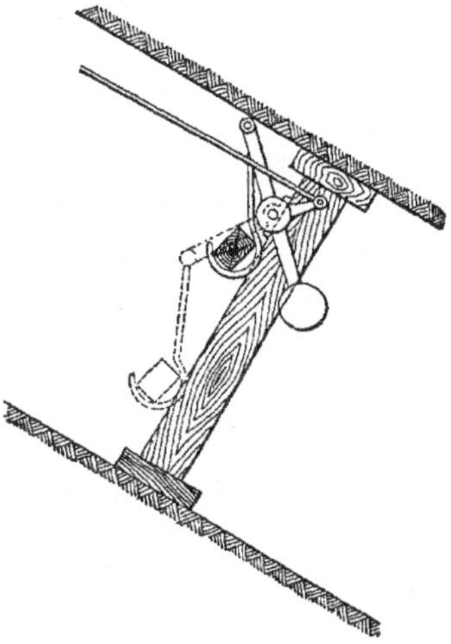

FIG. 203 a. FIG. 203 b.

BUFFER BARS FOR INCLINE SKIPWAYS.

Further, cables are frequently broken by winding the cage up to the sheave, and a force strong enough to break the cable is generally also sufficient to bend the rods and jam the moving parts of the safety clutch. Assurance should be made doubly sure by placing a set of chairs in the headframe so that when the cage is pulled to the sheave the chairs will be about 6 in. below the floor of the lowest deck (a separate set for each deck would be preferable). Chairs so placed are high enough above the collar to be out of the way of ordinary work. They are always held in the shaft by weights or springs and, being easy of access, are kept in first-class condition. Then when an accident occurs, if the dogs of the clutch fail to work, the cage cannot fall down the shaft.

Testing Safety Devices on Mine Cages.—At practically all the mines operated by the Oliver Iron Mining Co. in Minnesota and Michigan, the safety catches on the man cage are tested at least once every month. Fig. 204 shows the device that is used to test the cage. By using this the time of making the test is reduced to only a few minutes. A cable is tied across the shaft to the guides or runners about 30 ft. above the ground, to which is fastened a 6-ft. cable with a loop in each end. The lower end is attached to the top of the device at *A*. Another short piece of cable with a loop in either end is used to connect the tripper with the cage. The clutches *B* are drawn together by pressing down on

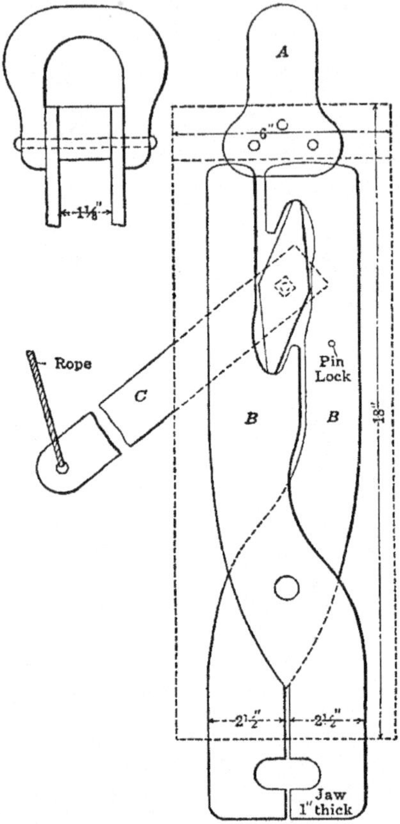

FIG. 204.—TRIPPING DEVICE USED IN TESTING SAFETY APPLIANCES ON CAGES.

the lever *C*, and a lock pin placed in as shown. This is to insure no tripping until all arrangements are complete. When all connections are made the main cable is given 5 or 6 ft. of slack, all of which is drawn to the cage by means of a small block and tackle. A good strong timber is placed across the top of the shaft to prevent the cage from dropping too far. The pin is then taken out and by means of a long rope, attached to the lever *C*, it is drawn up sufficiently to separate the jaws and allow the piece of supporting cable to drop out. At most the cage can only drop 3 or 4 ft. before it strikes the timber across the top of the shaft. In most cases the safety dogs will catch in the guide within less than 1 ft. If not, something is wrong and the difficulty is remedied.

Cage Landing Chairs (By W. F. Boericke).—In Fig. 205 a simple cage chair is shown, together with the "fingers" used for throwing the chair into position. It is, of course, necessary that the supports *A* and *B* of the chair should be out of the way as the cage goes up and down, an inch clearance on

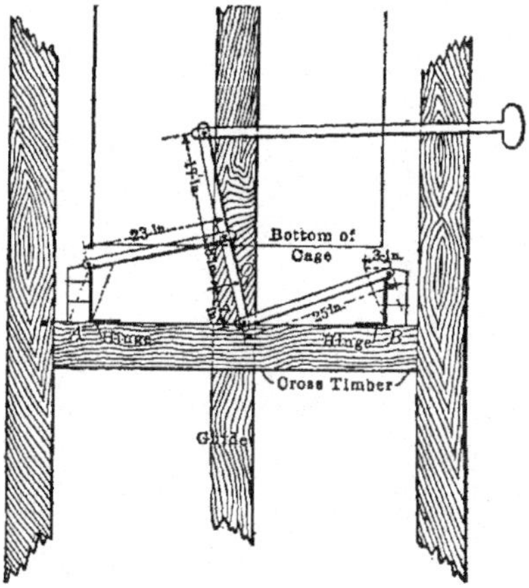

FIG. 205.—LANDING CHAIRS IN SHAFT.

either side is therefore provided. The supports are of 3×10-in. timber, shod with flat iron, and reinforced by bands of steel going completely around. The tops are sloped as indicated, to provide a flat supporting surface for the cage when the chairs are thrown in. The supports are securely hinged to the cross timber, and work with little friction. The fingers are made of 2-in. flat iron, and at the point *C* the device is bolted to the guide. The chair should be of such height that the car will be exactly level with the tracks on the landing, and roll off without trouble.

Improved Landing Chair.—An admirable type of landing chair, patterned after one in use for some years at several German mines, is installed at the Doe Run mines, Flat River, Mo. The prime feature of the chair is that it obviates the inconvenience and delay, and the consequent waste of time and power, attending the lifting of the cage from its supports by the hoisting engine before lowering as is necessary with the usual types of chairs. The details of construction are shown in the drawings given in Fig. 206. The upward projecting sides of the sole plate *A*, are bored to receive the two axles, *L* and *D*. To the former are rigidly attached the hand lever *H*, and the short lever K. On the shorter axle *D*, are fastened the two short links *E E*, through the ends of which passes the pin *B*. The pin *B* also passes loosely through the inner end of the chair piece *C*, and through the link *F*, the other end of which is loosely attached to the lever K, by the bolt *I*. The under side of *C* slides freely on the pillow block X. The axles *L*, of the two mechanisms on the opposite sides of the shaft collar are connected by keyed levers and link rods, so that they work in unison. The principal dimensions are as follows: *L*, 2 7/16 in. in diameter; *D*, 2 7/16 in.; *B*, 2 1/8 in.; *I*, 1 1/2 in.; center to center, *L–I*, 3 1/2 in.; *I–B*, 4 1/2 in.; *B–D*, 4 1/4 in.; length of *D*, nine inches.

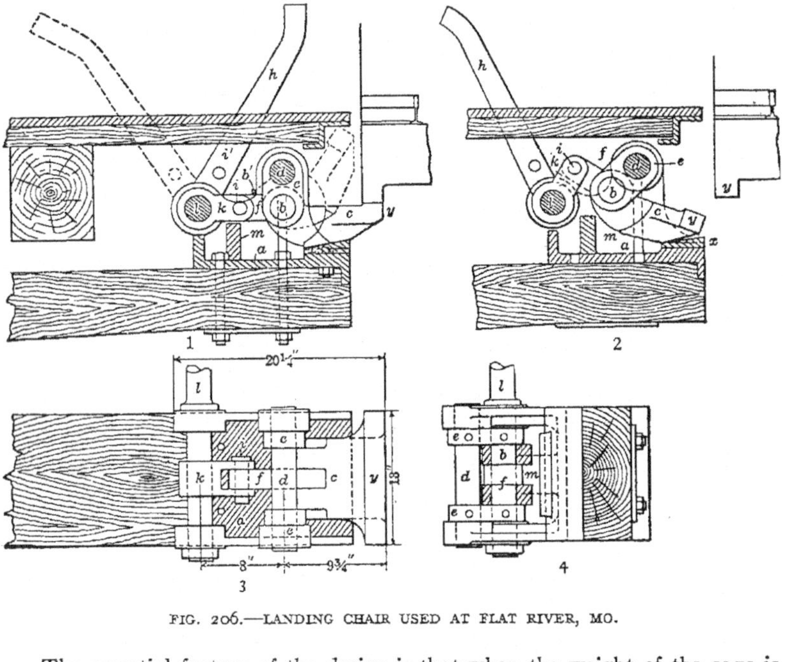

FIG. 206.—LANDING CHAIR USED AT FLAT RIVER, MO.

The essential feature of the design is that when the weight of the cage is resting on the outward end of the block *C*, the latter is held firmly in position by this weight; it cannot tip because the links *E E* lie in a vertical line from *B* to *D*, and it cannot slide backward because the link *F*, and the lever K, also lie in a straight line from *B* to *L*. In order to prevent a downward motion of the joint *I*, due to the weight of the hand lever *H*, a block *M*, is placed below the lever, K, to stop it at the horizontal position. When the hand lever is thrown backward, through an angle of 60°, the piece *C*, is obviously withdrawn, and takes the position shown in Fig. 2. After the cage has passed down the shaft, the lever is returned to its first position, putting the piece *C*, in place, ready for the next landing. Owing to the hinged construction at *B*, the catching of the cage by *C* is automatic, as the piece simply lifts when the cage comes into contact with it, and falls into place when the cage rises above it. The total friction of the apparatus, most of which occurs at the rubbing surface *Y*, is so small that with a load of 5 tons resting on the chair, a force of less than 45 lb. at the handle will suffice to operate the mechanism. One of these chairs is placed on each side of the shaft, and by means of a connecting rod both are operated by the same lever. As used at Flat River, the beam upon which the chair is mounted, is cushioned by a spiral spring placed beneath.

Chairs on the Cage.—In the effort to do away with the chairs in a shaft, many devices have been tried. An apparatus similar to the one described here

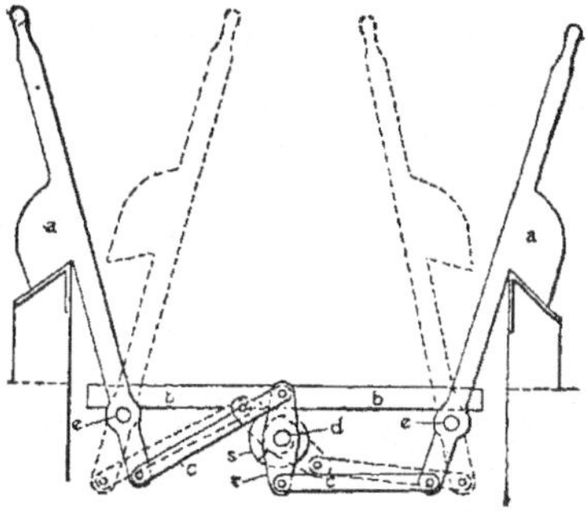

FIG. 207.—CAGE CHAIRS.

can be made in any mine shop and will give satisfaction. Referring to Fig. 207, the levers *a a* carrying the chairs are four in number and are set at each corner of the cage floor *b b*, just within the side rods. As shown, they are connected by the rods *c c* to rocker arm *r*, and are held clear of the shaft when the cage is in motion by the powerful spring *s*. The shaft *d* and bars *e e* extend across the cage and the arrangement is duplicated on the other side. Thus a movement of any lever actuates all. The chairs rest on blocks with sheet-iron caps placed at each corner of the shaft compartment. The height of these blocks is, of course, such as will insure alignment of the cage rails with those on the station. The mechanism can be operated from the cage or station floor.

Landing Chair for Cage (By C. L. Severy).—The cage chair shown in Fig. 208 was designed at the Poderosa mine, Collahuasi, Chile, to overcome the difficulties in the use of the ordinary kind where the timbers of the shaft were continually shifting due to bad ground. Since this chair was put into use no trouble whatever has been experienced, for as long as the cage will pass through the shaft the chairs are ready for use. The advantages of this style of chair are many. Only one chair is needed for each cage. They are quicker to operate in changing levels and in case of accident with men aboard they can be seated on any set in the shaft. This feature is convenient in shaft repair work. They add little more weight to the cage and are so simply and easily operated that boys of 12 years who can hardly reach the handle operate them here. The chairs can be easily attached to any cage.

The construction of the chair is simple, being a system of levers to throw out four iron rods, one at each corner of the cage, for it to rest upon, these levers being operated by another upright lever extending above the top of the car, on

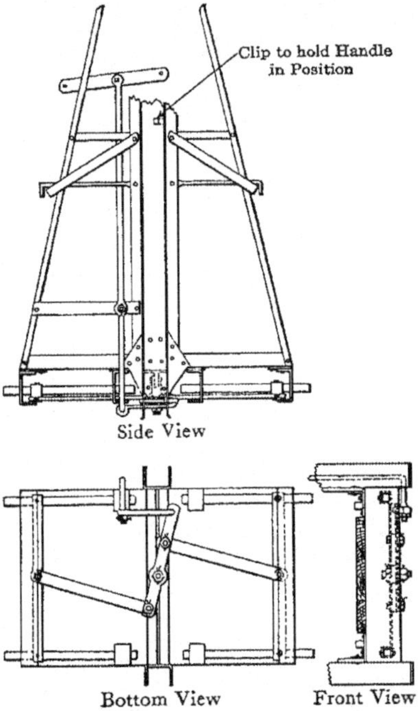

FIG. 208.—CAGE FITTED WITH LANDING CHAIRS.

the cage. The four rods for seating the cage are of 1 1/4-in. square iron and upset at the proper point for a pin run through holes in channel irons as guides, the channel irons being riveted to the bottom of the cage. The center pivot for the levers is riveted to the middle of the T-iron on the under side of the cage. The long arm of the operating lever is made of 1×5/8-in. steel. The handle is also made of steel so as to insure strength and to give it the necessary spring to hold it behind a pin at the side of the cage to keep the chair from jarring open while the cage is in motion.

A Safety Gate for Cages (By R. B. Wallace).—In vertical shafts an open cage is often used for hoisting men in preference to one inclosed with sheet metal. Fig. 209 shows a gate for an open cage which gives a full opening and which can be used where there is not sufficient room for an up- and down-sliding gate. The bars of the gate are hinged on one side and fold up to a vertical position where they are held by a hook that fastens into one of the handles of the gate. The top bar is 1 in. square while the other cross bars measure 1 × 1/4 in. These are loosely riveted together. Two channels *C* made from plate and bolted

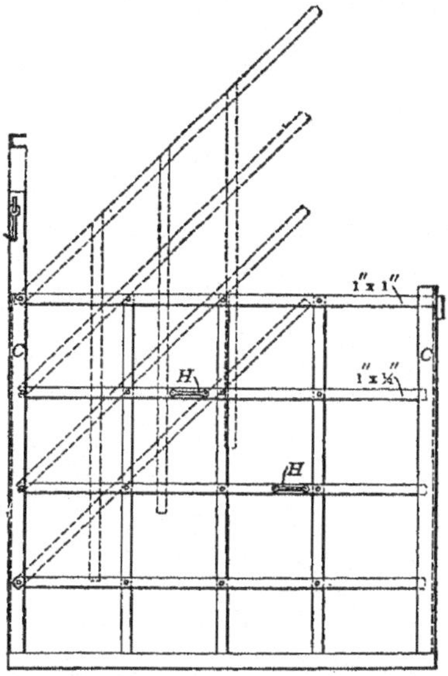

FIG. 209.—A FOLDING SAFETY GATE.

or riveted to the sides of the cage hold the gate in position. Two handles *H* are placed in convenient positions on the cross strips for the cage conductor to grasp when raising the gate.

A Safety Device for Cages at the Chapin Mine.—Many accidents in mine shafts are due to carelessness on the part of the gate tenders in failing to close the bars to keep the mine car in its place. Moreover, it is possible for this bar to be knocked out of place by a projecting timber or otherwise, resulting in the car rolling off the cage and catching in the shaft timbers. The device shown in Fig. 210 has been used many years at the Chapin mine, Iron Mountain, Mich., where it has proved a success in preventing accidents of this kind. It consists of a drop track directly under the wheels. This track can be thrown into position only by setting the cage on both chairs. The instant the cage is lifted from the chairs the track drops down 2 1/2 in. so that it is impossible for a car to roll off, even should the bars not be in place. The iron bar *a* used to support the drop track is 2×4 in. and as long as the cage. The drop section of the rail is

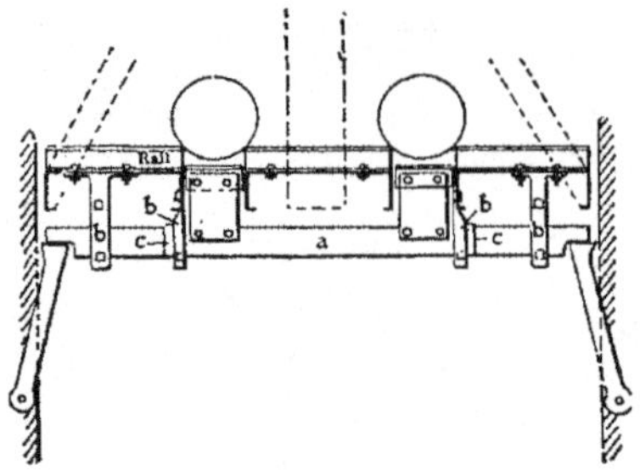

FIG. 210.—SAFETY DEVICE FOR CAGES.

10 in. long. The bar is supported in a slot *b* that allows 2 1/2 in. play and is secured by an angle-iron lug *c* so that it cannot slip endwise. It is operated entirely by the contact of the cage with the chairs at the desired landing.

Shaft Gates.—The type of shaft gates shown in Fig. 211 gives satisfactory

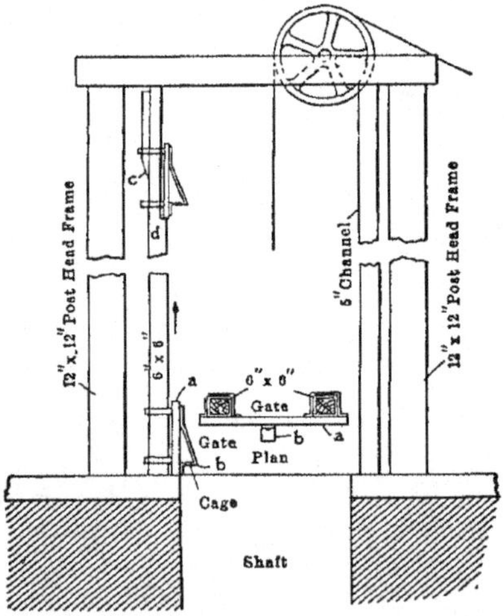

FIG. 211.—OPERATION OF SHAFT GATES.

results and is of simple and inexpensive design. The gate *a* is built with iron guides as shown in the plan, which allow it to travel up and down the guide posts. When the cage comes out of the shaft mouth, it strikes the iron *b* on the back of the gate, lifting it out of the way until the cage is again dropped. In case of overwinding, the gate is diverted by the guide at the point *c*, and swings clear of the cage into the position marked *d* on the drawing. This same arrangement may be used on both sides of the shaft so that as the loaded car is run off the empty may be run directly upon the cage and lowered, thus saving considerable time. The advantage of these gates raised by the cage lies in the fact that the necessity for a top man to watch the shaft is eliminated, and an additional guard is afforded against danger from persons falling down the shaft, as it is difficult to lift the gates without raising the cage. Where there is little danger of overwinding, the guides for the gates may be made of channel irons or straight posts without the notching shown at *d*. The gate itself may be made of any desired dimensions or construction.

Anaconda Gates (By F. L. Fisher).—In several of the Anaconda company's mines at Butte, where ore is hoisted in skips, it has been found necessary to

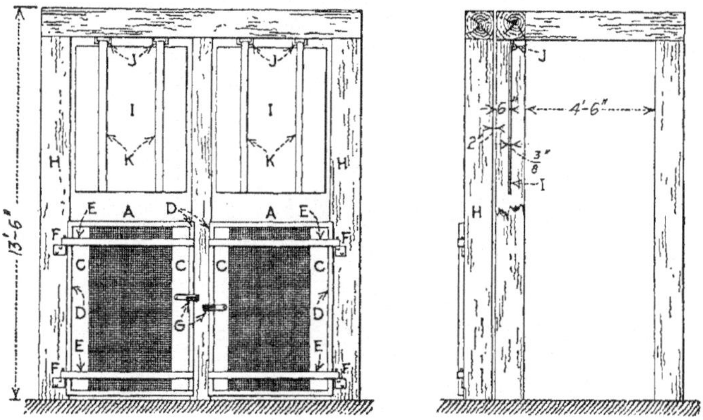

FIG. 212.—STATION GATES IN A BUTTE MINE.

protect the openings from the shaft to the stations from occasional falls of small quantities of ore that are inevitably spilled in loading. To insure against accidents from this source, swinging gates have been devised and are installed at the stations. The main gates *A*, in Fig. 212, are made of two wide vertical strips of 1/8-in. iron sheeting *C*, and a central screen of 1-in. mesh, No. 8 iron wire, all bordered by a 2-in. angle iron *D*, and crossed horizontally by two strips of 3/8-in. iron *E*. The doors are hung on hinges *F*, and locked with lift-latch *G*, which can be operated from either side. They are attached to the outward faces of the first station sets *H*, which are separated from the shaft timbers by a space of 2 in. Above and back of these doors is a second set of doors *I*, hung from the center of the cap at the top of the shaft-station sets by hinges *J*, so that the doors can be swung in and out of the shaft, to facilitate the unloading of long timbers from the cages; the doors being simply swung up and away from the shaft as the timbers are dragged through. The bottoms of the upper doors are 12 in. above the tops of the lower, and 22 in. back toward the shaft. They are made of 1/8-in. iron sheeting, strengthened by two vertical strips of 3/8-in. iron *K*, and of their own weight they tend to deflect any falling pieces of ore to the turnsheet below, where the lower gates prevent them from bounding into the station. The opening between the gates, and the wire screens on the lower gates are for the purpose of admitting light to the shaft.

Guards at Shaft Stations.—At the stations of the inclined shafts in the Michigan copper district many different kinds of guards have been tried. At some shafts a wire rope or a light chain is used having a hook at its end that fastens to a ring carried by a post at the hanging-wall side of the shaft. It is

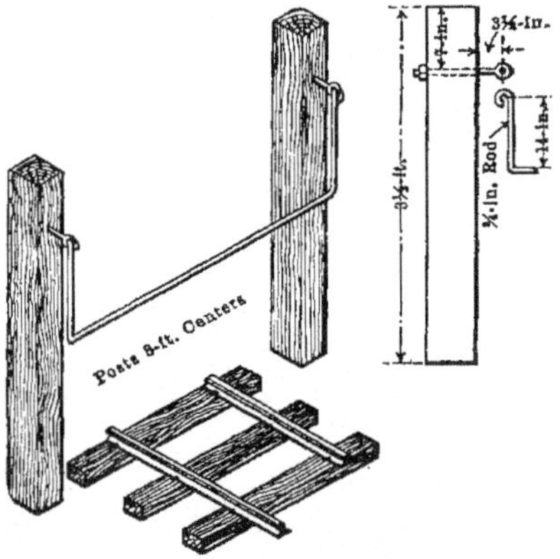

FIG. 213.—A SIMPLE IRON SHAFT GUARD.

an unhandy device as the men have to go clear over to the foot wall, stoop down and pick up the chain each time to put it across the shaft opening. At other inclines the shaft bar that is in common use at vertical shafts has been copied. Owing to the length that this guard arm must have at the inclines, and its weight, a counterweight is sometimes arranged to make the lifting of the arm easier. A nail driven into the post that carries the stop on which the guard arm rests holds the arm out of the way, when it is raised and a car is being dumped.

A guard arm is handier than a chain, but the best form of bar for the inclined shaft stations consists of a 3/4-in. iron rod that is bent at each end so as to have arms 14-in. long with eyes in them that hook into an eye-bolt that is fastened into the two posts of the fence that is built around the shaft at the station. These posts are about 3 1/2 ft. high and the eye-bolts are placed about 7 in. from the top, and extend out from the posts about 3 1/2 in. Therefore when the guard is turned up so that a car can be dumped into the skip it rests against the top of the posts and will not jar down easily. As soon as the car is dumped the bar is given a push so that it swings down to prevent any car from being pushed into the shaft. This guard is illustrated in Fig. 213.

NOTES ON MINE TRACK AND SWITCHES

Mine Track (By Alvin R. Kenner).—A poorly laid mine track is the cause of much trouble and the additional time required to lay it properly will be saved many times over by the avoidance of derailed and overturned cars. When laying short lengths of rail for a temporary track in the face of a drift tunnel or crosscut there is usually a tendency upon the part of the tracklayer to put the ties in carelessly with the intention of fixing them properly when the permanent rails are placed. When this time comes it is too much work to dig out each tie and place it properly. If the ties are placed permanently the first time a much better track will result, especially if short pieces of rail are dispensed with and full length rails used instead, in the manner here described.

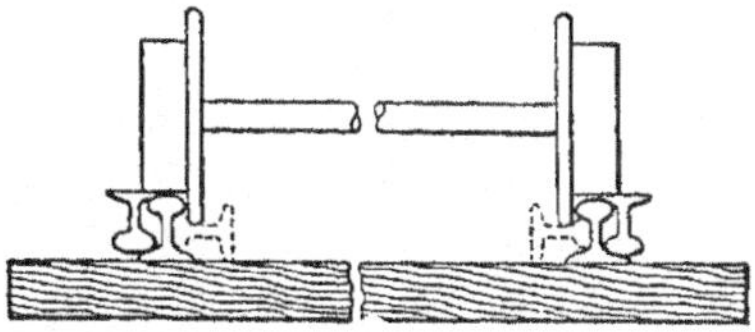

FIG. 214.—EXTENDING TRACK WITHOUT USING SHORT RAILS.

Lay the ties permanently as near the face as advisable and place two full length T-rails as shown by the full lines in Fig. 214, on the outside of the last set of permanent rails. Drive a track spike at both ends of the lap of each pair of overlapping rails to hold the new rail against the old, and the track is ready for the car. As the heading advances slide the rails forward. When the ends are reached, turn over and spike them in place. After the face is advanced a few feet two new rails are placed as before.

A decided advantage of this method is due to the stability of the rails near the face due to the fact that they are fastened to the track behind. This permits advancing the ends much nearer to the face than when short rails are used, and the car can always be brought into close proximity to the muck pile. Instead of track spikes being used to hold the new rails in place, a block may be nailed on a tie and a wedge driven between the block and the rail. The block being on the outside of the rail need not be removed. In spiking the new rails to the ties, spikes should only be driven on the outside and when the rails are turned over, these will be found to be correctly placed for spiking down one side of the base of the rail and need only to be driven down.

This method is even more useful in the sinking of an inclined shaft or in advancing the waste-dump track than in driving headings. In sinking an inclined shaft with a skip, a source of annoyance is the false track necessary to bridge the distance from the last set of timbers to the bottom of the shaft. A long set of rails will work better than a false track and has the added advantage of being readily extended below the water level and remaining in a fixed position. In unwatering and retimbering an old inclined shaft this was of great assistance making it possible to lower the water sufficiently to put in a new set of timbers, as readily as though the track were already in place. In hoisting water from a sump where only two shifts could be used in sinking, the engineer on the third shift was annoyed by the skip jumping the false track and tearing out timbers. By adopting this method the trouble was eliminated.

In advancing a dump track, lack of stability is often the cause of the car going over the dump. To avoid this the new rails are not only firmly fastened to the old track but the rear ends of the new rails are so placed that the outer ends cannot give downward except by bending. A support at the outer end of a set of rails placed as suggested will give dumping room for a considerable period of time, as the new rails are easily pushed forward any distance desired.

A similar system which is much used is indicated by the dotted lines in the illustration. This method of placing the rails, however, has a drawback in that muck accumulates rapidly in the groove of the new rail and causes annoyance

A New Track Spike.—A new track spike, the invention of W. H. Floessell

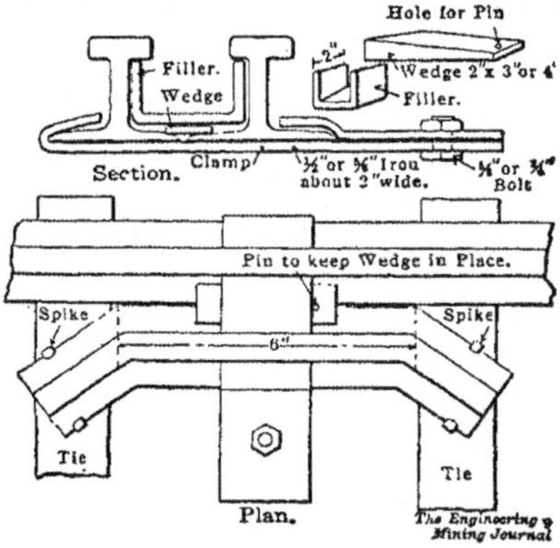

FIG. 215.—METHOD OF FASTENING A GUARD RAIL.

of Sydney, N. S. W., has been tested by W. H. Warren, professor of engineering at Sydney University and proved to have 1.29 times the holding power of a black-iron spike of square section and 1.30 times that of one of circular section. The spike is made by twisting a bar of square, hexagonal or octagonal cross-section, so as to form a helix of large pitch, and a square head is forged at the top. The original bar can be of fluted or plain section. The spike is driven into a hole bored to a depth of 4 in. in a railway sleeper; the spike revolves as it is driven into the hole.

Short Guard Rail and Fastening (By G. M. Shoemaker).—A satisfactory guard rail and method of fastening is shown in Fig. 215. A shorter piece of rail may be used in making it than with the older method. The purpose of a guard rail is to guide the flange of the wheel away from the point of frog and 6 in. of rail should be sufficient to do this, 3 in. on either side of the point of frog. With the old method it is necessary to use a much longer piece to get sufficient spiking surface to make the rail fast. In the old method, unless notches are cut in the bases, it is necessary to drive the spikes between the outer rail and the guard to prevent the base of each from meeting. By this method the ball of each rail is brought closer together thereby making the liability of derailment less, especially so in the case of a bent axle or a wheel which has become wobbly from wear on bore or axle.

Mining Track Frog.—The cheap, simple but effective device shown in Fig. 216 is used in some of the Utah mines for guiding the wheels of mine cars to the rails when passing from a smooth floor or from a turntable. It is composed of a piece of 2-in. plank, 15 in. wide (if used with 18-in. track)

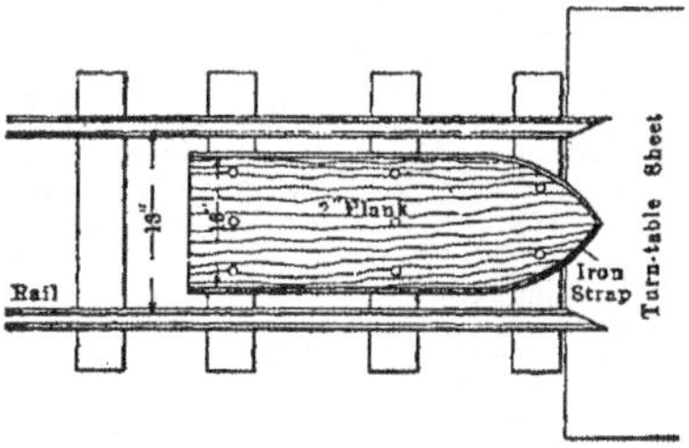

FIG. 216.—TRACK AND FROG.

and 2 1/2 ft. long. One end is cut into a rounding point and the edge of the plank is protected by a piece of strap iron, 2 in. wide, nailed on. The device is then spiked down between the ends of the rails, with the point toward the turning place, leaving enough space between the edges and the inner sides of the rails for the flanges of the wheels to pass. Unlike most other devices for this purpose, this contrivance does not get battered out of shape, does not require a blacksmith to make it, and never fails in its duty.

Mine Track Switches.—At many mines it is the practice to order standard switches direct from supply companies, while at others the switches are made at the mine, usually by the blacksmith, who makes them according to his own ideas, guided by the data supplied by the foreman of the track-laying crew. Standard switches are expensive when bought from supply houses. Made-at-the-mine switches are comparatively cheap, but unless well made they cause much trouble from cars being derailed. At the mines of the Tonopah Mining Co., the engineering corps designed the switches illustrated in Fig. 217. These are made at the company's shops according to standard specifications and are suitable for tracks where the tramming is done by hand. The car is made to take the turn by shifting the rear end in the opposite direction. The switches are so placed that only the empty cars need be so shifted; the loaded cars, running in the opposite direction, take the straight-away track.

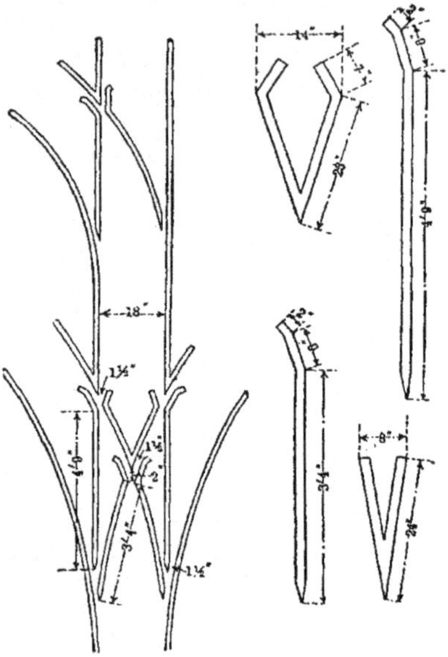

FIG. 217.—ONE- AND TWO-WAY MINE SWITCHES.

Calculating a Crossover Switch.—The following formulas for calculating the lengths and distances required to lay a crossover switch between two parallel tracks in a mine were published in *Coal Age*. The data usually given are: The frog number, n; the gage of the track, g; and d, the track centers. The data required are found by the following formulas: The chord of lead rail, $c=2\,ng$; the radius of lead rail, $R=nc$; the frog angle, $\sin. 1/2a=\frac{1}{2n}$; the length of lead rail, $L=\frac{a}{180\pi}R$; the length of the follower, $l=\frac{R-g}{R}L$; the length of straight track, $r=\frac{d-g(1+\cos. a)}{\sin. a}$; the lead of switch, $x=R\sin. a$; the frog distance apart, $y=r\cos. a-g\sin. a$; and the distance between switches, $D=2x+y$. The letters in the formula refer to the dimensions specified in Fig. 218.

Gravity Tram Switch (By B. A. Statz).—While operating a small milling plant near Kelly, N. M., I had trouble with a surface tram connecting the mine and mill and having a total length of 2300 ft. The tram line was single track, with a 30-in. gage, having a switch or turnout in the center; the brake was of a drum type, so arranged that each car had to pass, coming and going, on the same side of the switch. The grade of the tram line ranged from 15 to 40°. The

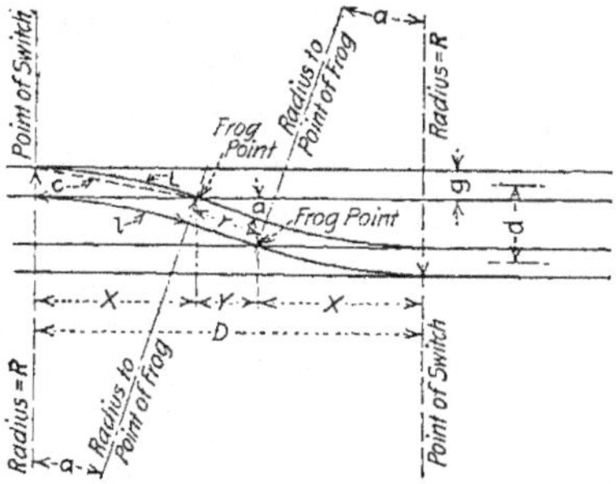

FIG. 218.—CALCULATING A CROSSOVER SWITCH.

cars used held two tons of material and had straight bodies. When the cars were running at fairly good speed the front wheels of the cars would be 1 in. above track when on the steepest grade, hence when the cars came to the lesser grade the front wheels nearly always missed the track, causing delay. To overcome this I put in the switch, shown in Fig. 219; the switch was cast in a local foundry. This arrangement worked admirably in conjunction with cars designed so that the top line was level when the car was on the steepest grade.

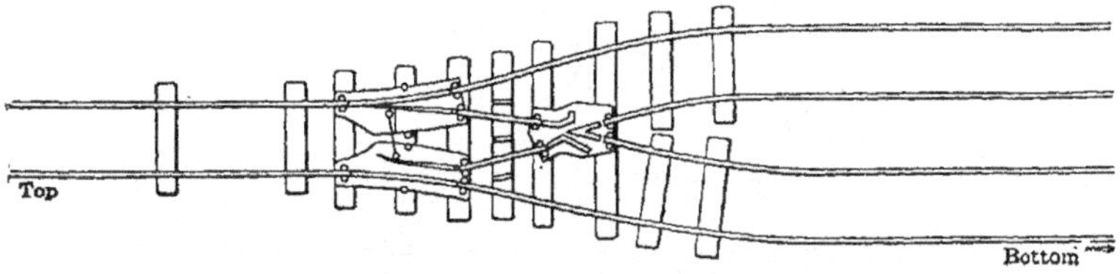

FIG. 219.—SWITCH FOR GRAVITY TRAM.

A switch such as is shown in the accompanying drawing was put in at both ends of the turnout. The tongues of these switches were made of wrought iron and coupled together. These slid over a cast-iron bed plate. The switches were set opposite from one another, of course, so that the car would pass on the switch. Consequently, the car coming up would throw over the tongues of the switch at the head of the turnout, while the car going down would throw over the tongues of the switch at the bottom end of the turnout. So, when the car at the top was loaded and the car at the bottom dumped, and they began their journeying again, the switches were set so that each car traveled along the same side of the turnout that it had taken before.

A Double Gage Turnout.—The turnout shown in Fig. 220 is used, states Frederick MacCoy (*Eng. News*, Jan. 25, 1912), wherever 36-in and 4 ft. 8½-in. tracks are employed together at the Esperanza mine in the El Oro district, Mexico. A motor car runs on the narrow-gage tracks, drawing either narrow- or standard-gage cars. One lever is used to operate both switches. Frogs *A* and *B* are standard while *C* is a crossing frog.

An Automatic Switch.—The accompanying illustration, Fig. 221, indicates the design of a switch operated by gravity instead of a spring. It can be

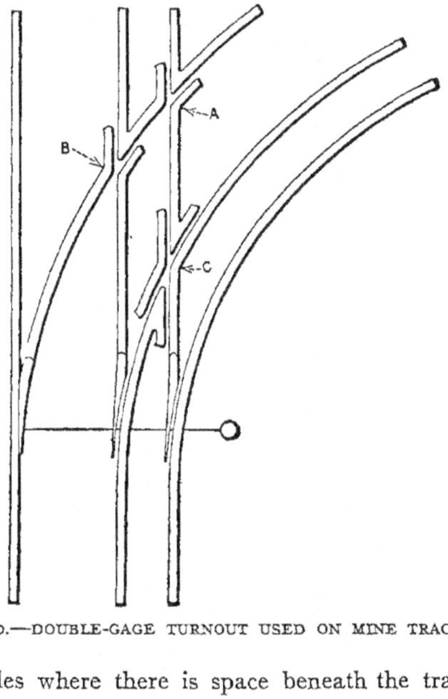

FIG. 220.—DOUBLE-GAGE TURNOUT USED ON MINE TRACKS.

easily used on trestles where there is space beneath the tracks. It consists simply of the ordinary two short rails fastened together as in the case of a spring switch. Beneath the connecting bar is a small lug which is engaged by an L-shaped lever. One arm of the lever is 2 ft. long, and the other 1 ft. At the angle point it is fastened to a post or beam. The short arm has a wide face which engages the lug on the bar above, and by means of a small weight on the end of the lever it throws the switch. It works satisfactorily, and has been in use by the Cleveland-Cliffs Iron Co. for a number of years. It is arranged so that it is opened by the loaded cars and then closes, thus throwing the empty cars on another track upon their return.

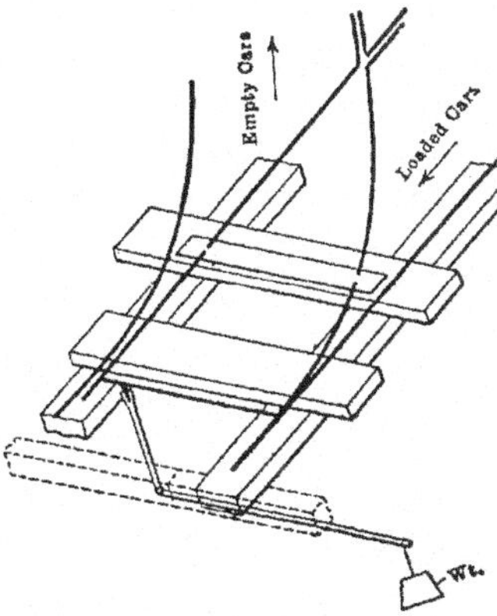

FIG. 221.—GRAVITY SWITCH FOR ORE CARS.

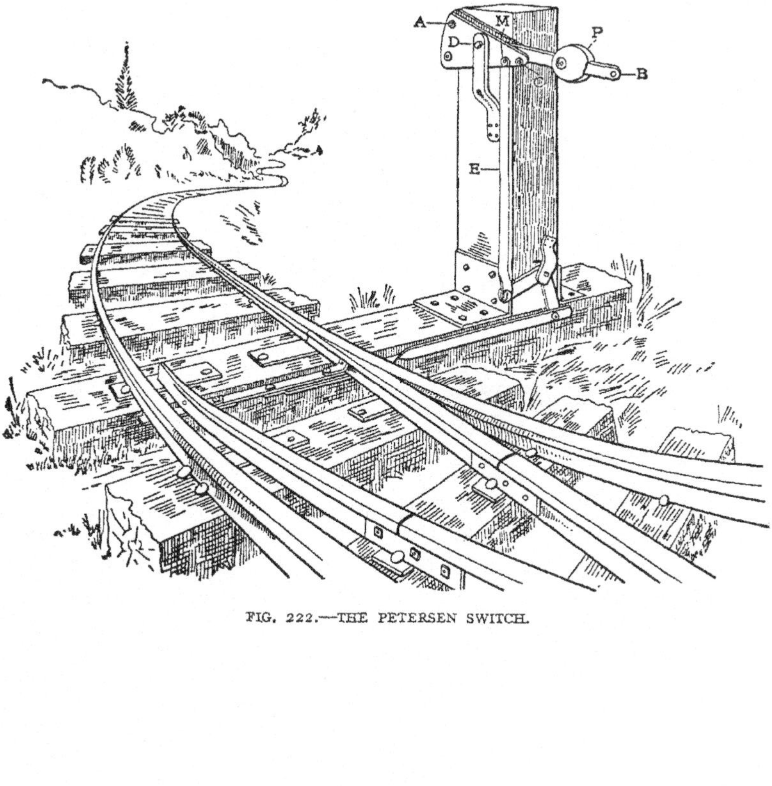

FIG. 222.—THE PETERSEN SWITCH.

A Convenient Switch-throwing Device.—The switch-throwing device shown in Fig. 222 is in use at the Homestake mine, and is made at Newport News, Va., by Peter H. Petersen, a former employee of the Homestake company. The function of the apparatus is to permit the switch being set in either position: When trailing a closed switch no adjustment is necessary, the action being automatic; and when facing it, the required adjustment can be made by the engineer from the train. The device consists of a triangular lever box composed of two plates spaced by roller-shaped fillers on the bolts, and pivoted at *D*. The switch rails are connected to this lever box through a bell crank and bar *E*, which is pivoted at the bolt *M*. The gravity lever *B* with weight *P* is pivoted between the plates at *D*. When *B* is in the position shown, it rests on the bolt *C* and holds the switch points against the right-hand rail, thus keeping the left-hand track open. When *B* is thrown over center, it rests on the bolt *A*, raising *E* and holding the points in the opposite position.

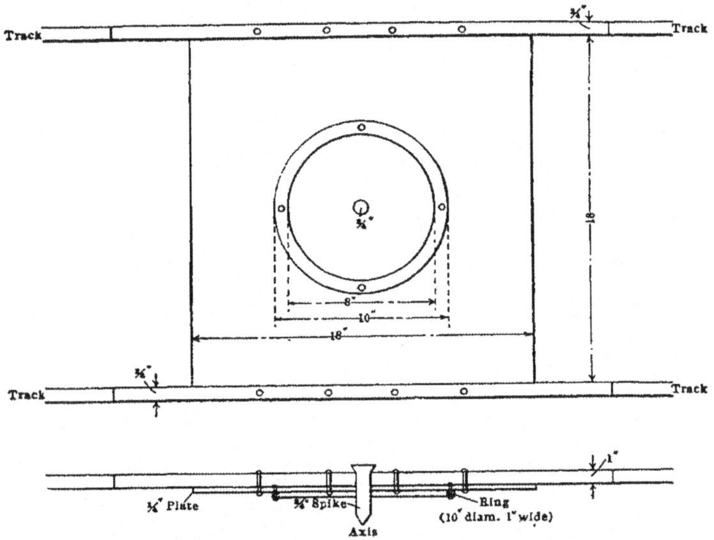

FIG. 223.—TURNTABLE USED IN HIGHLAND BOY MINE.

When trailing a closed switch, no matter for which track it is closed, the switch points will open to let the cars pass through and adjust themselves again to their former position after the last car has passed over them. The engineer or motorman can throw the lever to set the switch for either track he desires while his locomotive is passing over the switch points. When facing a switch, should it be placed wrong, the engineer drives to the throw, adjusts it to suit, and then moves off the switch points, whereupon the weight sets the switch to its new position. Or a rope may be attached to *B*, run through a pulley directly over the pivot *D* and extended to any point along the track, permitting the engineer to set his switch as he approaches it. Thus no switchman or extra trainman is necessary.

Turntable for Mine Cars.—A turntable of simple construction and requiring no bed other than an ordinary tie is shown in Fig. 223. In place of switches or iron plates such small turntables are used at tunnel crossings, in the Highland Boy mine of the Utah Consolidated, Bingham Cañon. The turntables act quickly, are easy and cheap to build and keep in repair, and save space at the tunnel junctions. A piece of 1/4-in. iron plate is riveted to two 3/4×1-in. iron

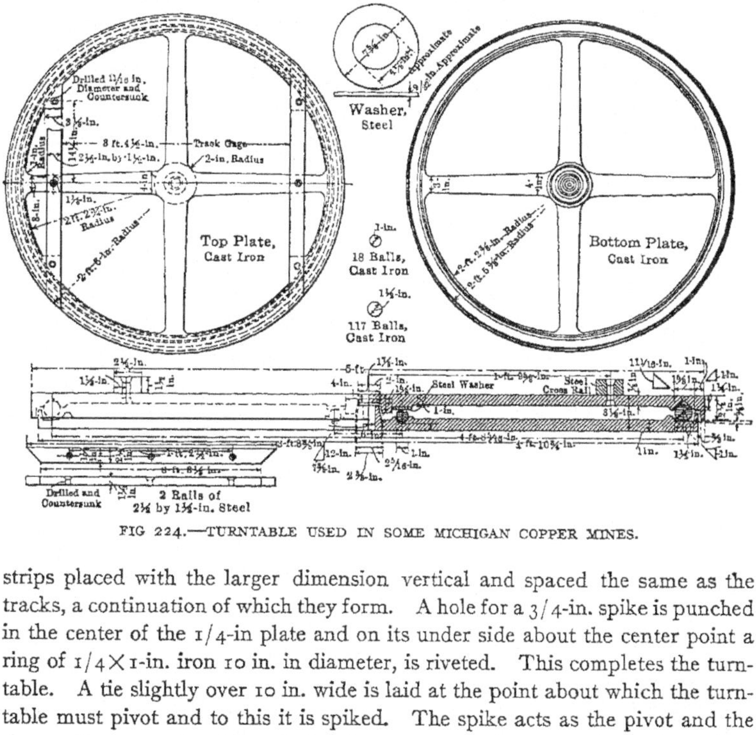

FIG 224.—TURNTABLE USED IN SOME MICHIGAN COPPER MINES.

strips placed with the larger dimension vertical and spaced the same as the tracks, a continuation of which they form. A hole for a 3/4-in. spike is punched in the center of the 1/4-in plate and on its under side about the center point a ring of 1/4×1-in. iron 10 in. in diameter, is riveted. This completes the turntable. A tie slightly over 10 in. wide is laid at the point about which the turntable must pivot and to this it is spiked. The spike acts as the pivot and the ring on the underside of the 1/4-in. plate serves as a bearing on the surface of the tie. A plentiful supply of grease is provided at this point to keep the table turning easily. There is practically no opportunity for dirt to get on this bearing surface, so little attention is required for the device.

A Ball-bearing Turntable.—Turntables are generally used at the shaft stations in the inclined shafts of the Lake Superior copper country, as the tram cars generally hold from 1 3/4 to 2 1/2 tons. At the Tamarack, however, where the shafts are vertical and the cars hold 2 1/2 tons as loaded, turnplates are used. These are somewhat thicker and hence more rigid than turnsheets, the usual substitute for turntables in other districts. The turntables are not especially rapid in operation as, owing to the weight of the loads, speed is not so important as in Western mines where ton-cars are used. The turntables are put in the main tracks on the hanging-wall side of the shaft and are used so as to get the cars on the short tracks that lead right up to the edge of the plat. Some of these turntables are made with ball bearings, while others have two flat bearing rings. The turntables with the balls are stiff at first, but after the balls have worn smooth they are superior to those with bearing rings. If anything, the top part of the turntable could be cast a little heavier as occasionally an arm breaks. The gage shown is 3 ft. 4 in. which is the standard of many mines in the copper country. Fig. 224 shows the turntables used in the Wolverine and Mohawk mines; the Calumet & Hecla company uses a similar table on tracks of 4-ft gage.

XI

PUMPING AND DRAINING

Operation of Pumps—Air Lifts and Eductors—Mine Drainage

OPERATION OF PUMPS

A Useful Pump Formula (By A. Livingstone Oke).—Some years ago, while in charge of the work of unwatering a mine in Portugal, I noticed the following simple relation between the tons of water delivered per hour by the pump and the diameter in inches of the pump plunger, or piston: Tons per hour equal the plunger displacement in cubic feet per hour times the weight of a cubic foot of water divided by the number of pounds in a ton.

$$T=\frac{d^2\times 22\times 100\times 60\times 62.5}{4\times 7\times 144\times 2000}=d^2\times 1.023$$

or only 2.3 per cent. more short tons than the square of the plunger diameter in inches. For the long ton the value is $d^2\times 0.924$, or 7.6% less than the square of the plunger diameter in inches.

These factors are based on the assumption that the piston speed is 100 ft. per minute, which is that usually adopted in ordinary reciprocating steam and other pumps. In any case the formula is easily applied by multiplying the speed and dividing by 100. It will be seen then that the square of the diameter of a pump plunger expressed in inches is nearly the same as the short tons it will deliver in an hour, neglecting slip. In dealing with long tons, this amount should be reduced by one-tenth, thus an 8-in. pump will deliver $(8\times 8)-6.4=$ 57.6 long tons, which, in most cases, will be rather over, than under, the actual amount on account of slip in the valves. In short tons it is quite close enough to say that it is simply the square.

This formula is applicable to pipes when the rate of flow per minute is known. Thus a 4-in. pipe, through which the water is flowing at 400 ft. per minute is delivering $4\times 4\times 4=64$ tons of water per hour. In approximations where the spouting velocity and nozzle diameter are known, the values obtained will be, of course, a little high, but the formula affords a means whereby a rapid calculation gives a quantitative approximation. I have often found this formula surprisingly useful when examining mines where numerous small pumps are in use, and also in rapidly approximating the capacity of pipe lines.

Unwatering Flooded Mines (By D. Lamont).—It often occurs in opening an old mine that a considerable quantity of water has to be removed. I propose to give a few details and hints, gleaned from actual experience as to the plant required for this work, its installation and working. I do not propose to deal with elaborate and costly installations, such as have been used in some cases, but confine myself to the style of plant in more common use, and which, in nine cases out of ten, would be used in a medium undertaking by the average engineer with an eye to economy in first cost.

Before definitely settling on the size and capacity of the pumps required, the size of the shaft and available space must be considered. It is also necessary to ascertain the amount of water the mine is producing, and add a percentage to allow for extra water by seepage from the surface during heavy rains or melting snow.

Many mines have an adit level communicating with the shaft as low as the contour of the country will permit. The amount of water flowing from the adit is generally a fair guide to the amount of excess water that the mine is yielding. This may be measured by an ordinary weir.

In ordering a pump, a good margin must be allowed on its capacity for the excess water. Of the different types of sinking pumps little need be said, as all have their particular merits, and an engineer or pumpman will generally swear by the particular type of pump with which he has had most experience. I consider it a good policy to give an experienced pumpman his choice of pump.

Most sinking pumps can be driven with either steam or air, or both together. The steam heats the air, and increases its efficiency considerably. If the distance between the boilers and pump is not too great, the combination of air and steam prevents freezing of the exhaust, which is often a great trouble in pumps using compressed air only. Compressed air is expensive, as it involves the use of steam or other power to work the compressors. The losses in efficiency through friction in pipes, leaks, etc., are also considerable, and although compressed air is a boon in a mine, it is not always convenient in the initial stages of unwatering the mine. Steam power is most favored to begin with, as fuel for boilers is obtainable in most parts of the world.

The boiler ordered, should be a little in excess of the actual horsepower required, and of a type suitable for transport if the mine is situated at a distance from the rail or waterway. The boilers should be placed as near to the shaft as space and solid ground will allow.

If the mine has been shut down for a good many years, it is possible that the shaft timbers have rotted, or fallen in, and it is always safe to begin by putting in a good collar set, well spread, and carrying two or three sets down on hanging bolts. The collar set should be placed a little above the ground level, and the ground sloped outward, to prevent water from entering the shaft. A temporary headgear should then be erected over the shaft to carry the weight of the pump. A small steam winch should be rigged in line with the pulley for use in lowering and raising men and materials.

A crosshead is useful, the light timber guides for which should be carried down as the work of unwatering proceeds. A signal line should also be fitted in the shaft, and a code of signals arranged. If the sides of the shaft are in good condition it may not be necessary to carry down the timber sets, and, in that case, the only timber required would be the chain-block timbers, and bearers for the pump hangers, and cross timbers to carry the guides for the crosshead. These should be wedged into hitches cut in the wall. Cross timbers should also be placed every 50 ft. or so, to carry the weight of the steam and water pipes. The pipes are supported by clamps. It is a good plan to arrange platforms and ladders in the pumping compartment for executing repairs, and to serve as an exit for the men, in the event of any accident.

A hand crab-winch should be well anchored at the surface. The pump should be hung on this with a flexible wire rope passing over the pulley, and lowered through the hoist compartment. When the pump has been lowered into position and hung with a set of chain blocks in the pumping compartment, the rope should be passed over the other pulley and down the pump compartment and secured to the hanger chain by a strong shackle. In this way the pump is always in hand, and, in the event of water rising in the shaft, it is generally possible to lift the pump out of the water.

Sometimes, even in the case of steam pumps, a pump can be made to work under water and clear itself. I recall a case in point where a steam pump was covered with 6 ft. of water. When steam was turned on it started easily, and lowered the water to the previous level. This was the Tangye Cameron pump with a capacity of 15,000 gallons per hour working against a head of 400 ft., the exhaust being led to surface. The exhaust is sometimes carried into the water. This arrangement has a tendency to heat the water and any escaping steam makes it uncomfortable for the men. Suction condensers take up too much room under the pump, and interfere with its efficient working. I have always found it the best plan to carry the exhaust to the surface, although it entails a little more work and extra piping.

Sinking pumps are generally fitted with heavy hangers and hooks to take the timbers top and bottom. With heavy heads, however, the vibration of the pump is sometimes so great that it is necessary to supply extra support in the shape of an extra timber from the opposite wall.

The suction hose is generally a great source of trouble, and it is not always convenient to use an iron-pipe suction, as, in the event of encountering débris, it is essential that the suction may be shifted. Rubber suction hose as supplied by the makers should not be put into a shaft without being protected with tarred rope, or wound with light chain of about 3/16-in. link. If tarred rope is used, the end should be passed through the bight at each turn. It should not be pulled too tight on account of subsequent shrinkage of the rope in the water. Securing the rope at each turn in this way prevents it becoming unwound should it be cut in any part. A foot valve and strainer should be used with a strong rope attached, the end being secured near the pump platform, this greatly facilitates the handling of the suction hose.

It is always best to lubricate the cylinder and slide valves from the boiler room, or, at any rate, from the surface. For this purpose a one-pint sight-feed lubricator should be fitted on the main steam pipe. A 1/2-in. or a 3/4-in. valve should be placed on the lower end of the steam pipe, near the pump, to blow out any water when starting. A check valve should be placed in the water column immediately above the air vessel, or, failing this, a small pipe connection and valve to empty the column when it is required to open the water end for repairs.

All bends or sharp angles should be avoided in the water column. At the top of the shaft or wherever the water is delivered a T should be placed so as to give free exit to the air.

A spare pump of a similar type should be kept in working order at the surface ready to lower in case of a bad breakdown. Metal valves and seatings are not suitable for gritty water and in the case of rubber-composition valves no time should be lost in turning or changing them if they are in any way leaky. It is poor economy to continue pumping with defective suction or delivery valves. A good stock of these should be kept on hand, and as many spare working parts as possible.

As the water is lowered, the different working levels should be thoroughly explored to see that no bodies of water have been held back by falls of ground or other causes. Water so dammed is liable to break away later and cause damage, besides endangering the lives of the men in the shaft.

It is wise to keep under the head specified by the makers and when this limit has been reached the pump should be securely fixed near a level in which a tank should be made either by damming a portion of the level with concrete or by cutting out the floor or side. Another pump should then be installed to continue the work deeper. The steam piping should be large enough to supply the number of pumps considered necessary to unwater the mine, and it is better to put this in at the beginning and save the trouble of changing later on. A book should be kept by the pumpmen in which should be noted: The running time; stoppages; causes; and the depth the water is lowered in each shift.

Another system of unwatering old workings is to sink a shaft in virgin ground to a depth below the level of the bottom of the old workings, and tapping the water by a drill hole which is plugged with a special form of plug and valve. The water is then under control, and can be drained into the sump of the new shaft and pumped to surface. This system requires a large outlay of capital, but, is often advisable, especially in the case of extensive and dangerous workings,

The Sinking Pump and Its Troubles (By M. T. Hoster).—In sinking wet shafts, both vertical and inclined, one of the most important problems confronting the engineer is to keep the bottom of the shaft so free from water as to enable the miners to do their work efficiently. The sinking of many shafts, especially in remote districts where a few days are required to get supplies, is accompanied by far too many temporary shutdowns on account of not being able to keep the water down low enough for the men to work.

Sinking pumps and bailing skips, or buckets, are the most common means of keeping the water low. For prospecting a new property, sinking a compartment shaft, or where but little water is encountered, the bailing skip has the advantage over the sinking pump for the first few hundred feet; but for single-compartment shafts or where considerable water is met with, the pump is by far the better, especially as the shaft gets deeper.

For use in inclined shafts the pump rests on its lower side; therefore it will be found best to take off the hanger irons. If they are simply turned so that the bent arms stand up, trouble will be encountered later in case the cylinder head must be taken off. For vertical shafts the pump is hung from the ring or, in timbered shafts, from the hangers. If a hose is used for suction it should be tightly wound with 1/4- or 1/2-in. rope to prevent its being cut. The foot valve at the end of the suction hose should always be protected by a good strainer; it takes but little time to wrap some mosquito wire around this, and such a strainer may save much trouble by preventing chips of wood or the like from getting into the water valve chest.

Inside the valve chest are two sets of valves of two valves each; the upper are the exhaust valves, the lower the suction valves. These are made of soft-rubber set in brass seats and mounted centrally on brass valve rods. Each valve is forced against its valve seat by its own spring. The priming valves in the cover act as a bypass over the exhaust valve beneath them. The trouble experienced with these sinking pumps can generally be found in this valve chest or in the suction hose and its foot valve.

In case of trouble, hold open one of the suction valves and pour water down the suction hose. If it fills up and remains full it is evident that all is in good condition below. If the water leaks out examine the entire pipe; the smallest leak will cause much trouble. Hold the foot valve above water and pour water in above to ascertain whether a new foot valve is required. Foot-valve leather should always be kept on hand.

If the suction valve and foot valve are in good condition the trouble is probably in the valve chest. Be sure that the gasket under the cover is good or else air will leak in. If the priming valves leak when closed, the pump will churn its water from the exhaust-valve chamber to the suction-valve chamber and will not throw water up the shaft. To detect this, close the priming valves and try to blow through them; if they leak, new gaskets (preferably leather) may remedy the leak or else new valves are required. I think it best if these valves are kept tightly closed at all times and the pump primed as will be described later.

A sinking pump always handles muddy, gritty water which is bound to cut the priming-valve seats if the valves are opened to let the water through. The soft-rubber valves often become worn or cut along the edges, letting the water pass back between the valves and their seats; hence very little water will go up the shaft. To remedy this, pull out the valve rod and take the valve and cap out. The valve can be turned over on its cap, but after both sides are worn a new valve is needed.

The dirt and grit in the water will in time wear the valve rods and caps also, the result being that the caps become too loose on the rods (Fig. 1) and water will rush back past the valves through the space *a*. (References are to Fig. 225.) In one case where the sinking pump would not work, nearly the whole pump was overhauled and still it would not throw water. Finally new rods and caps were put in and the pump has worked well ever since. Only under severe conditions will new valve seats or springs be needed. The packing boxes should be kept in good condition.

The air (or steam) end of the pump will probably never give much trouble, but to take out the reverse valves at times and blow air (or steam) through will

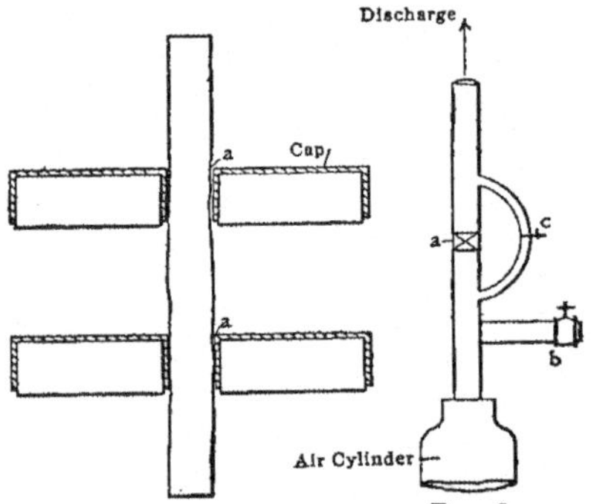

FIG. 1 FIG. 2

FIG. 225.—VALVE AND COLUMN PIPE OF SINKING PUMP.

be of benefit. An inexperienced pumpman will often blame his troubles on the air end and thereby make a big mistake. The following repair parts should always be kept on hand and should be bought with the pump; foot-valve leather, two priming valves, four valve caps, two valve rods, good sheet-rubber packing (1/8-in.), packing for the boxes (air and water).

The column pipe should be fitted up as is shown in Fig. 2, with a check valve *a*, priming pipe and valve *b* and bypass *c* around the check valve. The check valve relieves the pump valves of the back water pressure, and the bypass is advantageous for starting the pump. To start pumping, open both *b* and *c*, running the pump slowly. The water column being full of water, enough water will flow through *c* to prime the pump without the priming valves being open. After the pump starts throwing water out at *b*, close *c* and keep pumping into the sump, through *b*, until all air has been pumped out. Then close *b* while pumping and the pump will take its head slowly. Should the column pipe be empty, water can be poured in at *b*.

Unwatering a Mine with Electric Turbine Pumps (By Percy E. Barbour). —The unwatering of the Columbus Consolidated mine, at Alta, Utah, in the summer of 1911 was an interesting work. This shaft is very wet, and has been flooded several times, and on one occasion was unwatered at an expense of about $30,000. On the last occasion the expenditure of such a sum was prohibitive, yet the mine had to be unwatered. Bids were asked for from contractors, and E. G. Stobel, of Salt Lake City, who had just returned from a six months' trip to European turbine-pump plants, undertook the contract for about one-fifth of the previous cost, his price to cover the cost of the pumps, which he designed and had built under his direct supervision in Salt Lake City.

The mine is opened by a tunnel, at the end of which is an incline shaft dipping about 45° to the 400-ft. level. Down to the 150-ft. level the shaft is a single-compartment 5×6 ft.; from the 150- to the 400-ft. level the shaft has two 5×6-ft. compartments. Between the 100- and the 150-ft. levels, the shaft had squeezed until the maximum opening left was 38 in. The pumps had to be built to pass through this small space.

Two specially designed, German-type, high-efficiency turbine pumps were used, one a single stage and the other a two-stage pump. The former was direct connected to a Westinghouse, vertical, 75-h.p. alternating-current motor running at 1720 r.p.m. The other pump was direct connected to a 50-h.p., horizontal Bullock motor, transformed to a vertical, and run at 850 r.p.m. Two pumps were used in order that the old motors on hand at the mine could be used, and thus reduce the cost of the pumping installation.

The single-stage pump was used as a sinker, and was set, together with its motor and starting box, on a steel frame on trucks to run down the skipway in the shaft, and was handled to the 200-ft. level by the hoisting engine and cable. Below this level a chain-block was used, and the pump was lowered at 5-ft. intervals. When the water had been pumped out to the 200-ft. level, the two-stage pump was installed as a station pump, and thereafter the single-stage sinker discharged into the suction of the two-stage station pump. A sollar was built over the shaft here, and mining on this level immediately resumed.

The suction pipe for the sinker was 8 in. diameter and 16 ft. long, equipped with the usual foot valve and strainer. The discharge connection was an innovation. To lessen the time and difficulty of making changes in the piping, the pump was connected to the water column by a short length of 6-in. rubber hose, guaranteed to withstand a pressure of 150 lb. per square inch. This required piping to be done only every other time the pump was lowered.

The main water column was 7 in. diameter, and the average rate of discharge was 1200 gal. per minute, by weir measurement. The total static head was 290 ft.; the friction head was 35 ft. The average vacuum was equivalent to 9 in. of mercury. The altitude of the mine was about 8000 ft. and the corresponding barometric pressure was equivalent to 19 in. of mercury. The average efficiency of the single-stage sinker was 73% and of the two-stage station pump

77.5 %, both high efficiencies for that altitude. The work was successfully done without any delay, except the burning out of one motor armature, which got wet. The pumps were so small and so easily handled that only two men per shift were employed during sinking for all labor required.

Not the least interesting feature of the work was the number of pumps recovered as the water was pumped out. At the 200-ft. level, four air-lift pumps of various types, two electric-sinkers, and one steam-driven standard sinker were recovered. At the 300-ft. level, one triplex power pump with 75-h.p. motor, one duplex steam pump, three centrifugal pumps, one belted and two direct

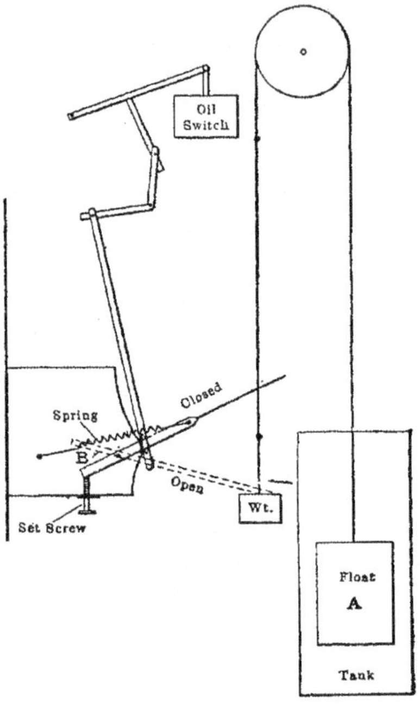

FIG. 226.—ARRANGEMENT OF AUTOMATIC CUT-OFF.

connected, three steam sinkers and two air lifts were recovered. At the 400-ft. level were two electric station pumps, one triplex power pump, three large sinking pumps, and two air lifts; 25 pumps in all.

The cheap and easy solution of this problem rested solely on the minute details given to the design and construction of the interior of the pumps, which gave them large capacity and high efficiency. The reverse nozzle used reduced the velocity of discharge from the pump vanes without the loss of head, due to the usual excessive eddy currents and friction; these particular features are original with the Germans.

An Automatic Cut-off for Electric Pumps.—The device shown in Fig. 226 has been in use by Witherbee, Sherman & Co., at Mineville, N. Y., more than three years and has operated satisfactorily. It is used on a three-plunger electric pump for handling mine water intermittently from a large sump. A float *A*, an 18-in. cube, is constructed of sheet copper and is placed in a tank connected with the sump. A cord is attached to the float and operates over a pulley with a weight suspended on the other end. The cord has two clamps securely fastened, one above and the other below a slotted lever arm. As the water rises in the sump the float in the tank will also be elevated. The rope passing through the slotted lever arm will, when the clamp comes in contact with it, carry the arm down to a point past the center, when the tension spring *B* operates instantly, and thus closes the circuit, which starts the motor and pump. As the water is lowered, the cord is reversed and the lever is driven past the center when the spring again acts automatically and opens the switch. The scheme here used is in connection with an oil switch, yet could be used with an air switch when using only a small current, or with carbon contacts. However, for heavy current the oil switch should be used. The levers are adjustable so that the desired motion is given. The advantage of this cut-off is that it acts automatically and very little attention other than an occasional oiling need be given the pump. The first cost of the equipment is much less than for the solenoid type. The mechanism is simple and can be inclosed so that the little moisture that reaches the parts does not injure the apparatus.

Pumps for Fire Protection (By A. W. Newberry).—At the Belen mine of the Sierra Mining Co., Ocampo, Chihuahua, Mexico, a simple and effective means of fire protection was made necessary by the presence of a large quantity of dry pine timber in one of the main haulage ways. A Smith-Vaile, $8 \times 5 \times 10$-in. duplex pump, situated on the 950-ft. level, handled all the mine water from a main sump. The water level was kept within 4 ft. of the station, so as to provide sufficient water in case of fire. The pump, which was supplied with air at 90 lb. pressure, could raise 50 gallons per minute against a head of 185 ft. To turn this water into the mine, a 2-in. pipe was provided, so as to connect the discharge end of the pump with the main air line to the drills, as shown in Fig. 227. The valve shown in this line at *A* was kept closed, and the valve in the main air line *B* was left open to allow the passage of air. In case of fire, the valve *B* could be closed; the release at *C* opened to bring the air in the line down to atmospheric pressure, and the pump started, if not running, by opening the valve *D*. If necessary, the pump could be quickly primed at *E*. As soon as the valve *B* was closed, the pumpman signalled the men at the working face by striking the pipe. These men then proceeded to disconnect the pipe at a point as near as possible to the fire, unions being provided every 120 ft. for this purpose. Over each union was hung a hose connection, and one 50-ft. armored hose was kept in the station for use in case the hose in the face could not be reached. As soon as the hose was connected, the pipeman signalled by striking the pipe twice, and the pumpman opened the valve *A*, allowing water under 185 ft. head to flow through the air line. This arrangement obviated the necessity of a valve in the discharge column, and allowed all excess water to escape through the latter

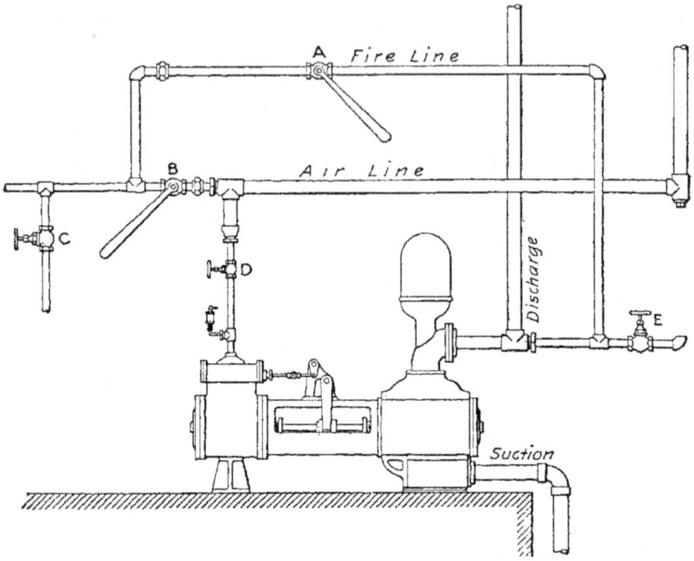

FIG. 227.—PUMP CONNECTED TO AIRLINES FOR FIRE SERVICE.

Repairing a Cracked Pump Cylinder.—A cracked pump cylinder was repaired, states F. H. Coleman, in *Power*, by drilling a hole at one end of the crack and tapping for a 3/8-in. plug. Another hole was then drilled close to the first, the drill extending into the first plug about 1/8 in. The second hole was then tapped and plugged, and the method continued until the crack was covered. It was then smoothed off with a file and calked. Straps 3/8 in. thick used with iron cement further secured the job. The pump has been running for some time with a 110-lb. pressure without signs of leaking.

Air Escape on Small Pump Columns.—At some of the mines in the Joplin district the pumps are operated intermittently. The water is pumped into the column pipe in the shaft and at the surface is discharged into a pipe through which it flows to the mill pond. Where the difference in elevation between the collar of the shaft and the water in the pond is only a few feet and the pipe is only about 3 in. in diameter it is the custom to use a tee at the top of the pump column. The side leg of this tee is connected to the surface pipe, and a short piece of pipe is screwed into the upper leg. The length of this short, vertical extension of the column above the tee is sufficient to prevent any water being discharged over its top. The object of using the tee and the short vertical extension of the column is to permit air to escape at the top of the column which otherwise would have to flow through the surface pipe.

Concrete Water Column.—At one of the shafts of the Continental Zinc Co., Joplin, Mo., the water contains so much sulphuric acid that an iron water column will not last more than 30 days. The quantity of water to be pumped is about 800 gallons per minute. The water contains 1/4 oz. sulphuric acid per cubic foot and 3 1/4 oz. soluble sulphates. A hole 16 in. in diameter was drilled from the surface with a Star No. 23 drill, to a depth of 208 ft. A 10-in. casing was then inserted, leaving an annular space of 3 in. between the pipe and the solid ground. This annular space was then filled with neat cement and allowed to set, making a solid concrete tube from the surface to the pump station. The pump used was a 12×24-in. Scranton lined with brass. A flexible joint with brass discharge connected with the bottom of the cement column. The cement column was used for several months until a cave in the mine destroyed it.

Expansion Joint for Pipe Lines (By C. L. Edholm).—At a Western mine, a U-shaped pipe is used in a long pipe line for steam to take up the expansion and contraction caused by changes in temperature. The bend is supported by iron posts in such a manner that the pipe can slide freely over the supports in contracting or expanding. The device not only takes care of all the expansion in the pipe line, but also reduces the vibrations that, if not checked in some way, would quickly cause leakage at all the joints in the line.

[This device is a very common and extensively used joint for taking care of expansion in pipe lines. A similar device consists in bending a pipe into a complete circle of as great a radius as practicable. The circular pipe has an advantage over the U-pipe in that it can be hung on a single support; the resistance to the flow of fluid is probably greater in the U-pipe than in the circular pipe.—EDITOR.]

Utilizing Water in Mines.—Where it is necessary to drop water a hundred feet or more to a pumping level, it can be utilized to furnish power by conducting it through tight pipes instead of allowing it to find its own course through stopes, raises, etc. One Cœur d'Alene mine uses such power to drive a fan, and another drives a generator which supplies electric light to six stations of the mine.

Pump Station at Leonard Mine, Butte.—The pumps for handling all of the water from the Boston & Montana company's mines, and from some of the mines of the Butte Coalition company, are stationed on the 1200-ft level of the Leonard mine. The new pump station is situated about 150 ft. to the south of the No. 1 Leonard shaft, the old pump station being close to the shaft. The No. 1 Leonard shaft serves as an airway. Between the No. 1 and the No. 2 shafts, the latter of which is the main hoisting shaft of the mine at the present time, separate parallel drifts are run for haulage-ways and for carrying the air, steam and water lines, and electric cables.

The pumping equipment in the new station comprises three 600-gal., five-throw, electrically driven pumps, one of which is of Aldrich and two of Nordberg build. The pumps are 7×12-in. size and are each driven at 60 strokes per minute by two 150-h.p., 440-volt, 180-amp. motors run at 495 r.p.m. In the old station there is a 1500-gal. auxiliary steam pump and a smaller one with a capacity of about 600 gal. per minute.

The construction and timbering of the station is particularly interesting. It is built with an idea of providing ample space and in such a manner as to assure permanence. Provision against the crushing of timbers and caving of the roof

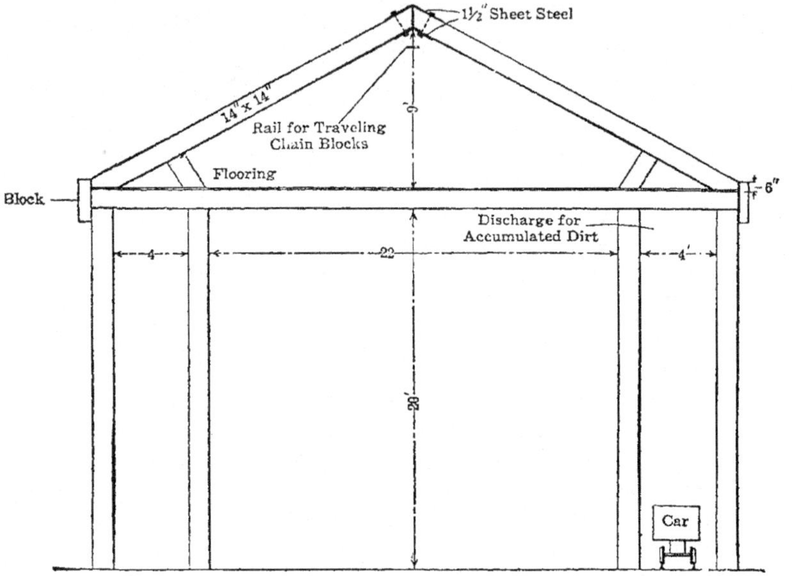

FIG. 228.—TIMBER SET IN PUMP STATION, LEONARD MINE, BUTTE, MONT.

was especially necessary, for as stated, the pumps in this station handle practically all of the water from a number of the mines of two of the large Butte companies, hence, the stopping of the pumps for any length of time would entail a large loss.

The station is cut out about 35×75 ft. and 30 ft. high in the clear. Fig. 228 shows the scheme of timbering employed. The caps reach entirely across the station and are supported at their ends by posts and also at points 4 ft. from the ends by auxiliary posts. Thus along each side of the station there are two rows of posts. Angle braces are used above the inner row to give added support to the roof of the station, which is trussed, the peak being 9 ft. above the center of the caps. (Above pumps, caps are cut out and horizontal angle braces used.) At the point of their abutting, the caps are held firmly in place by bolts through cover plates of 1 1/2-in. sheet steel. The plate over the joint of the roof members extends 16 in. down each side. The station sets are constructed throughout of 14×14-in. Oregon fir timber; sets are 5 ft. center to center.

By giving the station a peaked roof considerable extra excavation is necessitated and additional timber is required. The added cost this entails, is, however, more than counterbalanced by the safety from caving that is assured by the additional strength given to the timber framing. In a similar station with a flat roof it was necessary to clear out the caved material above the caps each year. The station described has already stood for three years and it has not yet been necessary to clear away any débris; flooring is, however, built over the caps to catch any material that caves, and an opening is left between the rows of posts along one side for the discharge of any caved material. Tracks are laid along this aisle for cars to handle the dirt.

Under the floor of the station is built a small concrete-lined tank, 3×20 ft. and 8 ft. deep, about which the pumps are grouped. The pump plungers all draw directly from this small sump, and a feeder 20 in. wide and 8 ft. deep connects this to the main tank, which is excavated in the solid rock to one side of the station and has a capacity of 25,000 gallons. The concrete lining of this larger tank is 2 ft. thick. A great saving in non-corrodible piping is effected by having the pump plungers draw directly from the small pump tank and at the same time the pumps are seated upon firm foundations. It is figured that the additional cost necessitated by building the station as described will be more than overbalanced by the saving effected in its upkeep.

AIR LIFTS AND EDUCTORS

Notes on the Pohle Air Lift[1] (By W. S. Anderson).—The proportions of piping, submergence, etc., in the Pohle air lift may be arrived at mathematically, basing the calculations on the expansion of the air bubbles as they rise to the discharge where the head due to the sum of the water pistons above diminishes, and on the usual formulas covering capacities of air and water pipes; but for practical purposes in the case of small wells reliable results may be derived from the following:

It is found in practice that the length of pipe submerged below the normal pumping level after the water has fallen should range from 30 to 60 per cent. of the total length of pipe from the bottom end to the top of discharge, the greater ratio of 60% generally giving the greatest efficiency. The air pressure should vary from three-quarters of the pressure due to the total lift for shallow wells to six-tenths for deep wells. It should be such that it will about equal atmospheric pressure at the point of discharge, unless high velocity is desired at the expense of economy. That is, it should equal the head in feet multiplied by 0.434 plus a few pounds for friction, this latter depending on the length of pipe and friction

[1] Reprinted from *Power*, December, 1909.

head. In this connection it should be remembered that where a central air pipe is used in drilled wells, on account of there not being room for the air pipe outside of the water pipe, the friction head is greater. If the compressor is close to the well, the reservoir should be proportioned to reduce all impulses, though in the case of a long air line this will to a great extent take care of itself.

In the case of a large well casing, where the inflow into the well is small, or where the desired amount of water is small in proportion to the capacity of the pipe, the air pipe, if central, should purposely be made large enough to prevent blowing, the most desirable proportion, as indicated by practice, being about 1:2 for small wells and 1:3 where the diameter of water pipe is over 2 1/2 in.; or the area of the water pipe is about six times the area of the air pipe for average cases. If the air is taken from shop air mains, where variable service causes a large variation in air pressure, and the air is not run through a reducing valve, the sudden expansion of high-pressure air at the discharge is likely to cause trouble, if the discharge is into a covered or practically closed tank. This may be relieved by putting a large-sized safety valve at any convenient bend.

Assuming 60% submergence, the volume of free air for maximum economy may be roughly taken from the following: For lifts of 25 ft., 13 cu. ft. of air for 100 gallons of water; for lifts of 75 ft., 28 cu. ft. of air per 100 gallons of water; for lifts of 100 ft., 36 cu. ft. of air per 100 gallons of water; for lifts of 150 ft., 57 cu. ft. of air per 100 gallons of water; for lifts of 200 ft., 75 cu. ft. of air per 100 gallons of water.

With side inlets and 3/4-in. air pipes and 1 1/2-in. water pipe, the ordinary capacity will be about 25 gallons per minute, and for larger wells the capacity may be assumed as varying nearly but not quite with the square of the diameter of the water pipe. For example, a 6-in. water pipe will ordinarily give a flow at fair economy of from 400 to 425 gallons per minute. In the case of central air pipe with 3 1/2-in. casing, the output should be about 110 gallons per minute, or 230 gallons with 5-in. pipe, and other sizes in proportion. Generally speaking, from 10 to 14 gallons per square inch of area of water pipe, after deducting the air-pipe area, will be about right, the smaller figure being used for short lifts and the larger figure for lifts of 100 ft or over.

The matter of economy is generally misunderstood and most of the claims of high efficiency should be taken with rather a liberal allowance. While under favorable conditions it may be possible to approach 70 or 80% efficiency figured from the discharge back to the indicated horsepower of the compressor, most of the tests of ordinary wells where extreme submergence is not possible seem to indicate efficiencies between 20 and 35%. It should also be understood and remembered that the air lift is not well adapted to forcing water horizontally, and horizontal discharges of over 40 or 50 ft. should be avoided, except by discharging into a tank with a gravity flow the rest of the distance.

Unwatering Shaft by Compressed Air (By Louis Boudoire).—A simple air lift can be quickly set up to unwater mine shafts. Fig. 229 shows the arrangement of the piping. In the mine where it was used 40-mm. and 90-mm. pipes were at hand and, as the necessity of unwatering a certain shaft on the property was urgent, no time was spent in an effort to improve the efficiency of the appliance by tapering the ends of the pipes. Air was delivered at an effective pressure of 65 lb. per square inch; the vertical length of the 90-mm. water pipe was 40 m., and its horizontal length 300 m. The results were as follows: With a submergence of 30 to 35 m. and a lift varying up to 5 m. the output was over 200 liters per minute; with a submergence of 14 m. and lift of 21 m. the output was 50 liters,

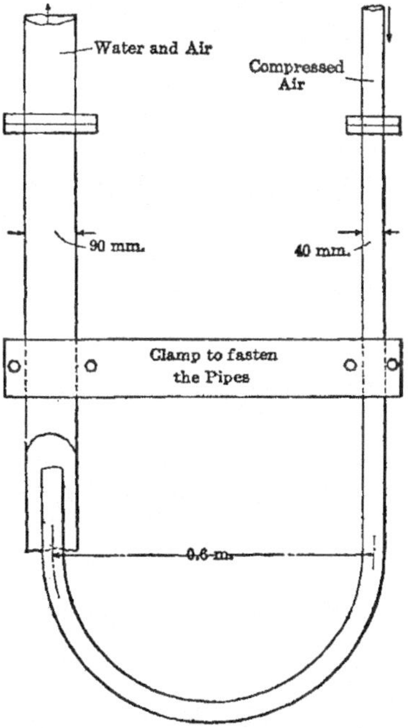

FIG 229.—AIR LIFT FOR UNWATERING SHAFT.

and with a submergence of 11 m. and a lift of 24 m., it was only 30 liters. The output decreases, therefore, as the height which the water must be lifted increases. Although the efficiency of the system is not high, it presents valuable advantages for emergency use, as it can be quickly installed, does not require any attention, oiling nor, as does a pump, adjustment for every 7- to 8-m. variation of the head under which it is operating. In the case of a deep shaft this appliance might be used to assist the sinking pump which would then have to be lowered only for every 25- or 30-m. reduction of the water level.

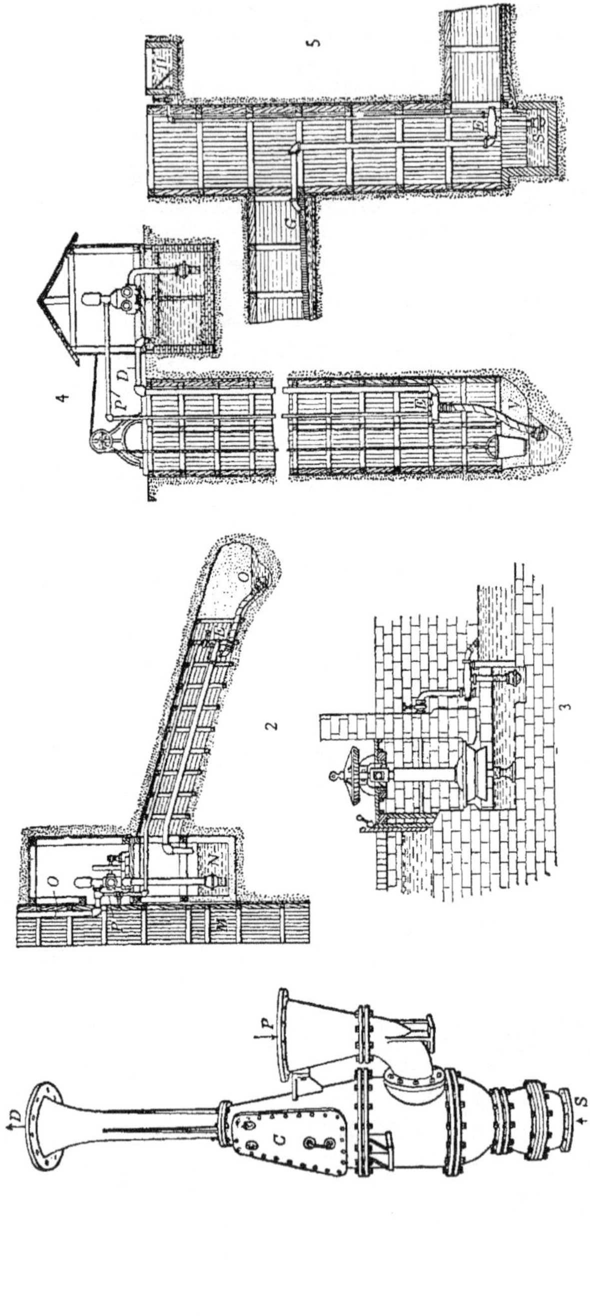

FIG. 230.—TYPES OF MINE EDUCTORS.

Mine Eductors (By Oskar Nagel).—The lifting of large quantities of water, a problem of great importance in mining, is mostly performed by means of pumps. In places, however, where the space available is insufficient for the installation of a pump, the water-jet eductor, Fig. 1 (reference is to Fig. 230), is the proper machine to use. The pressure water enters the eductor at *P*. Passing through a nozzle, it creates a vacuum, and raises the water by suction at *S*, discharging the entire volume of liquid at *D*. These eductors are designed to raise water by means of high-pressure water, and are used as follows:

(1). Water collecting at a considerable altitude is used to raise water which has collected further down, both being discharged at a medium level, thus permitting flow off through horizontal openings down a hillside, or to a pumping engine.

(2). In a similar manner the water from a condenser of an underground pumping engine may be raised with advantage.

(3). Even if the pressure should have an available head of but a few yards, it is possible to effect a considerable suction which is particularly useful in tunneling.

Fig. 2 shows an installation and illustrates the simple manner in which an eductor may be installed in the smallest possible space. *E* is the eductor, *O* the suction hose, *P* the main pump. Fig. 3 shows how in a turbine installation the fall of a river may be utilized by means of eductors. Even with a low fall the eductor retains its capacity for high suction up to 16 ft. and above.

The advantages of eductors for sinking shafts have caused their application in connection with high-pressure pumps. Fig. 4 shows such a plant. *D* is the discharge, *E* the eductor, *P* the pressure line from pump creating water pressure. This method has the following advantageous features: It is easy to handle the apparatus, as only piping of small diameter and small weights is to be taken into consideration; the small space required, and, above all, the free working space; positive working, due to the absence of moving parts; the apparatus works with equal sureness in case the water does not flow sufficiently to the shaft.

Fig. 5 shows the eductor in a shaft taking the head-water from the surface and discharging into an upper gangway. From the flume *H* on the surface, the water flows to the eductor *E* and lifts the water from sump *S*, discharging into the gangway, or passage *G*.

These mine eductors are used by: Thomas Shelton (Engelbach Machinery Co.), Leadville, Colo.; Compañia de Santa Gertrudis, S. A., Pachuca, Hidalgo, Mexico; Benito Juarez Mines Co., Salinas, Mexico; Arizona-Parral Mining Co., Denver, Colorado.

MINE DRAINAGE

Draining a Shaft through a Drill Hole (By Lucius L. Wittich).—By installing a deep-well pump in an 8-in drill hole, sunk 7 ft. from the shaft, J. M. Short, operator of the Geronimo mine, on a lease of the Connor Estate at Ellisville, west of Joplin, Mo., is able to dispense with the pump that he has been using in sinking the shaft, thus reducing the cost of sinking from $25 to $17.50 per foot, a saving of $7.50 per foot. As the shaft was 70 ft. deep when the pumping arrangements were changed and as the shaft must be sunk to a depth of 126 ft. to reach the ore, as indicated by the drill hole on which the shaft is being sunk, 56 ft. remain to be sunk. The total saving will be $420. Deducting from this, $150, the cost of sinking the 8-in. drill hole and installing the deep-well pump the net saving in money will be $270. Of course this does not include the cost of the drill rig and pump which were on hand. Had the system been adopted earlier the saving would be much greater.

But the most important feature is in the saving of time in sinking. Before the drill-hole pump was placed in operation, the shaft was being sunk at the rate of 5 ft. per week; now it is being sunk 5 ft. in 16 hours. Previously it was necessary to remove the pump after loading each round of holes and to replace it after blasting, or, at least, it was necessary to guard the pump with heavy timbers, and even with this precaution the machinery was almost invariably damaged by the blasts. On this account it was impossible to use heavy blasts when the pump was left in the ground. Now it is possible to use full charges of dynamite, and the working ability of the pump in the drill hole near-by is not impaired.

In following a system of this kind, adopted for the first time in the Joplin district by Short, it would seem necessary that the shaft should be either sunk through open ground, permitting free drainage of water to the drill hole, or that two drill holes should be sunk, one squarely in the center of the shaft and the other near-by for the pump. The latter is the system followed by the Geronimo company. The shaft is being sunk through hard limestone beneath which occurs the open ground in which the ore is found. Both holes penetrate the open ground, the deep-well pump having been installed to a depth of 150 ft. Care is taken to keep open the drill hole in the bottom of the shaft, for through it the water from the shaft seeps into the lower open ground, thence across the interventing space of 7 ft. to the second drill hole, whence it is pumped out. In this manner the shaft, formerly so wet that the miners were forced to wear rubber boots, is now dry and the excavated rock is handled with much greater ease.

Draining with Well Points.—It is often the case in placing concrete piers, foundations and retaining walls, that there is a heavy influx of drainagewater that rapidly fills the pit dug for foundations, or accumulating behind the forms for retaining walls, rises to exert a considerable hydrostatic pressure that may warp the forms either prior to or after filling with concrete. In laying concrete foundations supported on all sides, the inflowing water is not of much moment as the concrete may be filled in to displace the water, and if there is no heavy upward flow that carries away the fine cement, a substantial foundation is obtained. But there are many instances when ground water is detrimental to both the stability and facility of making the structure, more particularly in the case of retaining walls. In laying brick or stone masonry in pits, such as elevator pits, the inflow of water must be controlled and usually the desire is to draw the water from back of the wall being built, *i.e.*, between the wall and face of earth which the wall is to support. The problem of draining, while building a wall that will later be impervious to water, may often be satisfactorily accomplished by using well points, driving them down back of the wall. A number of such points can be attached, preferably by a flexible coupling such as a short length of hose, to a main leading to the suction chamber of a pump. The well points are made of 1-, 1 1/2- or 2-in. pipe of suitable length, to one end of which a solid iron or steel driving point has been welded. Above the point the pipe is drilled with many small holes for a foot of its length to make a strainer. In driving the pipes, a steel or iron bar, just a trifle smaller than the diameter of the well point and a foot longer, which at a point about 14 in. from the end has been upset enough to make it a little larger than the pipe, is placed in the pipe, and the point is driven by sledging the upper end of the bar. When the point is driven to the desired depth the bar is withdrawn. When first used the pump draws considerable sand through the pipe but gradually the gravel and bits of rock too large to pass the strainer form a coarse filter that not only prevents much sand from entering the pipe but gives freer passage for the entering water.

Draining an Ore Chute (By Arthur O. Christensen).—At the Enterprise mine, Gunnison County, Colo., a raise is being driven from No. 3 to No. 2 tunnel, 800 ft. above. The raise is a three-compartment incline following the vein. The manway is in the middle with skipway and chute on either side. The whole is cribbed with 8-in. timber. At every 100 ft. a level is driven on the vein in each direction. As the work progressed more or less water was encountered on the levels. This flowed to the stations, and as no provision had been made for carrying off the water, it found its way into the chute. The quantity of water was not large, but was enough to keep the dirt in the chute at just the proper consistency to stick there, the dirt being largely of a clayey character. The objection offered to making any provision for carrying off the water from the levels was that there was no room in the raise to put any conduit for it. The ladder and suction pipe filled all the space which could be spared in the manway.

At intervals of about a foot holes were bored from the manway into the chute. By the use of drills and a bent rod it was possible, with sufficient labor, whenever the chute became "hung up," which was the greater part of the time, to get it down by working through these holes. This was expensive and laborious. At each level a length of the suction pipe was removed and a T inserted, having a 5-in. side opening. A section of 5-in. pipe was carried through the back of the raise into the station, entering it a little below the bottom. This was connected to a trough, as shown in Fig. 231. The water from the levels is run into the trough and carried by it into the suction line, through which it falls into a barrel provided at the bottom of the raise.

By having the water fall through the airway the flow of air has been noticeably increased, and the water, instead of being trammed out with the dirt, now helps to run the compressor which supplies power for the mine. Comparatively little difficulty is now experienced by the dirt being hung up, as the only water entering the chute is what runs in from the breast of the raise. What hang-ups do occur are readily taken care of by providing the following arrangement in the partition between the manway and the chute:

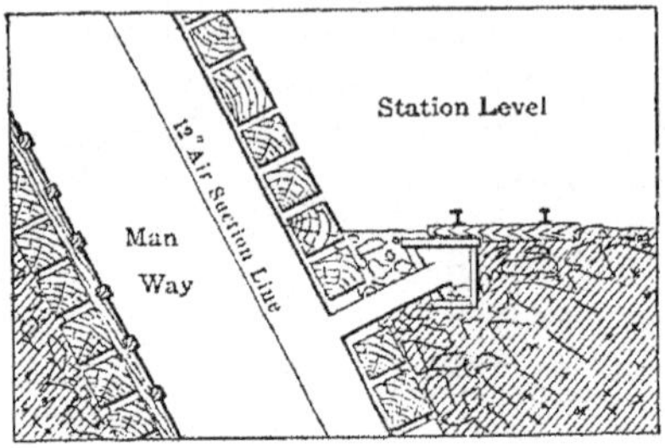

FIG. 231.—DRAINAGE SCHEME FROM STATION TO SUCTION PIPE.

Slits 2×4 in. are cut between alternate timbers. Over the slit is nailed a piece of inch board about 5×8 in., by driving one nail at the top so the door thus formed can be swung up when it is desired to look in or to poke hung-up dirt. These doors keep small rocks and dirt from flying into the manway. The timbers now being framed are made with a groove across the middle of one edge 2×2 in. on the manway side and 6×4 on the chute side, as shown in Fig. 232.

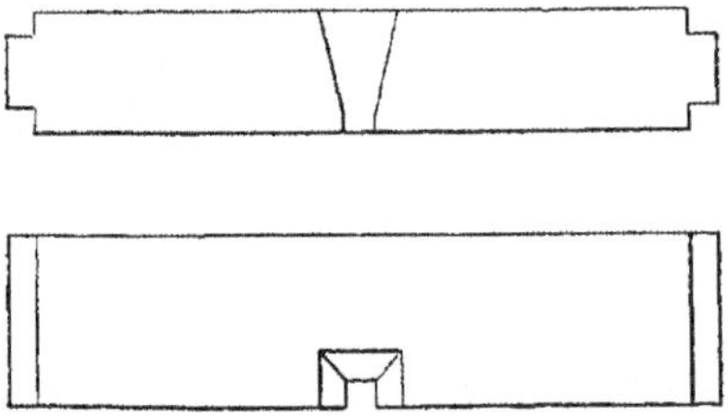

FIG. 232.—METHOD OF CUTTING TIMBERS.

When these partition pieces are put in they form a slit 2×4 in. on the manway side and 6×8 in. on the chute side, the slits being between every other partition piece. By having openings this size and shape a hang-up is easily pried down by working from the manway.

The chute in question is 2 ft. 8 in. × 3 ft. 8 in., cribbed solidly, but not lined. The dip of the raise varies between 60 and 70°. This chute seems to be too small, for even dry dirt, after piling up in it, has to be barred down occasionally.

A chute of this dip should be lined on the bottom to prevent wear and hanging of the dirt. Planks 2 in. or 4 in. thick, laid longitudinally, or still better, sheet iron, make good lining when well supported from underneath.

Draining Gravity Planes.—The best plan to drain a gravity plane and keep the track ties in place is to lay a line of tile-pipe with properly cemented joints on one side of the track. All the water that can be diverted from the head of the plane should pass down this drain. Several Y's should be put in the line and 6-in. branches run from these under the track. These branches should be cemented on the under half. The upper half should be left uncemented so as to drain the roadbed and the ditches in which the branch pipes are laid covered with broken stone.

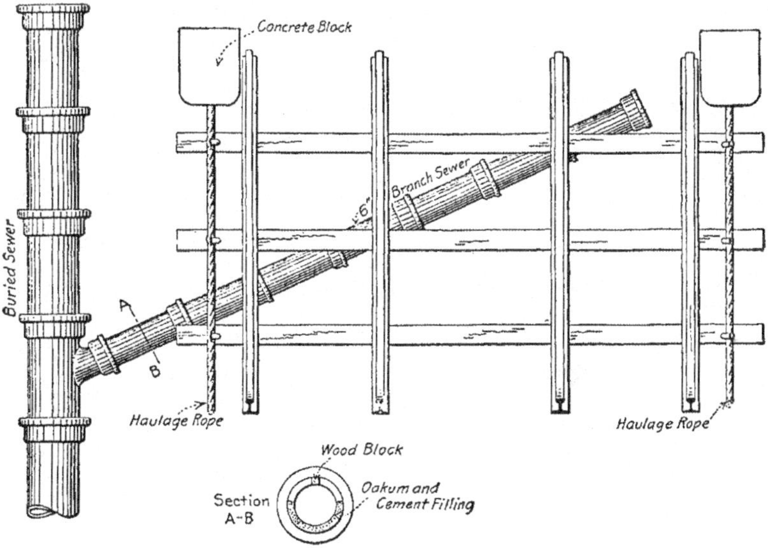

FIG. 233.—DRAINING A GRAVITY PLANE WITH SEWER PIPE.

Some planes are laid with anchored wood-sills into which the cross-ties are notched. The notching results in speedy rotting of the timber and the expense of such a method is considerable. The use of long ties common to both tracks as shown in Fig. 233 is recommended by *Coal Age*, but perhaps the best security and the cheapest is the use of a haulage rope spiked to the end of ties and secured to a heavy timber, masonry or concrete support at the head of the plane. If the plane is long, two or three of these may be necessary so as to divide up the expansion from heat. Old, discarded ropes will do, and a heavy coat of tar will prevent them from rusting away; this coat being applied before and after their attachment to the ties.

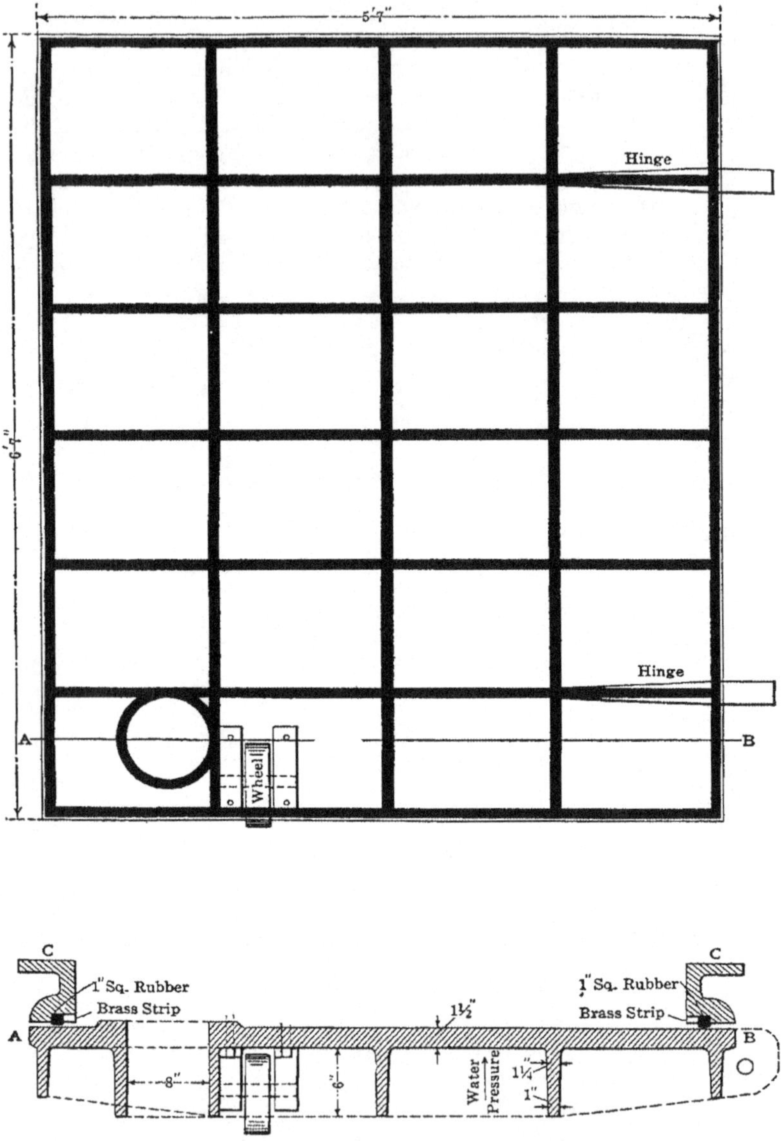

FIG. 234.—CAST IRON GATE FOR MINE DRIFTS.

Gate for Controlling Mine Water.—A gate used for controlling water flow in mines is shown in Fig. 234. This particular gate was used at Flat River, Mo., to protect pumps and mine workings from sudden flows of water. The gate is of cast iron, 1 1/2 in. thick, with longitudinal and cross ribs 6 in. high and 1 1/4 in. thick. The door is 5 ft. 7 in. wide, and 6 ft. 7 in. high and has an 8-in. opening in the lower left-hand corner to which a pipe and valve may be attached. Near this opening is a small wheel to support a portion of the weight of the door, and thus relieve the hinges of such a large burden. The door closes against a cast-iron frame *C* which is securely cemented into the wall of the drift. A groove is cast in one side of the frame, and a piece of 1-in. square rubber packing is inserted and held in place by a small brass strip. The door closes against the rubber packing, and is so hung that the pressure of the water from within will hold it shut. In order to open the gate it is necessary to open the valve and allow the water to flow out, thus relieving the pressure.

Stopping the Flow of Water from a Drill Hole.—Several years ago a bore-hole drilled from the No. 7 plat of the Armstrong shaft, in the Great Fingall mine, Western Australia, encountered a flow of 6000 gallons per hour of salt water, at a depth of 500 ft. The bore-hole inclined 27° from the horizontal and attained a total depth of 747 ft. When drilling operations were concluded it was decided to control the flow of water, and the following method, described by G. C. Klug in the *Journal* of the Western Australia Chamber of Mines, was employed:

A piece of 1 1/2-in. steam pipe 5 ft. long was tapered for half its length from its full diameter at the middle to 1 11/16 in. at the end, the metal at this point being 1/32 in. thick; the other end of the pipe was threaded to receive a 1 1/2-in. plug-cock. The tapered end was forced into the bore-hole and held in position by means of clamps and bolts which had previously been cemented into holes drilled into the surrounding rock as shown at *B* in Fig. 235. This proved effectual in sealing the bore-hole itself, but when the water was shut off, leakage of water took place through fissures in the surrounding country. Owing to the gradual increase of water finding its way into the lower levels, it was found necessary to collect as much water as possible in the upper levels; and in order to assist in this, the plugging of the bore-hole was decided upon.

A double-ram Cameron pump having cylinders 7-in diameter, rams 4-in. diameter and 6-in. stroke, was erected close to the bore-hole, and the delivery pipe was connected up to the 1 1/2-in pipe at *C*. An ordinary cyanide case was used for a suction sump, and into this a water pipe was led. Compressed air at a pressure of 80 lb. per square inch was used to drive the pump. The suction sump was filled with water, the plug-cock *A* opened, and the pump started. The sump was kept three-quarters full of water, and pine sawdust was gradually added to the water, which was kept well agitated. When about 1 cu. ft. of sawdust had been pumped into the hole, it was noticed that all leakage from the face of the rock had stopped. As the sawdust was intended only to temporarily stop the larger cracks, the pump was stopped, the sawdust removed from the sump and clean water pumped for a few minutes into the hole. Portland cement was now added to the sump in such quantities as to form approximately a mixture of 25% cement and 75% water, and this was pumped into the bore-hole in the same manner as the sawdust, cement being added as re-

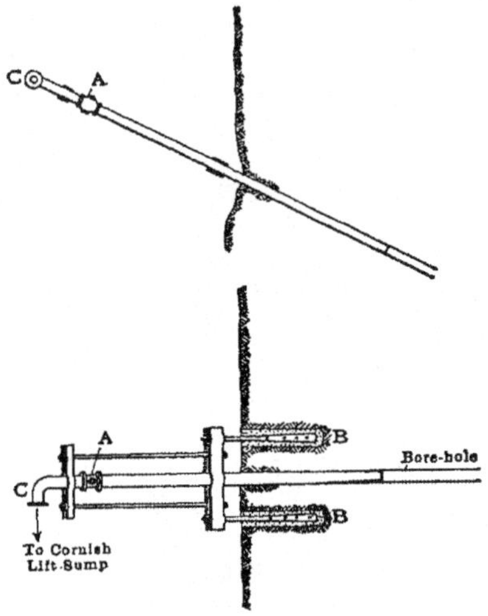

FIG. 235.—DRILL HOLE AND FITTINGS.

quired. At first, the pump worked quite freely against the pressure encountered; but as the hole filled with cement, the pressure gradually increased until it was necessary to assist the pump by barring the fly wheel round. When one cask of cement had been used it was found impossible to force any more into the hole. The cock *A* was then closed and after standing four days, it was opened, when it was found that the water had been completely stopped, and there is now no leakage whatever through the surrounding country.

XII

VENTILATION AND COMPRESSED AIR

Approved Practice in Ventilation—Devices for Improving Ventilation—Theoretical and Practical Considerations in the Use of Compressed Air.

APPROVED PRACTICE IN VENTILATION

Ventilation for Transvaal Mines.—Treating the subject of mine ventilation, the Mining Regulations Commission of the Transvaal makes recommendations for the sectional ventilation of the mines. The ventilating currents from downcast intakes should be split at the entrance of every working drift, such entrances being provided with brattices so constructed that the openings for the passage of air can be varied as required. After passing through the workings air should be led as directly as possible to the main return airway.

Recognizing the insufficiency of the ventilation in most of the Transvaal mines, it is recommended that in all portions of a mine or workings where the natural ventilating current is insufficient, suitable mechanical appliances for ventilation be erected and operated. The courses for the supply of air to all working places, and of foul return air from such places, should be kept separate and disused drifts, stopes, etc., where possible, should be completely closed in.

It is further recommended that plans and sections of every mine be kept at the mine office and these drawings show airways, direction of air currents, position of brattices, etc., drawings to be posted to date at intervals of not more than 3 months. Not less than once every 3 months, chemical determination of the following samples should be made at each mine: Air 100 ft. from the face of all drifts; 50 ft. from the face of all winzes and shafts; from the bottom of upcast shafts; from all stopes connected by only one drift.

With regard to the subject of the local ventilation, the Mining Regulations Commission recommended as follows: (1) That the use of mechanical appliances is indispensable for adequate ventilation of certain sections of a mine outside of the circuit of natural ventilation; (2) that every working place where rock drill ares used be furnished with suitable arrangements for laying and removing dust, smoke, gases, etc., and that no man shall return to a working face until the air is free from noxious gases caused by blasting; (3) that the intake pipes to compressors be led outside of the engine room to where the air is of suitable degree of purity; (4) that the lubricating oil used in compressors have a flash point of not less than 600° F.; (5) that periodical inspection by a responsible mine official be required for air cylinders of compressors; (6) that when mechanical ventilation is not provided, all compressors be kept running for at least 2 hours between shifts at not less than 20 lb. pressure except when necessary to stop for repairs; (7) that the vicinity of the collar of downcast shafts be kept clear of all cinder heaps, and as far as possible of smoke.

Carbon Dioxide Criterion for Ventilation.—The Mining Regulations Commission of the Transvaal has made a number of excellent recommendations for the bettering of underground conditions. The legal maximum for noxious carbon-dioxide is fixed at eight parts by volume in 10,000 of air; in addition four parts representing innocuous CO_2 present in the atmosphere, three parts where candles or similar iiluminations are used, and five parts in order to meet the difficulties of practical administration in regard to possible innocuous gas from country rock and other uncertain sources, are allowed. The total limit is, therefore, 20 parts of CO_2 per 10,000 of air. In the Lydenburg district, where there is geologically strong presumptive evidence of a production of ground CO_2, further investigation is recommended and an allowance of 1% maximum by volume CO_2 in the mine air is made.

It is stipulated in the recommendations that samples for testing purposes be taken not less than 1 hour after blasting. No allowance is to be made for the altitude of the land, as affecting the allowable CO_2 limit, as many samples will be taken at considerable depth. The maximum permissible amount of carbon monoxide, CO, in any part of a mine is not to exceed 0.01% and no practically determinable amount of NO_2 shall be permitted in any part of the mine.

The commission seems to recognize that the application of the existing Transvaal laws on the subject of mine ventilation is open to serious practical difficulties. The quantity standard (70 cu. ft. of air per man per minute) is judged as less satisfactory than one of quality. The quantity of carbon dioxide present is accepted as bearing a roughly constant proportion to the amount of impurity present and the carbon dioxide is considered the best criterion of the sufficiency of ventilation.

Lack of Oxygen in Hydraulic Air.—When the air from the hydraulic plant at Ragged Chutes, Cobalt district, Ont., was first turned on it was found that it was practically impossible to burn candles in the mines where it was used. It was claimed that this was due to the absorption of oxygen by the asphalt with which the inside of the pipes were coated, and that this effect would soon pass off. It was soon found, however, that hydraulic air contains an appreciably less percentage of oxygen than ordinary air, and analysis demonstrated that it contained only 17.7% oxygen, which is 3% lower than ordinary air. This is due to the oxygen going into solution in the water during compression, when a pressure of 130 to 135 lb. per square inch is maintained. The lack of oxygen does not apparently trouble the miners, but besides the difficulty experienced in keeping lights, the effect of the gases from exploded dynamite is much quicker and more serious than was found to be the case with air compressed by machinery.

DEVICES FOR IMPROVING VENTILATION

Wrinkles for Ventilating Mine Workings.—The schemes for aiding the ventilation of a tunnel or shaft in out-of-the-way places are interesting, and often effective. In a shaft the prospector often builds a fire at one side on the bottom. This is especially efficient in a two-compartment shaft that is timbered close to the bottom. A little sheet-iron stove with a long chimney to the surface is an "improvement" over the open fire. Sails are sometimes rigged to deflect the wind down a sheet-iron pipe to the shaft bottom. Ventilation at tunnel faces is helped by a simple sheet-iron pipe fitted with an elbow at the portal

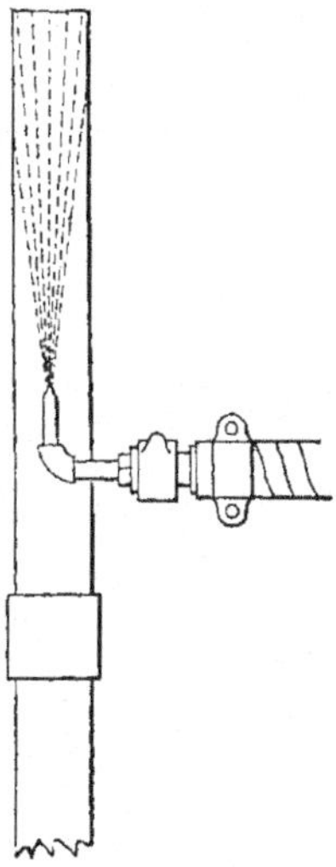

FIG. 236.—JET FOR VENTILATING BY COMPRESSED AIR.

and having a vertical length of pipe set in the elbow. A small stove may be inserted at the elbow, which draws air from the drift and uses the vertical pipe as a smoke stack. Small wooden fans are fitted with footpower attachments like that used on grindstones.

Ventilation by Suction (By Arthur O. Christensen).—The following method for sucking air through a pipe, although not new, may be novel to some and a suggestion to others. Fig. 236 shows an arrangement of the apparatus that has been found satisfactory in La Noria mine, Zacatecas, Mexico. A pipe about 2 1/2 in. in diameter or a wooden conduit 4 in. square, is laid in the working to be ventilated by suction. Into the lower end of the pipe a 3/8-in. pipe is inserted and bent or fitted, as shown in the illustration. This is coupled direct to the compressed air line. When the valve is opened the jet of compressed air, rushing into the larger pipe and parallel to its direction, creates a strong suction in the pipe or conduit, producing the ventilating current.

In operation it is better to use the full air pressure available, and cut down the amount used by the size of the nozzle rather than employ a large nozzle and only partly open the valve, as is sometimes done. Where the larger pipe is not over 4 in. in diameter, a 3/8-in. pipe tipped with a 1/4-in. nozzle is large enough for a 30-lb. air pressure. For higher pressures and where it is necessary to be economical, an aperture of from 1/8 to 1/4 in. may be used. The nozzle should be placed in the pipe or box at a point so situated that the jet issuing from it will be spread out to fill the pipe before leaving it. The higher the air pressure used the farther back the nozzle need be placed. I have found that placing the nozzle 12 to 18 in. from the outlet of the pipe was about right. If the nozzle is put farther back than is necessary the pipe ahead of it hinders the rush of air, and thus impairs the suction efficiency.

In the case of an opening where it is desired to blow air into rather than draw it out, about twice the amount of air can be carried in by laying a second pipe line and putting such a jet on the inner end. This arrangement not only secures the benefit of the compressed air, but also makes it suck in an equal amount of air through the second pipe line. Of course, this is suitable only for such distances within which it would pay to lay the second pipe line in order to save compressed air.

Ventilating with Compressed Air.—The practice of turning compressed air into a ventilating pipe to induce an air current is general in the Cœur d'Alene mines. This is undoubtedly the simplest method of ventilating drifts when compressed air is at hand and power to operate a fan blower is not available. On the 1200-ft. level of the Hecla mine at Burke, air is drawn in this manner 500 ft. from the face of a drift to the shaft. Twelve-inch pipe is used and a piece of 3/4-in. pipe turned up at the end serves as the air nozzle. The air current is in this instance sucked 500 ft. through the fan pipe, the air jet being introduced into the fan pipe about 15 ft. above the bend at the shaft.

A different scheme is used on the 1600-ft. level of the Mace mine. Here the air jet is applied within a few feet of the suction end of the fan pipe. In this manner a current of air is forced 400 ft. to the shaft through 8-in. fan pipe. The nozzle is, however, different in this case, being made of 1/2-in. pipe bent in circular shape so as to just fit around the interior of the fan pipe. The coil is drilled with a number of 1/8-in. holes on the side opposite the suction end of the fan pipe. It is claimed that this acts as a more efficient nozzle and requires much less air than does turning in the air in a single jet. These nozzles may be used at a number of places in the fan pipe if one will not draw a current of air sufficient for proper ventilation.

Scheme for Ventilating the Working Face.—In order to spray muck piles at the face of ill-ventilated drifts and assist the ventilation in a less wasteful manner than by "blowing out" the drift with air from the drill-supply line the apparatus shown in Fig. 237 has been adopted at several large mines. A pipe connection is provided from a water-supply barrel to the air line *a*; this air line is enlarged at *b*. To spray the muck pile the valve *A* is closed and water run into *b* by opening valve *B*. The air is then turned on, after *B* has first been closed again. In this way much less air is consumed than by the ordinary method of blowing out, and the water absorbs the gases contained or held by the muck.

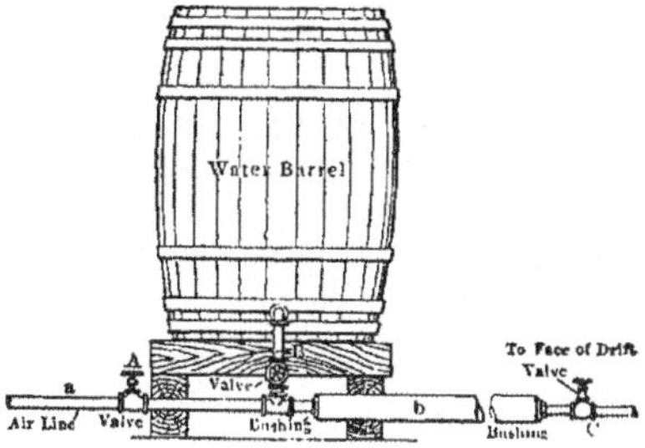

FIG 237.—WATER AND AIR-LINE CONNECTIONS FOR SPRAY.

A Hydraulic Air Blast.—A hydraulic air blast is easily rigged up where water under high head is available, and serves satisfactorily for affording ventilation, supplying the blacksmith shop, etc. Three holes are bored into a tight, strong barrel, one in the top and two on the sides, as indicated in Fig. 238. Into the one in the head of the barrel a funnel is inserted and fitted tightly. Pipes are tapped into the other holes and preferably some sort of valve or spigot arrangement provided on each. A smaller pipe connected with the water supply opens into the funnel, the end of the pipe being set a couple of inches above the throat of the funnel. On turning on the pressure water, air is entrapped and forced into the barrel. The lower pipe serves for an outlet for the water and the upper one as an air discharge. By regulating the valves on the discharge pipes so that the water is let out as rapidly as it enters, and setting the end of the pressure-water pipe at the proper height above the throat of the funnel, a strong air blast can be maintained. The amount and pressure of the water admitted regulate the amount of blast obtained. On the 800-ft. level in the Pittsburgh mine near Nevada City, Calif., such an arrangement is used with great success, water being taken from the pump column to operate the blast.

Wing Sail for Ventilating Shafts (By A. O. Christensen).—After sinking our shaft 60 ft. the gas from the powder smoke became so bad that work had to be suspended until we installed a "wing sail." The apparatus consisted of a canvas tube, held open by hoops made of willow branches, which was run down the shaft, and to the upper end of which a wing sail was attached. Fig. 239 shows the features of this arrangement. The sail is held in place by ropes stretched to nearby trees. Auxiliary ropes fastened to the middle of these can

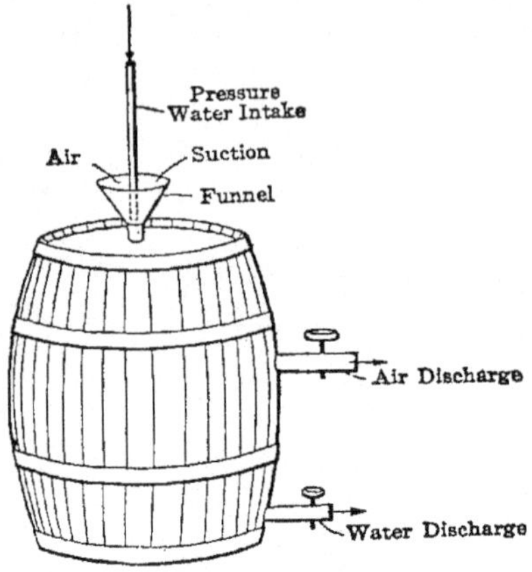

FIG. 238.—CONVENIENT AIR BLAST.

shift the direction of their pull, thus altering the position of the sail without necessitating more than two hitching posts. The sail can be turned to catch wind from any direction. It is necessary to set the sail so that it will draw air out of the shaft. When first tried, the air was blown down but it was found that

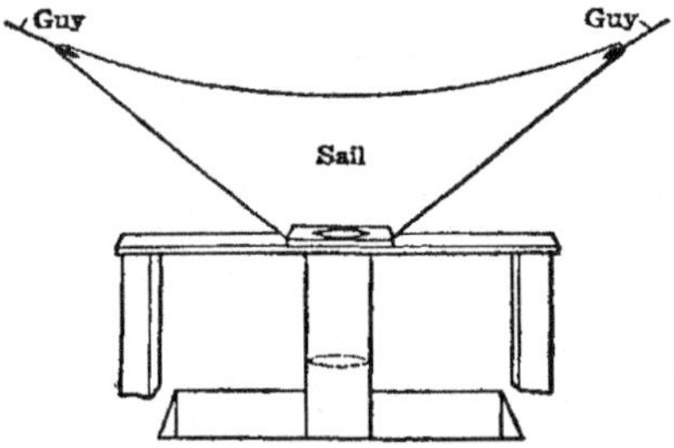

FIG. 239.—SAIL FOR SHAFT VENTILATING.

the heavy gases remained at the bottom while the fresh air merely worked to the surface again. When sucking air out, the draft through the bag is strong enough to carry the heavy gases up without trouble.

Ventilating Stopes in Bisbee (By F. W. Holler).—Ventilation in the Bisbee mines is natural as far as possible. Most of the shafts are connected on the different levels, and usually the levels are cool enough for comfort and the air is good. Levels are 100 ft. apart and are connected in many places by raises which are put up for prospecting purposes as well as to help the ventilation. Before doing extensive stoping, a raise is put through from one level to the next. Then stoping is started from this raise, keeping the latter in the corner of the stope which will vary in size from a square section four sets on a side to one seven sets on a side. In some cases the raise ventilates the stope naturally. In other cases the air in the raise may be good, but a set or two away it may be just the opposite. In this event special methods of ventilation are necessary, and several of these follow: The manway set of the raise is covered over with plank on the working floor of the stope, and the floor is removed in one of the sets in the far corner of the stope, thus forcing the air to travel across the working floor, down into the far corner and back to the raise on the floor below. In this way two floors of a stope can be ventilated if there is a current of air in the raise. When there is not a current of fresh, cool air in the raise, small centrifugal blowers run by electric motors are used to blow air from a main air passageway to the stope.

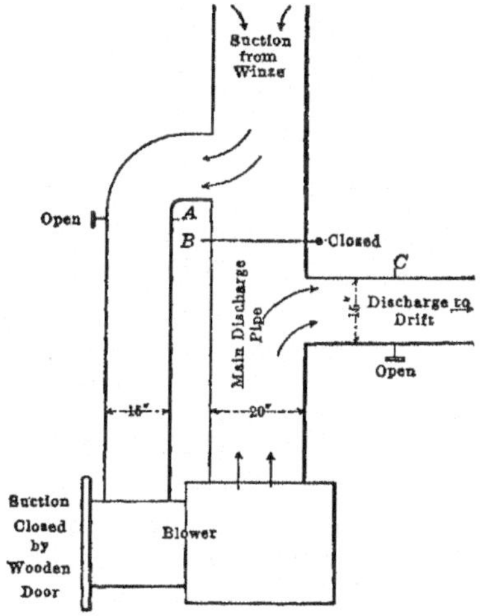

FIG. 240.—PIPE ARRANGEMENT FOR FAN BLOWER USED ON COMSTOCK LODE.

Six- or eight-inch galvanized pipe is used to conduct the air. Occasionally, compressed air is used in stopes where it is not deemed advisable to put in blowers. Results obtained are not good considering the power used and the quality of the air, but by using jets with the compressed air the results obtained are better; however, they do not compare with those obtained by using centrifugal blowers.

Piping Arrangement for Fan Blower.—In Fig. 240 is shown a simple piping arrangement for reversing the air current from a fan blower. The scheme is employed on the 2000-ft. level of the Union mine, at Virginia City, Nev., where a Sturtevant, multivane blower is used to supply air to a winze from which levels are being opened. The main discharge of this blower is 20 in. in diameter and the fan is run at 1120 r.p.m., being belt connected to a 20-h.p. motor. The power consumption is about 16 h.p. Ordinarily the fan is used merely to blow fresh air down the winze through the 20-in. main-discharge pipe. After blasting it is, however, necessary to draw the foul air and gas from the winze. The 20-in. pipe then acts as a suction pipe, the air current being drawn (into the blower) through the parallel length of 15-in. pipe and discharged through the 20-in. pipe and connecting 15-in. pipe. A wooden door or gate is used to close the suction end of the blower and the gates *A*, *B* and *C* in the pipes control the air current. The sketch shows the blower drawing air from the winze and discharging it into the drift. After clearing out the winze the door is removed from the suction of the fan, valves *A* and *C* closed, *B* opened, and fresh air is blown into the winze. This is a much simpler arrangement than is usually seen and requires a minimum amount of pipe. The wooden gate to close the suction end of the fan can be quickly constructed of a few nails and some plank. It is much quicker and more economical to draw out bad air than to force it out by blowing in fresh air. In the winze mentioned, no time has to be lost between shifts even though the temperature of the air would quickly rise to above 120° F. if artificial ventilation were not resorted to. By this arrangement it is possible to deliver the gases directly to an upcast air current instead of allowing them to mingle with the air currents about the winze station.

Mine Ventilation through a Drill Hole.—In underground operations it is necessary to have two openings in order to insure good ventilation. The second opening is generally made by sinking a new shaft. In the case cited here, the ore could be handled readily through one shaft, and a churn drill hole was used for the second opening. The apparatus is a fan about 2 ft. in diameter with a horizontal bottom discharge 8 in. in diameter. To this nozzle is fastened a short piece of canvas air pipe slightly larger than the casing of the drill hole with which it connects. The fan is belt-driven by an 8-h.p. upright engine. The engine obtains its steam from the boiler at the shaft several hundred feet distant. The apparatus is in an open field in the southwest part of Joplin, with no protection from the weather.

Ventilation by Drill Holes (By W. F. Boericke).—In the shallow zinc mines of Wisconsin, drill holes, aside from their primary purpose of serving to prospect the ground, are of considerable use later in ventilating the underground workings. The holes are usually put down with churn drills and are seldom much over 125 ft. deep, with a diameter of 6 to 8 in. The drilling cost is usually about 60 to 80 cents per foot, depending upon the amount of drilling.

Where the holes are at different elevations, a small current of air usually passes down one and up the other. This can be augmented by erecting a high standpipe above one, thus increasing the draft. If one of the holes is wet, as frequently happens, the dripping water aids considerably in catching the air and carrying it down, on the familiar principle of hydraulic compression. Sometimes a small fan and motor forces a strong current down, or a suction fan may be used. Occasionally a sail is rigged to deflect the wind down the hole.

A more effective means than the last is employed at the Ross mine, Linden, Wis. The device is simple and inexpensive, and can be made by the mine blacksmith. It consists of several lengths of ordinary 7-in. galvanized stove pipe joined together and projecting 6 ft. above the ground. The bottom piece is

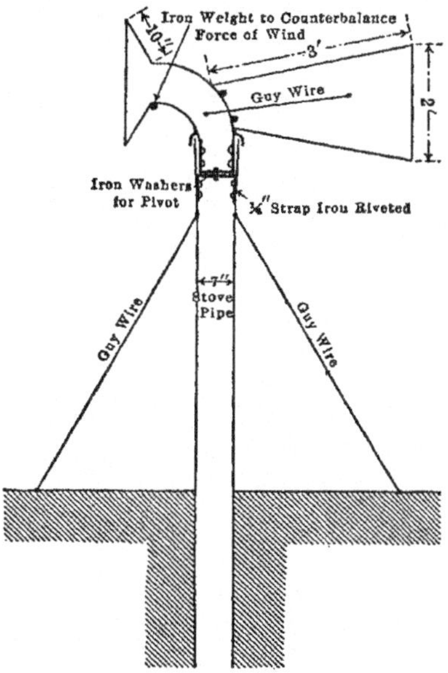

FIG. 241.—STOVE-PIPE VENTILATOR FOR DRILL HOLES.

sunk down the drill hole through the soil until it strikes rock. Guy wires hold it firmly in a vertical position. The top section consists of two pieces of pipe, one flared slightly so as to fit easily on the other. Strap iron, bent into angles, is riveted as shown in Fig. 241, and iron washers, with a bolt slipped through the strap irons, allows one to turn freely on the other. The top piece has an ordinary stove-pipe elbow securely fitted to it, which in turn is fastened to the funnel, a wide concave piece dimensioned as shown. The fan-shaped piece on the rear is, of course, to turn the device so as to face the wind at all times. A piece of bar iron is suspended from the front of the mouth of the pipe in such a way as to counterbalance the force of the wind when blowing against it.

Self-acting Mine Doors.—A device, used in a German mine, by which a door across an airway can be opened automatically by an approaching car or trip is illustrated in Fig. 242. The rail *G* is supported horizontally, at about 2 1/2 ft. above the ground, by two stulls on one side of the track, in such a way that the end of the rail toward the approach of a car is closer to the track than the other end. A slotted shoe *B* slides on this rail. Fastened to it is one end of a rope which passes around suitable pulleys, the other end being fastened to the outer edge of the door. A counterweight *g* is also connected to the sliding shoe to assist its return movement, if the usual pressure is insufficient. A

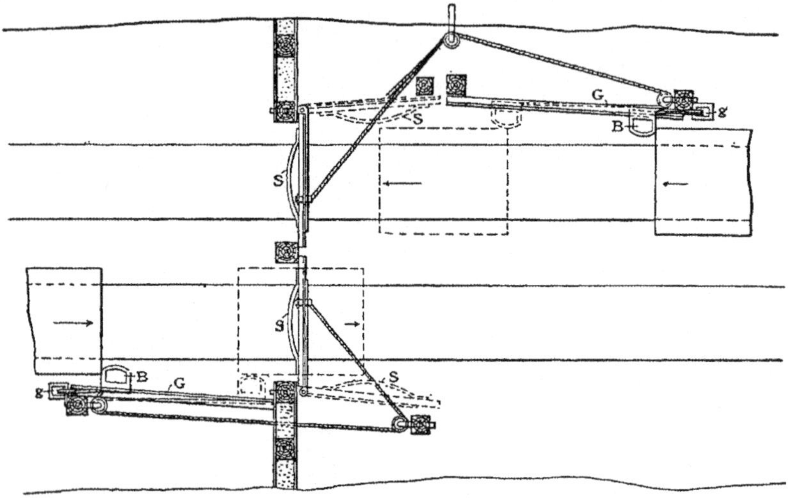

FIG 242.—SELF-ACTING MINE DOORS FOR DOUBLE TRACK DRIFT OR TUNNEL.

car coming in the direction of the arrows strikes the shoe *B*, and by pushing it ahead opens the door; by the time the door is open wide the shoe has traveled sideways far enough to allow the car body to pass it, but the door is prevented from closing by the springs *s*, which rub along the side of the car. When the car has passed, the natural weight of the door, which is purposely hung out of plumb, assisted by the wind pressure and the counterweight, causes it to close.

A Mine Air-door (By P. L. Woodman).—Details of a mine air-door and of an opening and closing device used in the motor-haulage drifts of the Copper Queen mine are shown in Fig. 243. The doors are opened and closed without stopping the train. When a train approaches, the motor or end car pushes the doors which are free to swing in either direction; upon opening, they are caught and held by the latches set in both walls and on each side of the door set. The releasing lever is operated by a cord a hundred feet or more in length conveniently hung at the roof of the drift. The motorman simply pulls the cord in passing and the levers release the doors allowing them to swing shut.

Starting a Ventilating Fan Automatically (By S. A. Worcester).—The Conundrum gold mine at Cripple Creek, Colo., now being operated under a lease to me, is ventilated by a system of my invention, with a large fan operated by a 15-h.p., three-phase induction motor. The motor is started from 1 to 2 hours before the shift goes to work, so that no gas will remain in the mine at "tally." For the first 2 or 3 weeks this starting was done by a miner who went to the mine early for this purpose. Later I devised and put in use the arrangement shown in Fig. 244, which saves several dollars each month, besides being accurate and reliable.

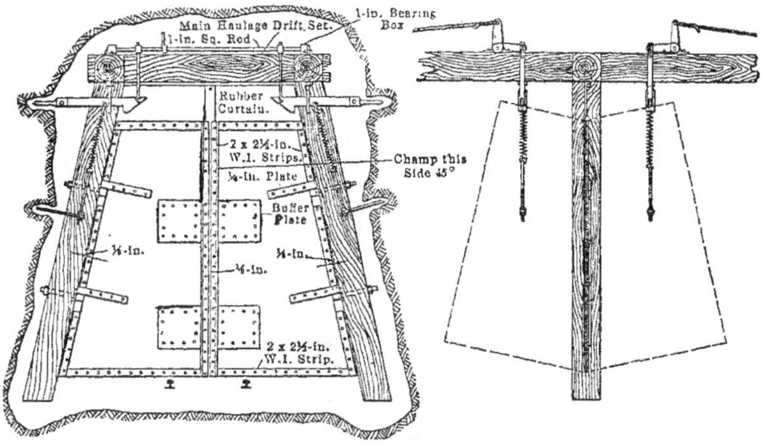

FIG. 243.—DETAILS OF MINE AIR-DOOR AND CATCHES.

The starting box *A* is the ordinary starting compensator used with induction motors, and has three "on" positions and the "off" position. The one-day weighted clock *B* is wound by pulling down the weight chain *C*, thus raising the weight *D*. The marks on the wall indicate the travel of the weight per hour and show how far the weight should be raised to start the fan within a given length of time. When the motor is stopped, the starting lever *E* is set as shown, in the "off" position, and is held in this position by the releasing lever *F*. The releasing lever has a bucket *G* suspended near its outer end and with its bottom a little below the surface of the water in the can *H*, which is an ordinary square 5-gal. oil can, with the top cut out. The bucket is made from a piece of 6-in. galvanized air pipe with a wooden plug for a bottom; a hole about 1/8 in. in diameter is bored through the bottom. The bail *K* of the bucket is hooked and hung on the trigger *L*.

When the clock weight *D* descends and lowers the long arm of the trigger, the bucket is unhooked and drops, carrying down the releasing lever *F* far enough to allow the starting weight *M*, which is fast to the handle *E* and moves with it, to drop one notch, bringing the compensator to the first "on" position. The bucket now sinks slowly as the water enters through the small hole in its bottom, requiring 18 seconds to lower the releasing lever so as to pass the second step of the weight *M*, and 12 seconds more to release the third, or full-speed step, 30 seconds being required to bring the fan to full speed. The water has a little oil on its surface to prevent evaporation. The operation of this arrangement is independent of manual skill and care and assures an easy and reliable start, with no danger of throwing the belt off or burning out fuses.

The fan draws air from the surface through a long tunnel. It is situated in a short crosscut from the tunnel to the hoist shaft and about 150 ft. below the

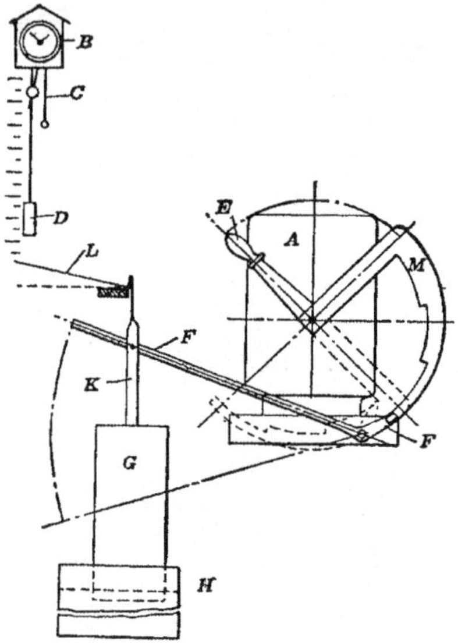

FIG. 244.—AUTOMATIC STARTER FOR VENTILATING FAN.

underground electric-hoist station. The air current is forced directly down the main hoisting shaft. The engineer visits the fan usually once each day, to see that the oil is feeding properly, and no further attention is required, except stopping and setting the starter for the proper time.

Before this ventilation system was installed the mine, which has about 3 miles of workings, was often entirely filled with mine gas, from the seventh level to the adit-tunnel entrance, a vertical distance of about 800 ft. The seventh level was inaccessible in even the most favorable weather and the gas zone was more than 150 ft. deep in all ordinary weather. One or more men had been killed in this mine by the gas which contains, by government analysis, 10% of carbon dioxide. The mine had been practically abandoned for 5 years on account of the gas. The ventilation is now perfect in all parts of the mine, and completely independent of weather conditions. The fungus or mold which was at first found throughout the mine, has all dried up and disappeared, and the air is cool and pleasant; candles will burn in all parts of the workings.

(By J. H. Dietz).—A method similar to that described by Mr. Worcester was employed at the coal mine of the Laning-Harris Coal & Grain Co., at Wellington, Mo. The fan at this mine is of the propeller type, belt driven by a 15-h.p. direct-current motor, and is placed directly in the air course, 1500 ft. from the mouth of the slope. The motor is operated by a type 70 Cutler-Hammer self starter, which replaces the elaborate mechanism Mr. Worcester has attached to his ordinary starting compensator. The simple solenoid, drawing the starting switch slowly over the contacts as soon as the current is turned on, takes the place of the counter-weight, oil can, 6-in. pipe and system of levers, described in the above article. The fan can then be stopped or started from the engine room simply by opening and closing the switch, which is equipped with a counter-weight for closing, operated by a string attached to the winding stem of an ordinary one-dollar alarm clock. This enables the fan to be started from the engine room at any predetermined time, and makes a simpler, cheaper, more convenient and reliable arrangement, with the advantage that it can be purchased properly made and ready to install.

The fan was manufactured and the starting arrangement installed by the Eagle Foundry & Machine Co., of Fort Scott, Kan. In addition to the equipment described, the motor is supplied with a variable-speed controller, without release, so that the fan can be operated with a 50% variation in speed, depending on the weather conditions and the mine resistance. This fan requires no attention, except for oiling, and is equipped with special self-oiling boxes, so that one trip a week is sufficient attention for the entire equipment.

When the fan was installed, there was a delay in shipment of the speed controller, and the fan was connected direct and run at the normal speed of the motor. The fan gave so much air that it became necessary to cover one-half of the discharge opening with a temporary wood brattice to enable the miners to hold a light anywhere in the workings. The fan is now running at minimum speed, with capacity for 50% increase, to take care of the future growth of the mine.

THEORETICAL AND PRACTICAL CONSIDERATIONS IN THE USE OF COMPRESSED AIR

Volumetric Efficiency of Air Compressors (By F. D. Holdsworth).—The term "piston displacement" or "displacement capacity," commonly used by air-compressor builders as a measure of capacity, expresses the quantity of air which would be delivered by a compressor in which there were no losses to prevent a discharge of a cylinderful of air at every stroke.

Certain losses, however, are unavoidable. These include: The reduction of cylinder volume, due to the space occupied by the piston rod; clearance space at each end of the cylinder; leakage past the piston and the inlet and discharge valves; failure to completely fill the cylinder during the intake stroke, due to loss of pressure through inlet ports and passages; rarefaction of the incoming air by absorption of heat from the air passages and piston. These losses materially reduce the quantity of air actually delivered; and the ratio between the actual delivery and the theoretical displacement, expressed in per cent., is termed "volumetric efficiency." This is sometimes measured from indicator cards by dividing the length of the atmospheric line included within the boundary lines of the card by the total length of the card. Results obtained by this method are misleading, as they invariably indicate that a compressor is being operated at greater efficiency than actually is the case. Compressors showing efficiencies, as calculated from indicator cards, as high as 95% have been found to have efficiencies of 85% or even less when the actual air delivered was carefully measured. For instance, a compressor, through faulty design of its inlet valves or having insufficient inlet-valve area, might have its cylinder filled with air at 2 lb. below atmospheric pressure, which would cause a serious drop in volumetric efficiency, yet with leaky discharge valves or with considerable absorption of heat through contact with heated surfaces, the pressure existing in the cylinder at the completion of the intake stroke might be almost or quite up to atmospheric pressure. An indicator card taken under such conditions would naturally lead to the conclusion that the cylinder was practically full of air and that its volumetric efficiency was correspondingly high.

In order to determine with accuracy the quantity of air actually compressed and delivered, the quantity of air entering or leaving the compressor must be measured. The measurement of the entering air by a gasmeter, for instance, is a troublesome matter, owing to the large volumes to be handled and to the fact that the pulsations of the compressor might affect its accuracy. The calibration of such large meters is likewise difficult.

Measurement of the air after compression is, therefore, usually attempted. In small compressors the air is sometimes compressed into receivers of known capacity or is measured by allowing it to displace water in such receivers, the quantity of free air delivered being proportional to the amount of water displaced; but a much more convenient method and one which may be applied to compressors of large or small capacity, with equal accuracy and with inexpensive apparatus and preparation for test, is by discharging the compressed air through orifices of known area and determining the quantity of equivalent free air by calculation.

The apparatus for this method consists of a manifold, as large or larger than the discharge pipe of the compressor, with numerous branches having orifices of different diameters attached and equipped with valves. One that I use has eight orifices, ranging from 3/32 to 5/8 in. in diameter. These orifices are carefully reamed holes in steel plates, which, for the larger sizes, are 1/2 in. thick, and for the smaller, 3/8 in. thick. The back or pressure side of the hole has its approach rounded to a radius 1/16 in. less than the thickness of the plate, leaving the remaining 1/16-in. length of the hole cylindrical. After the hole is thus finished, its actual diameter is carefully determined by micrometer and its area calculated.

The rate of flow through this type of orifice is obtained by the use of Fliegner's formula, the accuracy of which I have verified by carefully conducted tests in forcing compressed air contained in receivers of known volume through these orifices by displacing it with water. It is better to use a number of small orifices, as the formula is known to give unreliable results on orifices much larger than 5/8 in. diameter.

The manifold is tapped for a pressure gage and a thermometer well. An accurate pressure gage should be used, as it will be noted from the formula that the quantity of air discharged is directly proportional to the pressure and any inaccuracies in determination of pressure will materially affect the results. A reliable thermometer should be used and for a two-stage compressor should have a scale reading not less than 300° F. Pressure and temperature are the only quantities required to be observed for use in the formula.

In preparing for a test, the main air line should be disconnected near the air receiver and the orifices attached at that point. All other outlets from the receiver or from the piping between the compressor and the receiver should be either blanked or protected by valves known to be tight, in order to insure that all the air furnished by the compressor will be discharged through the orifices. A thermometer should be placed in the path of the air entering the compressor as near the cylinder as possible. For determining the speed of the compressor, a revolution counter should be attached at some convenient point.

In making a test, the proper combination of orifices required to maintain the desired pressure is determined by experiment, and it is usually found necessary to run the compressor for about 2 hours, discharging through these orifices, before the pressure and temperature reach a maximum and remain fairly constant. When this point is reached, an observer, on signal, records the counter reading and another observer begins taking readings, at 1-minute intervals, of the pressure and temperature at the orifices. The observer at the counter should then record the temperature at the compressor intake. At the end of 10 or 15 minutes, sufficient pressure and temperature readings will be obtained and the observer at the counter will, on signal, again read the counter; the difference between the last and the first counter reading will give the total revolutions for the interval of the test. Knowing the displacement of the compressor per revolution, the total displacement for the test period will be the product of the total revolutions and the displacement per revolution. The quantity of air discharged through the orifices, determined by the formula, using the observed data at the orifices, divided by the total displacement, will give the volumetric efficiency.

If accurate results are desired, the barometer reading at the time of the test should be known, which may usually be obtained from the nearest Weather Bureau station. This, together with the temperature at the compressor intake, should be used in the formula given below for reducing the pounds of air per second obtained from Fliegner's formula to cubic feet of free air per minute.

Fliegner's formula is:

$$G=0.53\frac{AP}{\sqrt{T}}$$

in which G=flow in pounds per second; A=area of orifice in square inches; P=absolute pressure of the air behind the orifice; T=absolute temperature (F.) of the air behind the orifice.

The weight of a cubic foot of air is found by the formula:

$$W=\frac{1.325\ B}{T},$$

in which W=weight of 1 cu. ft. of air; B=barometer reading in inches of mercury; and T=absolute temperature (F.) at the compressor intake.

The following extract, from a test recently made in New York City on a two-stage, direct-connected, motor-driven compressor to determine its volumetric efficiency, will serve to illustrate the use of the formula. The compressor had a low-pressure cylinder 26 in. diameter, a high-pressure cylinder 15 1/2 in. diameter, each with a stroke of 18 in. The piston rods in each cylinder were 2 1/2 in. diameter. The orifices used were as follows: 2 5/8, 2 1/2 and 1 5/16 in. diameter, having a combined area of 1.083 in. The observed data were as follows: Revolutions per minute, 188; gage pressure at orifices, 97 lb.; barometer, 30.1 in., or 14.8 lb.; temperature (F.) at orifices, 251°; temperature (F.) at compressor intake, 41°. Then,

$$G=0.53\times 1.083\frac{97\times 14.8}{460\times 251}=2.407 \text{ lb.}$$

per second. Then the weight of 1 cu. ft. of air at the intake temperature was

$$W=\frac{1.325\times 30.1}{460+41}=0.0796 \text{ lb.},$$

whence

$$\frac{2.407}{0.0796}\times 60=1814.4 \text{ cu. ft.}$$

of free air per minute delivered by the compressor.

This quantity, divided by the actual displacement of the low-pressure piston, gave the volumetric efficiency in per cent. This displacement in this case at the observed speed, with allowance for the piston rod, was found to be 2070 cu. ft. per minute whence $\frac{1814.4}{2070}=0.8765$, or the compressor actually discharged 87.65 % of its displacement capacity.

Testing Air Consumption of Drills.—The apparatus herein described has been satisfactorily used in testing the air consumed by rock drills. As shown in Fig. 245, a vertical air receiver is connected to the compressor line by a 1-in. pipe, fitted with a globe valve, and a 1-in. outlet pipe is led from the receiver to the drill. Water gages are placed one above the other throughout the entire height of the receiver, and a pressure gage is supplied. At the bottom of the receiver a 3-in. pipe with globe valve, furnishes water, and between the receiver and intake valve a tee connection is provided as a discharge.

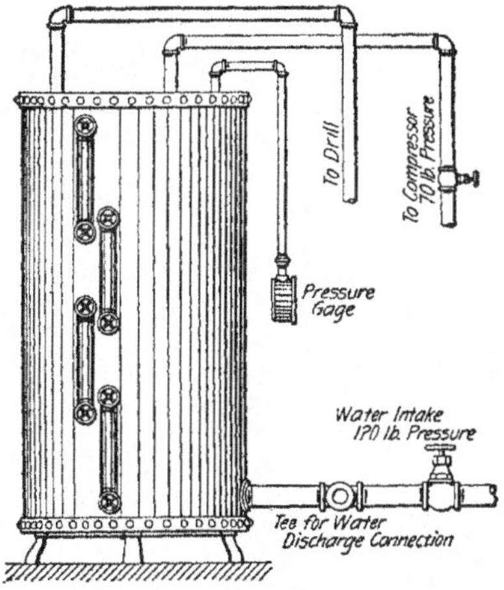

FIG. 245.—AIR TANK AND CONNECTIONS.

Before starting the test, water is allowed to flow into the receiver until it is just visible in the lower gage, and a chalk mark is made opposite this level. The receiver is then filled with air at the desired pressure, the air shut off and the test begun. As the drill takes air from the receiver, the operator maintains a constant pressure by regulating the water valve. Simultaneously with the stopping of the machine a chalk mark is made at the water level then shown by the gage and the water drained off in preparation for the next test.

The distance between chalk marks, depth of hole drilled and time of drilling are noted. The cross-section of the tank being known, the distance between the chalk marks makes it possible to find the volume of air used, at the given pressure, and from these figures the amount of free air can be computed.

It may happen, states *Coal Age*, that in spite of the control provided by the water supply, the air pressure will decrease during a test. In this event, the amount of free air used may be calculated as follows:

Let

P = Initial pressure,

P_1 = Final pressure,

B = Height above initial water level to top of receiver,

A = Cross-sectional area of receiver in square feet,

R = Rise of water in receiver, in feet,

H = Atmospheric pressure.

Then each foot rise of water in the tank, represents the consumption of A cubic feet of air and if the pressure had remained constant at P, the volume of free air used would have been given by $V=RA\times\frac{P+H}{H}$

However, the expansion of the air that remains in the receiver, from pressure P to P_1, represents the consumption of a volume of free air equal to $V_1=A(B-R)\left(\frac{P+H}{H}-\frac{P_1+H}{H}\right)$ and the total amount of free air used is then $V+V_1$. It may prove to be simpler to let V represent the initial amount of free air in the tank at pressure P, and V_1 represent the final amount of free air at pressure P_1. Then

$$V=AB\frac{P+H}{H}$$

$$V_1=A(B-R)\frac{P_1+H}{H}$$

and $V-V_1$ will give the amount of free air consumed during the test.

Proportions of Air-mains and Branches.—The accompanying table, showing the number of branch pipes of a given size that can efficiently be supplied with air from a main of given size, is taken from a bulletin issued by the Green Fuel Economizer Co. and is based upon the laws of the flow of air through pipes. The figures in the vertical columns to the left are the diameters of the mains; the numbers at the head of the vertical columns, the diameters of the branch pipes; the other numbers in the table show the number of pipes of the diameter designated at the top of the vertical column, equal to one pipe of diameter designated in the column to the extreme left on the same horizontal line.

For example, it is desired to find the diameter of the main equivalent to thirty 8-in. pipes. Follow down the vertical column for 8-in. pipes until the nearest number to 30 is found, then follow out horizontally to the left-hand column. The number there found will be the diameter of the main required, in this example 31 in. Conversely, the number of pipes of a given section that a given main can supply, can be determined from the table.

EQUALIZATION TABLE FOR PIPES

Diameter of mains	Diameter of branches															
	1	2	3	4	5	6	7	8	9	10	11	12	13	14	15	16
2	5.7	....	....	....	....	....	....	....	....	....	....	....	....	....	....	....
3	16	2.7	....	....	....	....	....	....	....	....	....	....	....	....	....	....
4	32	5.7	2.3	....	....	....	....	....	....	....	....	....	....	....	....	....
5	56	9.7	3.6	1.8	....	....	....	....	....	....	....	....	....	....	....	....
6	88	16	5.7	2.8	1.6	....	....	....	....	....	....	....	....	....	....	....
7	129	23	8.3	4.1	2.3	1.5	....	....	....	....	....	....	....	....	....	....
8	180	32	12	5.7	3.2	2.1	1.4	....	....	....	....	....	....	....	....	....
9	244	42	16	7.6	4.3	2.8	1.9	1.3	....	....	....	....	....	....	....	....
10	317	56	20	9.9	5.7	3.6	2.4	1.7	1.3	....	....	....	....	....	....	....
11	402	71	26	12	7.0	4.5	3.1	2.2	1.7	1.3	....	....	....	....	....	....
12	501	88	32	16	9.0	5.7	3.8	2.8	2.0	1.6	1.2	....	....	....	....	....
13	613	107	39	19	11	6.9	4.7	3.4	2.5	1.9	1.5	1.2	....	....	....	....
14	737	129	47	23	13	8.3	5.7	4.1	3.0	2.3	1.8	1.5	1.2	....	....	....
15	876	152	56	27	16	9.9	6.7	4.8	3.6	2.8	2.2	1.8	1.4	1.2	....	....
16	1026	180	65	32	18	11	7.9	5.7	4.2	3.2	2.6	2.1	1.7	1.4	1.2	....
17	1197	208	76	37	21	13	9.2	6.6	4.9	3.8	2.9	2.4	2.0	1.6	1.4	1.2
18	1375	239	88	43	24	16	10	7.7	5.7	4.3	3.4	2.8	2.3	1.9	1.6	1.3
19	1580	275	100	49	28	18	12	8.8	6.5	5.0	3.9	3.2	2.6	2.2	1.8	1.5
20	1775	312	114	56	32	20	14	9.9	7.4	5.7	4.5	3.6	2.9	2.5	2.1	1.7
21	1985	345	130	61	35	22	15	10	8.4	6.3	5.2	4.1	3.2	2.6	2.3	2.0
22	2250	398	145	71	41	26	18	13	9.8	7.2	5.7	4.5	3.7	3.1	2.6	2.2
23	2525	460	160	77	47	29	20	14	10	7.8	6.2	5.0	4.1	3.4	2.8	2.4
24	2800	493	180	88	50	32	22	16	12	8.9	7.6	5.7	4.6	3.8	3.2	2.9
25	3060	543	202	97	55	35	24	17	13	10	8.0	6.4	5.2	4.2	3.5	3.1
26	3425	590	219	108	62	39	27	19	14	11	8.6	6.9	5.7	4.7	4.0	3.4
27	3738	677	243	121	68	43	29	21	15	12	9.6	7.5	6.1	5.1	4.3	3.6
28	4100	725	265	129	74	48	32	23	17	13	10	8.3	6.8	5.7	4.8	4.1
29	4440	800	289	141	79	52	35	25	19	14	11	9.1	7.5	6.2	5.2	4.4
30	4898	864	315	151	88	56	38	28	20	16	12	9.9	8.0	6.7	5.7	4.7
31	5312	920	344	168	96	59	40	30	22	17	13	10	8.9	7.4	6.1	5.1
32	5631	1070	374	184	103	63	42	32	23	18	14	11	9.3	8.0	6.5	5.7
33	6154	1140	401	196	109	70	45	35	25	19	15	12	10	8.5	7.1	6.0
34	6675	1208	433	212	119	76	51	37	27	21	16	13	11	9.1	7.6	6.4
35	7075	1280	470	229	127	82	56	40	30	23	18	14	12	10	8.4	7.1
36	7735	1355	497	242	138	88	60	43	32	25	19	16	13	11	8.9	7.6
37	8265	1435	537	260	146	94	64	45	33	26	20	16	13	11	9.3	7.9
38	8715	1625	575	279	157	100	68	49	37	29	22	18	14	12	10	8.6

Compressor Precooler.—Compressor efficiency can be materially increased in warm weather by a simple and inexpensive precooler. It should be emphasized in the beginning that water or moist air should have no chance to mix with the compressor suction. Fig. 246 shows how the cooling nest of tubes should be connected by a wood or light iron conduit to the suction valves of the compressor. The nest of pipes can be made up of odd sizes of iron, or even tin or galvanized-iron speaking tubes can be used. If water is worth saving it can be pumped back at slight expense, though the flow should be regulated to just the quantity required to keep the pipes wet. These should be wrapped with thin cloth and should be placed in the open, preferably where prevailing winds or drafts will cause a maximum evaporation.

As an extreme example of the efficiency of this arrangement, the following may be stated: The difference between the temperatures of wet and dry thermometers at a large plant in California in summer has reached 40°, the difference between 110 and 70°. This difference in the temperature of compressor intake amounts to over 8%. The arrangement shown in the sketch precludes the saturation of the air with water, as recently complained of in South Africa, where precooling by spraying water directly into the intake was attempted. In fact, if the temperature of the water supply is below the

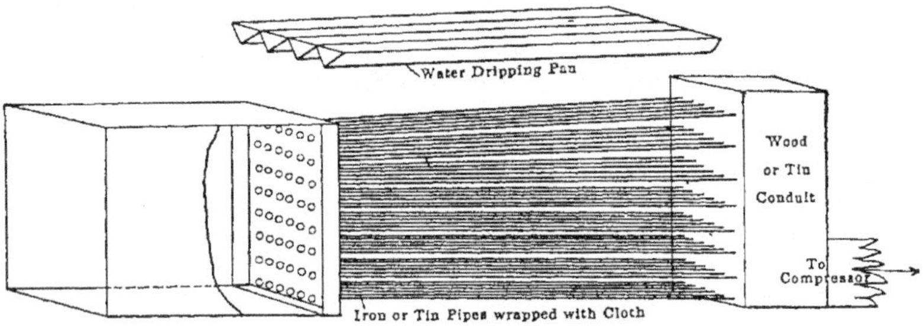

FIG. 246.—PRECOOLER FOR AIR COMPRESSORS.

dew point, and an abundance is available, water will actually be condensed out of the air as drawn to the compressor and can be drawn off from the bottom of the cool-air conduit.

Washing Air for Compressors.—The site of the compressor plant of the Penn Iron Mining Co. is on a sandy bottom-land at the base of a hill that furnishes more or less sand during windy weather. As the result of these conditions sand would get into the compressors and cause trouble. A scheme for washing the air was resorted to which is shown in Fig. 247. A concrete box 5 1/2 ft. wide, 5 ft. high and 16 ft. long was constructed. In the bottom of the box three 12-in. cast-iron pipes were laid, extending the full length of the box. The pipe is supported by the end walls. The lower side of the pipe is perforated with the sufficient number of 1/2-in. holes to allow the air to enter. Water is placed in the box so as to cover the perforations in the pipe 1 in. deep. This, however, is adjustable and can be arranged for any depth necessary. The box is covered with plank so that all the air must enter through the water-covered holes in the 12-in. pipe. The intake pipe from the compressor enters from the top of the box and does not extend to the water; therefore it does not take up much suspended water. However, the air does absorb a certain amount of moisture as it passes through the water. To reduce the amount of moisture in the air, the air is cooled as it passes from the first stage of the compressor to the second stage by passing through 150 ft. of water-cooled pipes. A small perforated water pipe is placed above the air pipe to furnish water for cooling the air. There is a trap between the two stages of compression to catch the condensed water. L. F. Armstrong, mechanical engineer for the company, designed this apparatus.

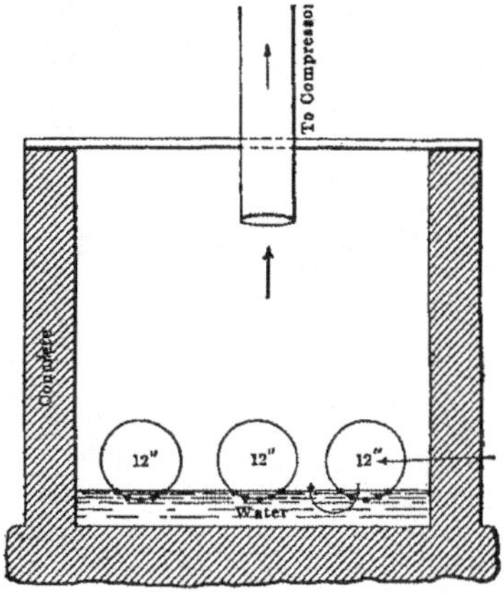

FIG. 247.—CONCRETE BOX FOR WASHING AIR.

Air Compressor Lubrication.—Explosions within the cylinders of an air compressor are usually caused by the ignition of inflammable gas, the presence of which is due to the use of too much lubricating oil of low flash point. The heat liberated from the air during compression may cause vaporization of the oil and the vapor mixing with the compressed air forms an explosive mixture that may be ignited at the temperature attained by the air in the cylinder. Excessive use of oil is open to the further objection that oil tends to cause sticking of the valves. Ordinarily, air cylinders and pneumatic tools require less oil than steam cylinders.

A lubricant that is free from the above-mentioned objections to the use of oil is soapy water, with which a small quantity of flake graphite has been mixed. The flakes of graphite remain suspended in the water until admitted to the interior of the cylinder, where they exhibit a tendency to attach themselves to the metallic surfaces, imparting a superficial glaze that is smooth, acquires a high polish and prevents actual contact of metal with metal. A small quantity of the mixture provides a safe and sufficient lubricating layer. As the soapy water may cause rusting, it is advisable to introduce a little oil into the cylinder when shutting down the compressor. The graphite is not affected by any degree of heat attainable in a compressor cylinder; it will not be carbonized or baked into a hard or gummy mass to interfere with the action of the valves, and under no conditions can it be volatilized.

Storing Compressed Air in a Natural Rock Receiver.—At one of the mines in the Rossland district there are two electrically driven compressors with a combined capacity of 7500 cu. ft. of free air per minute. These machines were described by C. Sangster in *Power*, Dec., 1909. Air from the compressors is stored in a crosscut the capacity of which is not less than 22,000 cu. ft. In free air compressed to eight atmospheres, it will hold 176,000 cu. ft., or the entire output of the compressors for 23 minutes. Allowing that one-third of this air is available at a working pressure, as cited, ten drills could be operated for 50 or 60 minutes after the compressor was stopped. The advantage of such a large storage is noticeable in the engine room. It tends to balance the rapid fluctuations in the load, the compressor and rope drive run more steadily and the unloaders cut out less often. The motors are not subjected to the strains of the load being constantly thrown off and on. In the mine, a hoisting engine or a group of drills may be thrown on or off without seriously affecting the air pressure. In short, it stores and restores the air, piling up a reserve when a machine is stopped and giving it back when a sudden call is made.

Using a Pump for Compressing Air.—It is occasionally desirable to use a pump as an air compressor where only low pressures are required, when the

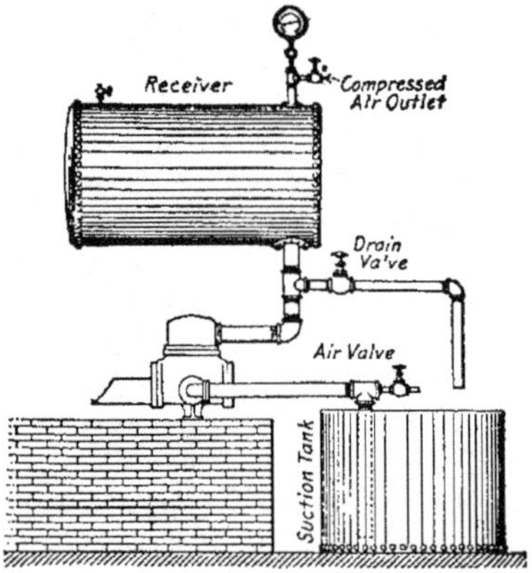

FIG. 248.—PIPING FROM PUMP TO TANKS.

work to be done is of only a temporary character and any makeshift will suffice. The scheme is shown in Fig. 248. The drain valve is closed. The pump is then slowly started and when primed the air valve on the suction line is opened just enough to prevent the pump from entirely "losing its water." By proper regulation of this air valve the pump will take in a large volume of air with each stroke and just enough water to keep the plungers and valves fairly well sealed. When a pressure of 8 or 10 lb. is reached the air valve on the suction line is closed, the pump takes water and the receiver is nearly filled. This forces the air out of the receiver and increases the pressure at the same time.

Should more pressure be desired the air-outlet valve is closed and the receiver is drained into the suction tank. The small valve shown on top of the receiver admits air when the receiver is being drained. The operation mentioned is then repeated. Incidentally it is not the most economical way of compressing air.

Reheating Compressed Air with Steam.—The practice of reheating compressed air by mixing it with steam is employed generally in the Cœur d'Alene mines of the Federal Mining & Smelting Co. Results obtained at these mines seem to indicate that this is the most economical and efficient method of getting the full measure of energy from the air. At the Mace mines, air at 90-lb. pressure for drills, and steam for the hoist were formerly conducted the 3000 ft. through the entry tunnel in separate pipe lines. The air is now compressed to 100 lb., mixed with superheated steam at the compressor house and piped into the mine in one line to supply both hoist and machine drills. The daily saving by this arrangement is figured at about $40, and besides an increase of 10 lb. in pressure is gained for drills. The steam plant formerly required 14 tons of coal per day, while from six to eight tons is all that is burned now. The boiler at this plant is rated at 80 h.p. and the capacity of the compressor is 4000 cu. ft. of free air per minute. No trouble has been experienced from either freezing or condensation.

Reheater for Air Hoist.—The six levels below the Sweeney tunnel, in the Last Chance mine at Wardner, Ida., are reached by an inclined winze containing two skipways and a manway. A compressed-air hoist operated in balance serves these lower levels with skips of 30-cu. ft. capacity. Trouble was experienced from the hoisting engine freezing, so that reheating of the air had to be resorted to. Gasoline torches playing on a coil of air pipe were tried, but with poor success, until finally the scheme now used was hit upon. A 3-in. perforated air line is run into the drum of a 40-h.p. fire-tube boiler, the air passing up through the heated water and being piped to the hoist from the steam chamber. A check valve is used on the air line before it enters the boiler, so that as the reheated air is drawn to the hoisting engine, more air is admitted to the boiler, but, at the same time, water is prevented from backing into the air pipes. The air pipe in the boiler is closed at its end and drilled with 1/4-in. holes, through which the air escapes. With the reheater, the air pressure is raised from 5 to 15 lb., depending upon the rate of consumption, and no trouble is experienced from the hoist freezing. Only a small wood fire is kept under the boiler, not more than 3/4 cord of slabwood being burned per 24 hours.

Electric Reheaters.—At the Bully Hill copper mines in Shasta county,

Calif., a novel type of reheater is used in connection with pumps operated by compressed air. The arrangement is an electrical resistance coil inclosed in a pipe through which the compressed air passes directly before being utilized. The arrangement was worked out by H. A. Sutliffe, electrician for the Bully Hill Copper Mining & Smelting Co., and has proved thoroughly satisfactory. The reheater consists of two principal parts, *i.e.*, an outer jacket and an inner length of pipe upon which is wound the resistance wire. The air line is bushed to the pipe jacket and through this jacket are tapped, as shown in Fig. 249, two 1/2-in. holes provided with insulated stuffing boxes through which the flexible lead wire is connected to the resistance coil.

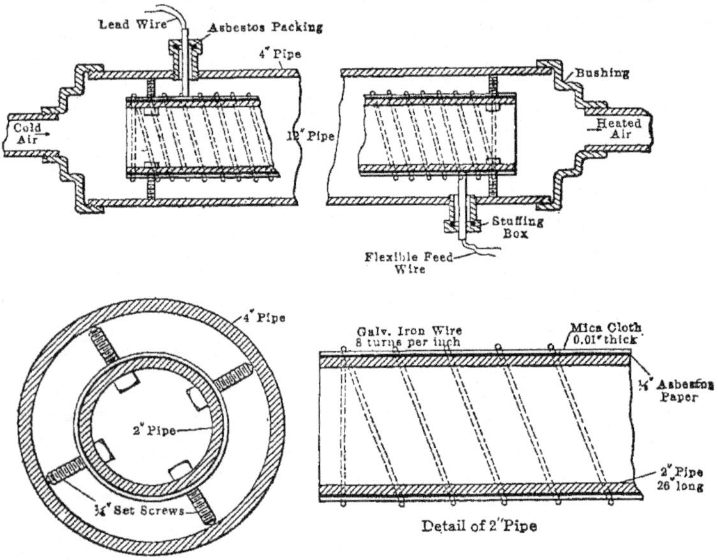

FIG. 249.—ELECTRIC REHEATER USED AT BULLY HILL, CALIF.

In the design shown the resistance coil is wound on a section of 2-in. pipe, 26 in. long, the jacket pipe being 4 in. in diameter. The central pipe is first wrapped with 1/8-in. asbestos paper, and this in turn covered with mica cloth 0.01 in. thick. Over this is wrapped a helix of No. 14 galvanized-iron, telephone wire pitched eight turns to the inch. At points 1 in. from either end, the central pipe is tapped for set screws, at four equally spaced points about its circumference. These set screws serve to keep the resistance coil from touching the outer pipe jacket. The wire coil is so wound as to not touch the set screws.

A reheater, as described, is designed for a 110-volt 40-ampere current and will use approximately 6 h.p., yet at the Bully Hill mine a saving of at least

$6 per month has been effected, it is claimed, by each reheater installed. The reheaters are credited with raising the available air pressure 5 lb. With the electric reheater it is well to have the valve controlling the air engine, pump, etc., for which the air is being heated, connected with a pilot light, so that when the engine is shut off, attention will be called to that fact at once and the reheater will be disconnected. If this is not done there will be danger of burning out the reheaters as they soon become hot enough to destroy themselves if allowed to run after the air is cut off.

Placing Air Pipes in Shafts.—Extensions of the compressed-air mains in the Champion shafts of the Copper Range Consolidated Co., in the Lake Superior copper country are made in 200-ft. lifts. The work is commenced by placing a tee in the shaft opposite the station at the bottom of the lift; then carrying up the piping for 200 ft. to connect with the air main already in position. The tee at the lower station is securely fastened by an iron yoke to a carrier timber. The pipes are then lowered and put together until the next level above the station is reached. There another tee is put in the line immediately above and below which yokes are used both to anchor the main and to prevent swinging at that point. When this tee has been put in place the other pipes are lowered and, by using a pipe of the necessary length, the top of the extension is brought to within 6 or 8 in. of the bottom of the main already in place. A flange is screwed tightly home on the upper end of the top pipe and yokes are attached to anchor the line in position.

The connection with the part of the main already in place is made after 1 p.m. on Saturdays at which time no drills are running in the mine so that the air can be shut off without interfering with mining work. Jacks are placed under the lower end of the extension pipe and the yokes are loosened so that all the weight comes on the jacks. The blind flange on the lower end of the part of the shaft main already in place is removed. The flange on the upper end of the extension is turned in the direction of unscrewing so that the bolt holes in upper and lower flanges are in line. The little loosening usually necessary does not cause leakage. Then the 200-ft. length of pipe is raised by the jacks until the flanges come tight together. The flanges are then bolted together, usually with a gasket between to make a tight joint. The yokes, of which there are two for each 100-ft. length of pipe, are then tightened to hold the main in position. During the operation, large wrenches are used to prevent turning of the upper part of the line when the extension is being raised. Making the connection takes an hour; two men are needed at the jacks, four at the yokes and two, with the boss timberman at the point where the connection is made.

A Method of Hanging Air Pipes (By Claude T. Rice).—The Copper Range company intends to connect its different properties in Michigan so that in case a compressor breaks down at one mine air can be delivered to it from another. Already the Trimountain and the Baltic mines are connected in this manner. The air pipe is carried above ground so that any leaks can be readily detected. Supports of the type shown in Fig. 250 are used. Timber legs have been used, this having been found a good way to employ old pipe. Any old piece of 2-, 2 1/2- or 3-in. pipe or old boiler tube 8 ft. or more in length was used for the legs. This pipe was bent to a radius of 12 in. at the middle and the legs given a spread of 1 in. in 4 in. Each piece of pipe was flattened at the top and a 1-in. hole punched through it to receive the 3/4-in. hanger bolt that holds the air pipe. These pipe supports are footed in concrete pedestals about 24 in. on an edge and buried in the ground. The under surface of the top of the pipe is set to grade so that a constant and standard length of hanger can be used in suspending the pipes from the supports.

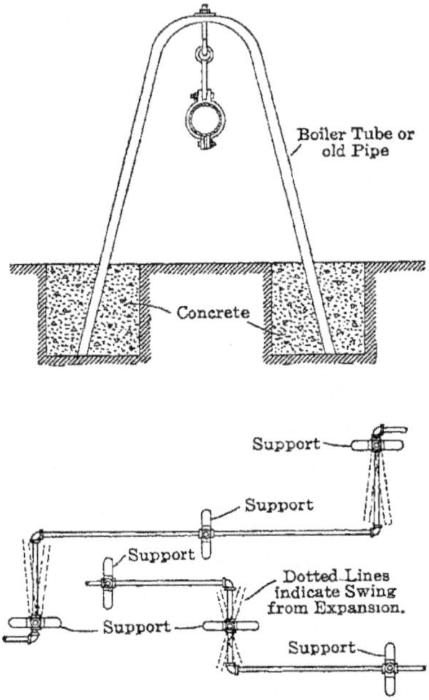

FIG. 250.—OLD-PIPE SUPPORTS FOR AIR MAINS.

These supports are placed along the line at a maximum interval of 25 ft. No expansion joints are used in the entire line which is about 1 1/2 miles long. Instead the expansion that would occur in the line with a change in temperature of 100° F. is figured for each straight stretch of pipe. This stretch of pipe is anchored at the middle, and the expansion is taken care of by the side swing of the pipes at the turns, or changes in direction. In case this is too much to be taken care of in that way with the ordinary course of the pipe a double-angle turn or reverse bend is put in long enough to allow the expansion to be taken up by the swing of the pipe at this turn without danger of loosening the joints. Often this turn is anchored in the middle so as to take care of the expansion of two stretches of pipe, but generally each turn is made to serve only one. In case that the stretch of pipe that must take up the expansion is short, the pipes are occasionally put in hot so that the end pipe of the expansion leg is pulled in the beginning in the opposite direction to that in which most of the expansion will occur. Owing to the pipes being suspended in swings at the supports, little resistance is offered to the movement of the pipes, and the expansion can be taken care of easily in this manner.

Stopping Leaks in Air Receivers.—A convenient method of stopping leaks around a loose rivet in air receivers or steel water tanks, or even for emergency

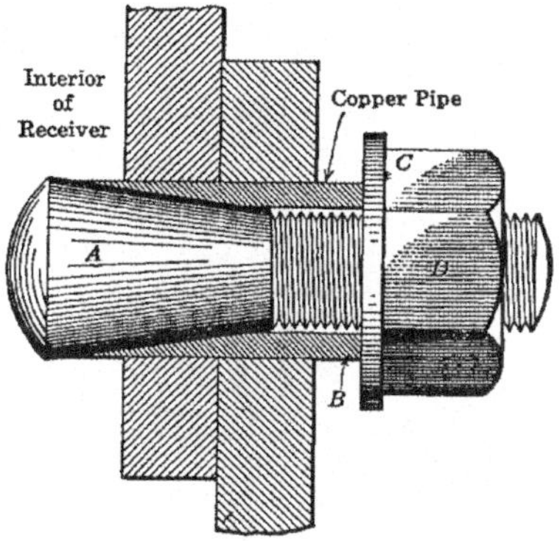

FIG. 251.—TAPER BOLT FOR STOPPING LEAKS.

repair work on boilers, and one which can be made entirely from the outside, is to use a taper bolt with copper sleeve, as illustrated in Fig. 251. The head of the faulty rivet is cut off, and the rivet knocked out of the hole or else the rivet may be drilled out. A taper bolt *A*, large enough to pass through the hole, and threaded up to 1 1/2 in. of the large end, is inserted in the rivet hole and a piece of copper pipe *B*, of the same internal diameter as the diameter of the bolt, is lipped over the threaded portion of the taper bolt; it should project 1/4 to 3/8 in. on each side of the plates. A washer *C* and nut *D* are then put on and the nut screwed up with a long-handle wrench. The bolt can be used for withstanding pressures up to 200 lb. per square inch with safety. The bolt may be used for repairing other small leaks such as may occur in pump columns by drilling out a hole at the point where the leakage occurs. A variety of uses for such a bolt will be suggested by the drawing.

Pipe Lines as a Factor in Rescue Work.—The introduction of compressed-air pipe lines into all the workings of a mine might be utilized to provide fresh air and even food to men imprisoned after explosions or through falls. This does not involve much expense, as mines are usually equipped with compressed-air apparatus, and the piping leading into the mine is of such a nature as to withstand considerable damage from the exterior. Telephone wires inserted within the air pipe might also serve a useful purpose in saving life.

Water in the Air Line.—One of the difficulties attending the use of compressed air arises from an accumulation of water in the pipe line. From time to time devices have been suggested to draw this water, some homemade and others of the workshop variety, but perhaps one of the most simple and ingenious forms has recently been developed in England. It possesses numerous advantages, among which may be mentioned automatic action, small size and light weight, which facilitates installation.

In Fig. 252 is shown the internal construction of this water ejector. It consists of a gun-metal cylinder with three apertures, two being threaded for pipe, the third being a small hole at the bottom, normally closed by a small

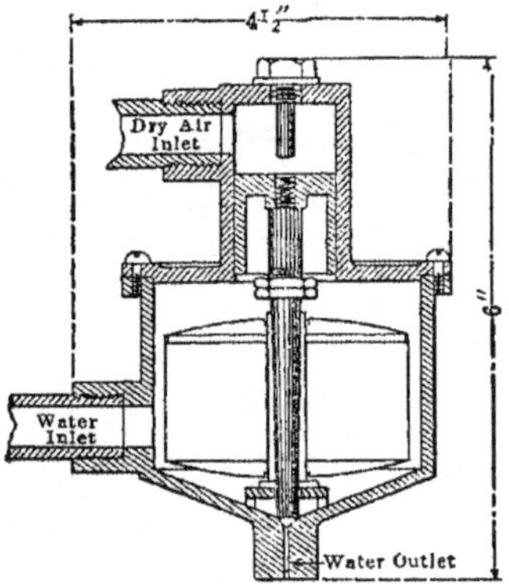

FIG. 252.—EJECTOR VALVE FOR AIR LINE.

cone-shaped valve. This valve forms part of a gun-metal spindle carrying a copper float and provided with a plunger, upon the top of which pressure is exerted by the air in the pipe line. This is balanced by an upward pressure on the under surface of the plunger from the air from the water inlet. Under normal conditions, the pressures on the piston and float are balanced. If, however, water enters the chamber from the water inlet, the float is lifted, its length of travel being limited by a stop, and the water outlet is uncovered. The pipe-line pressure from the water inlet thereupon blows the water out of the chamber, the valve remaining open until ejection is complete. When air blows out of the water inlet, the greater pressure on the top surface of the piston forces the valve down into its seat, and the emission of air is stopped. It is, therefore, impossible for air to escape after the accumulated water has been discharged.

In Fig. 253 the method of connecting the valve to the pipe line is shown diagrammatically. Where grit or other foreign substances are likely to get into the pipe line a strainer valve X is inserted at each of the points indicated.

Freezing of Compressed-air Pipe Lines (By Stacy H. Hill).—In northern countries great inconvenience is caused by the freezing of compressed-air pipe lines. The difficulty has been eliminated to some extent at permanent, well-regulated properties by burying the pipe. Even in these installations trouble is often experienced in the smaller lines to blacksmith shop or exploratory shafts, which may be at some distance. A method whereby this difficulty could be eliminated for all time would be acceptable and has been the cause of much study, as nothing disorganizes a force of men so much as the gradual or sudden loss of air supply.

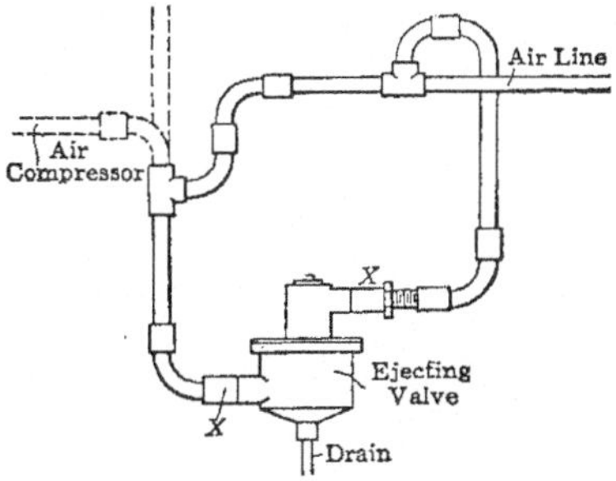

FIG. 253.—PIPING FOR EJECTING VALVE.

In the latitudes where weather from zero to 40° below is occasionally experienced, the pipes freeze from the circumference, gradually diminishing the pipe area until the passage is entirely stopped. As a preventive of this, the introduction of salt or sal-ammoniac has been found very effective up to the moment of entire blocking of the pipe. At a number of properties a barrel of salt is kept in the engine room and at regular periods, usually at the beginning and middle of each shift, several pounds are introduced into the pipe line just beyond the air receiver. The pressure is then increased by the air compressor and the salt blown through the pipe, flushing out the ice. The quantity of salt necessary is, of course, dependent upon the size of the plant. There is a record of one line transmitting approximately 1000 cu. ft. of free air per minute for 1700 ft., where 2 bbl. of salt were used in the course of the five winter months and completely did away with all freezing troubles.

In pipe lines where sags exist, no trouble is encountered by freezing in the dips, as the salt collects as brine, preventing freezing. For quick relief from a partly frozen line, the pipe is sometimes used as a part of an electrical circuit temporarily until the pipe is warmed sufficiently to blow the ice out.

Electric Heater for Air-line Drains (By G. C. Bateman).—When the Cobalt power companies first started to supply compressed air on a large scale to the mines, trouble was experienced in the winter by the water from the air collecting in the low parts of the pipe lines and freezing. In the pipe lines of the British Canadian Power Co., this difficulty was overcome by the use of an electric heater, designed by James Ruddick, the general superintendent for that company. The device, as shown in Fig. 254, consists of a heater which is placed in a small box built over the drain cocks in the pipe line. These drain cocks are placed wherever there is a drop in the line, so that the water will drain both ways to the cocks.

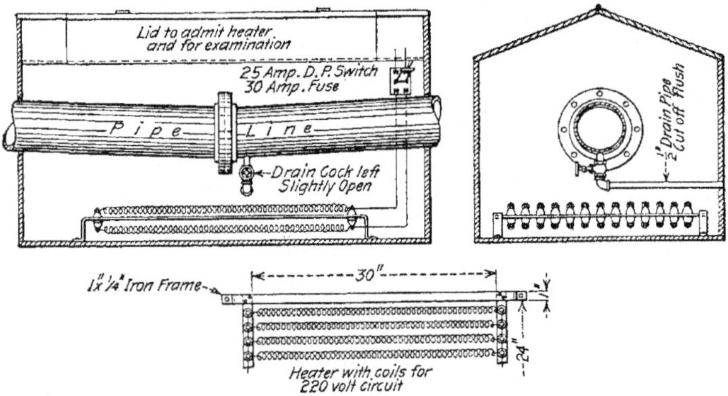

FIG. 254.—ELECTRIC HEATER USED ON COBALT AIR MAINS.

The heater, which is placed underneath the drain cock, is so designed that it fits snugly into the box, and a pipe leads from the cock to the outside of the box. This pipe is cut off flush with the box so that there is no danger of freezing on the outside. The frame of the heater is made of 1×1/4-in. iron, and on this frame insulation knobs 1 in. in diameter and 1 1/2 in. long are mounted back to back, as shown in the accompanying illustration, one bolt being sufficient for the two.

The coils are made by winding No. 14 galvanized-iron wire on a 7/8-in. rod, each coil consisting of 100 turns. The heater takes 220 volts, each coil being equivalent to 11 volts and 20 amperes. The coils are arranged so that there is at least an inch of space between them. There is a double-pole switch with 30-ampere fuses in each box and the wiring is mounted on cleats to prevent fire. Although the heater takes only 20 amperes the stronger fuses are used to take care of the current when starting, as the coils take more current when cold than when warm. The power required is 4 kw. for each heater.

The British Canadian Power Co. controls about 15 miles of pipe line between 6 and 10 in. in diameter, and several miles of smaller sizes. There are about 25 heaters in use, and since their installation no trouble from freezing has been experienced.

This device can be used advantageously wherever there is trouble from the freezing of surface lines, and as it is a simple device it can be built at the property. For pipes of smaller diameter than those mentioned, a heater of about half of the capacity of the one described can be used, in which case it would be necessary to have current at 110 volts, or less, and sufficient coils can be connected in series, to suit the voltage available. To increase or lower the voltage of the heater, it is only necessary to add or take out coils as the case may be. Should it be found that the heater does not warm up sufficiently, a coil or so can be cut out.

INDEX

E

F

G

H

I

J

K

L